Open-Channel Flow

M. HANIF CHAUDHRY

Professor of Civil Engineering
Washington State University,
Pullman, Washington

PRENTICE HALL, Englewood Cliffs, New Jersey 07632

Library of Congress Cataloging-in-Publication Data

Chaudhry, M. Hanif.
    Open-channel flow / M. Hanif Chaudhry.
        p.   cm.
    Includes bibliographical references and index.
    ISBN 0-13-637141-8
    1. Channels (Hydraulic engineering) 2. Hydrodynamics. I. Title.
TC175.C38 1993
627'.042--dc20   92-22564
                                                            CIP

Acquisitions editor: Doug Humphrey
Production editor: Jennifer Wenzel
Copy editor: Linda Thompson
Cover designer: Joe DiDomenico

Prepress buyer: Linda Behrens
Manufacturing buyer: Dave Dickey
Supplements editor: Alice Dworkin

The author and publisher of this book have used their best efforts in preparing this book. These efforts include the development, research, and testing of the theories and programs to determine their effectiveness. The author and publisher make no warranty of any kind, expressed or implied, with regard to the programs presented in the Appendices and stored on the diskette or the documentation contained in this book. The author and publisher shall not be liable in any event for incidental or consequential damages in connection with, or arising out of, the furnishing, performance, or use of these programs.

Printed in the United States of America

10 9 8 7 6 5 4 3 2 1

ISBN   0-13-637141-8

Prentice-Hall International (UK) Limited, *London*
Prentice-Hall of Australia Pty. Limited, *Sydney*
Prentice-Hall Canada Inc., *Toronto*
Prentice-Hall Hispanoamericana, S.A., *Mexico*
Prentice-Hall of India Private Limited, *New Delhi*
Prentice-Hall of Japan, Inc., *Tokyo*
Simon & Schuster Asia Pte. Ltd., *Singapore*
Editora Prentice-Hall do Brasil, Ltda., *Rio de Janeiro*

Contents

8 COMPUTATION OF RAPIDLY VARIED FLOW   203

9 CHANNEL DESIGN   231

10 SPECIAL TOPICS   250

Preface

Analysis of open-channel flow is needed for the planning, design, and operation of water-resource projects. The use of computers and the availability of efficient computational procedures have simplified such analyses as well as making it possible to handle complex and complicated systems. The main objective of this book is to present modern computational procedures for such analyses and other necessary up-to-date information on the topic. Although written mainly as a textbook, practicing engineers and researchers will find it useful as a reference.

Except for some limited material on erodible channels in Chaps. 9 and 17, the book deals with rigid-boundary channels without sediment deposition and/or erosion. Throughout the book, a very strong emphasis is given to the application of numerical methods suitable for computer analyses. A number of computer programs are presented in the appendixes and on a diskette. Several examples are included for illustration purposes. At the end of each chapter, up-to-date references and problems in both SI and customary units are given.

The book is divided into two parts: Chaps. 1 through 10 cover steady flow and Chaps. 11 through 17 deal with unsteady flow. Chapter 1 summarizes basic fluid-flow concepts and Chap. 2 presents application of momentum and energy principles. Critical flow and uniform flow are discussed in Chaps. 3 and 4. A qualitative presentation of gradually varied flow follows in Chap. 5, and several methods suitable for computerized analyses for the computation of these flows are presented in Chap. 6. Chapter 7 deals with rapidly varied flows. The material of this chapter is mainly empirical in nature; however, higher-order accurate finite-difference schemes for the analysis of these flows are presented in Chap. 8. A number of procedures for channel design are outlined in Chap. 9 and special topics related to steady flow are discussed in Chap. 10.

Unsteady flow is briefly described in Chap. 11, and equations describing one-dimensional unsteady flows are derived in Chap. 12. The numerical integration of these equations and the incorporation of boundary conditions are discussed in Chapter 13. Several explicit and implicit finite-difference methods for one-dimensional flows and two-dimensional flows are presented in Chaps. 14 and 15, respectively. The stability and convergence of the finite-difference schemes are described in Chap. 14. The finite-element method is presented in Chap. 16, and flood routing and aggradation and degradation of channels are discussed in Chap. 17. The listing of several computer programs written in FORTRAN for the analysis of steady and unsteady one-dimensional flows are included in Appendixes A through G.

For the last few years at Washington State University (WSU) we have used Chaps. 1 through 6, Chap. 9, and part of Chap. 10 for a 3-semester-hour senior-level undergraduate and graduate course and Chaps. 11 through 14 and part of Chap. 17 for a gradaute 3-semester-hour course. Although this sequence is recommended, other instructors may prefer to reduce coverage of Chap. 6 and instead add material from Chap. 7 and/or Chap. 11. The author required his students in each course to write four or five programs and do some parametric studies; however, others may prefer to utilize the programs given on the diskette. In addition, parts of Chaps. 3, 4, 6, 8, and 13 through 16 may be used in an advanced course on computational hydraulics or for directed studies, and Chap. 7 may be used in a course on hydraulic structures.

Any new book on the topic will benefit from the books by V.T. Chow and by F.M. Henderson; this text is no exception. The inclusion of contributions of several of my graduate students is thankfully acknowledged. Several colleagues reviewed the manuscript as it was being developed and offered suggestions for its improvement. Notable among these are (listed alphabetically) Professors Osman Akan of Old Dominion University, Dale Bray of University of New Brunswick, John Finnie of University of Idaho, Willi Hager of Swiss Federal Institute of Technology, Forrest Holly of University of Iowa, C. Sam Martin of Georgia Institute of Technology, and Marshall Richmond of WSU. Dr. Amgad Elansary of Cairo University, Visiting Professor, WSU, and my graduate student Walter Silva assisted in proofreading, preparing the index and some of the computer programs. The financial support and facilities provided by Professor D. Vischer of Swiss Federal Institute of Technology during author's sabbatical leave when the final draft was prepared are acknowledged. Dean Guenther of WSU Computing Center provided assistance in the preparation of TEXfiles, and Karen Parvin and Simin Heydari prepared the illustrations. The patience and understanding of my family during many hours needed for the preparation of this book are appreciated.

M. Hanif Chaudhry

To
Shamim

1 Basic Concepts

Contra dam, reservoir, and spillway (Courtesy of Verzasca SA, Lugano, Switzerland).

1-1 Introduction

Liquids are transported from one location to another using natural or constructed conveyance structures and flow passages. These flow passages may have cross sections that are open or closed at the top. The structures with closed tops are referred to as *closed conduits*, and those with open tops are called *open channels*. For example, tunnels, pipes, and aqueducts are closed conduits, whereas rivers, streams, estuaries, etc., are open channels. The flow in an open channel or in a closed conduit having a free surface is referred to as *free-surface flow* or *open-channel flow*. The properties and the analyses of these flows are discussed in this book.

In this chapter, commonly used terms are first defined. The classification of flows is then discussed, and terminology and properties of a channel section are presented. Expressions are then derived for the energy and momentum coefficients to account for nonuniform velocity distribution at a channel section. The chapter concludes with a discussion of the pressure distribution in a channel section.

1-2 Definitions

The flow in an open channel or in a closed conduit having a free surface is referred to as free-surface flow or open-channel flow (Fig. 1-1). Both of these terms will be synonymously used in this book. The free surface is usually subjected to atmospheric pressure. Groundwater or subsurface flows are excluded from the present

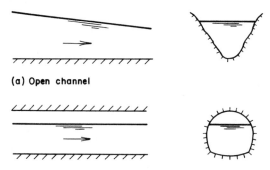

(a) Open channel

(b) Closed-conduit flow with free surface

Figure 1-1 Free-surface flow.

discussions. If there is no free surface and the conduit is flowing full, then the flow is called *pipe flow*, or *pressurized flow* (Fig. 1-2).

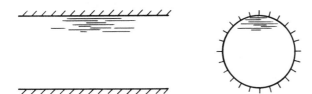

Figure 1-2 Pipe, or pressurized, flow.

In a closed conduit it is possible to have both free-surface flow and pressurized flow at different times. It is also possible to have these flows at a given instant of time in different reaches of a conduit. For example, the flow in a storm sewer may be free-surface flow at a certain time. Then, due to large inflows produced by a sudden storm, the sewer may start to flow full, thereby pressurizing it. Similarly, a closed conduit may have free-surface flow in part of the length and pipe flow in the remaining length. This condition usually occurs in a closed conduit when its downstream end is submerged (Fig. 1-3).

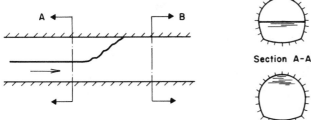

Figure 1-3 Combined free-surface and pressurized flow.

The photographs in Fig. 1-4 show unsteady flow in the hydraulic model (1:84 scale) of the tailrace tunnel of Mica Power Plant located on the Columbia River in Canada. The tunnel normally has free-surface flow. However, during periods of high tailwater levels, the tunnel may be pressurized following major load changes on the turbogenerators. The flow conditions shown in this photograph are for the reduction of flow in 8 seconds from maximum flow to zero on three turbines. Similarly, the photographs in Fig. 1-5 show free-surface and pressurized flow in a laboratory experiment.

The height to which liquid rises in a small-diameter piezometer inserted in a channel or a closed conduit depends upon the pressure at the location of the piezometer. A line joining the top of the liquid surface in the piezometers is called the *hydraulic-grade line* (Fig. 1-6). In pipe flow, the height of hydraulic-grade line above a specified datum is called the piezometric head at that location. In a free-surface flow, the hydraulic-grade line usually, but not always, coincides with the free surface (see Sec. 1-6). If the velocity head, $V^2/(2g)$, in which V = mean flow velocity for the channel cross section and g = acceleration due to gravity, is added to the top of

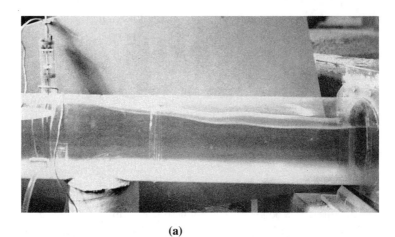

(a)

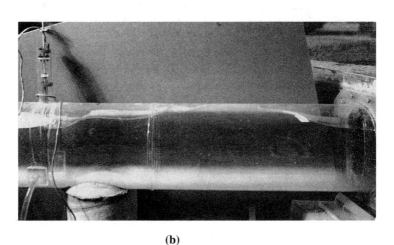

(b)

Figure 1-4 Transient flow conditions in the 1:84 scale hydraulic model of the Mica Tailrace Tunnel (Courtesy of B. C. Hydro and Power Authority).

the hydraulic-grade line and the resulting points are joined by a line, then this line is called the *energy-grade line*. This line represents the total head at different sections of a channel.

1-3 Classification of flows

Free-surface flows may be classified into various types using different criteria for their classification (Fig. 1-7). Definitions for each type of flow are given in the following paragraphs.

(a) Positive surge from downstream

(b) Positive surge from upstream

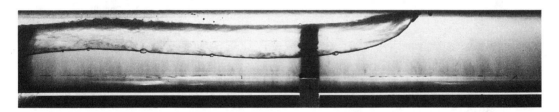

(c) Negative surge from downstream

(d) Negative surge from upstream

Figure 1-5 Free-surface and pressurized flows; flow is from right to left (Courtesy of Professor C.S. Song University of Minnesota, 1984).

Steady and Unsteady Flows

If the flow velocity at a given point does not change with respect to time, then the flow is called *steady flow*. However, if the velocity at a given location changes with respect to time, then the flow is called *unsteady flow*.

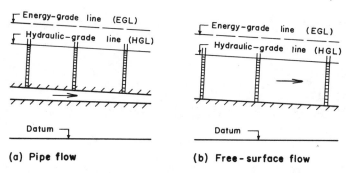

Figure 1-6 Hydraulic- and energy-grade lines.

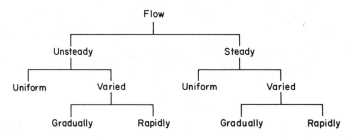

Figure 1-7 Classification of flows.

Note that this classification is based on the time variation of velocity v at a specified location. Thus, the local acceleration, $\partial v/\partial t$, is zero in steady flows. In two- or three-dimensional steady flows, the time variation of all velocity components is equal to zero.

It is possible in some cases to transform an unsteady flow into a steady flow by having coordinates with respect to a moving reference. This simplification offers a number of advantages, such as ease of visualization, ease in writing the governing equations, etc. Such a transformation is possible only if the wave shape does not change as the wave propagates. For example, the shape of a surge wave moving in a smooth channel does not change, and consequently the propagation of a surge wave in an otherwise unsteady flow may be converted into steady flow by moving the reference coordinates at the absolute surge velocity. This is equivalent to an observer traveling beside the surge wave so that the surge wave appears to the observer to be stationary; thus the flow may be considered as steady. If the wave shape changes as it propagates, then it is not possible to transform such a wave motion into steady flow. A typical example of such a situation is the movement of a flood wave in a natural channel, where the shape of the wave is modified as it propagates.

Uniform and Nonuniform Flows

If the flow velocity at a given instant of time does not change within a given length of channel, then the flow is called *uniform flow*. However, if the flow velocity at a time varies with respect to distance, then the flow is called *nonuniform flow*, or *varied flow*.

This classification is based on the variation of flow velocity with space at a specified instant of time. Thus the convective acceleration in uniform flow is zero. In mathematical terms, the partial derivatives of the velocity components with respect to x, y, and z are all zero. However, many times this strict restriction is somewhat waived by allowing a nonuniform velocity distribution. In other words, a flow is considered uniform as long as the velocity in the direction of flow at different locations along a channel remains the same.

Depending upon the rate of variation with respect to distance, flows may be classified as *gradually varied* or *rapidly varied*. As the name implies, if the flow depth varies at a slow rate with respect to distance, then the flow is called gradually varied flow, whereas if the flow depth varies significantly in a short distance, then the flow is called rapidly varied flow.

Note that the steady and unsteady flows are characterized by the variation with respect to time at a given location, whereas the uniform and varied flows are characterized by the variation at a given instant of time with respect to distance. Thus, in a steady-uniform flow, the total derivative $dV/dt = 0$. In one-dimensional flow, this means that $\partial v / \partial t = 0$, and $\partial v / \partial x = 0$. In two- and three-dimensional flow, the partial derivatives with respect to time and space of the velocity components in the other two coordinate directions are also zero.

Laminar and Turbulent Flows

If the liquid particles appear to move in definite smooth paths and the flow appears to be as a movement of thin layers on top of each other, then the flow is called *laminar flow*. In *turbulent flow*, the liquid particles move in irregular paths which are not fixed with respect to either time or space.

The relative magnitude of viscous and inertial forces determines whether the flow is laminar or turbulent: If the viscous forces dominate, the flow is laminar; if the inertial forces dominate, the flow is turbulent.

The ratio of viscous and inertial forces is defined as the *Reynolds number*, i.e.,

$$\mathbf{R}_e = \frac{VL}{\nu} \tag{1-1}$$

in which $\mathbf{R}_e$ = Reynolds number, V = mean flow velocity, L = a characteristic length, and ν = kinematic viscosity of the liquid. Unlike pipe flow, in which the pipe diameter is usually used for the characteristic length, either hydraulic depth or hydraulic radius may be used as the characteristic length in free-surface flows. *Hydraulic depth* is defined as the flow area divided by the water-surface width and the *hydraulic radius* is defined as the flow area divided by the wetted perimeter. The transition from laminar to turbulent flow in free-surface flows occurs for $\mathbf{R}_e$ of about 600, in which $\mathbf{R}_e$ is based on the hydraulic radius as the characteristic length.

In real-life applications, laminar free-surface flows are extremely rare. A smooth and glassy flow surface may be due to surface velocity being less than that required to form capillary waves and may not necessarily be due to the fact that the flow is laminar.

Care should be taken while selecting geometrical scales for the hydraulic model studies so that the flow depth on the model is not very small. Very thin depth may produce laminar flow on the model, even though the prototype flow to be modeled is turbulent. Such a model may give incorrect results.

Subcritical, Supercritical, and Critical Flows

A flow is said to be *critical* if the flow velocity is equal to the velocity of a gravity wave having small amplitude. A gravity wave may be produced by a change in the flow depth. If the flow velocity is less than the critical velocity, then the flow is called *subcritical flow*, and if the flow velocity is greater than the critical velocity, then the flow is called *supercritical flow*.

The *Froude number*, F_r, is equal to the ratio of inertial and gravitational forces. For a rectangular channel, it is defined as

$$F_r = \frac{V}{\sqrt{gy}} \tag{1-2}$$

in which y = flow depth. General expressions for F_r are presented in Sec. 3-2. Depending upon the value of F_r, flow is classified as *subcritical* if $F_r \leq 1$; *critical* if $F_r = 1$; and *supercritical* if $F_r \geq 1$.

1-4 Terminology

Channels may be natural or artificial. Various names have been used for the artificial channels: A long channel having mild slope usually excavated in the ground is called a *canal*. A channel supported above ground and built of wood, metal, or concrete is called a *flume*. A *chute* is a channel having very steep bottom slope and almost vertical sides. A *tunnel* is a channel excavated through a hill or a mountain. A short channel flowing partly full is referred to as a *culvert*. A channel having the same cross section and bottom slope is referred to as a *prismatic channel*, whereas a channel having varying cross section and/or bottom slope is called a *nonprismatic channel*. A long channel may comprise several prismatic channels. A cross section taken *normal* to the direction of flow (e.g., section BB in Fig. 1-8) is called a *channel section*. The depth of flow at a section, y, is the *vertical* distance of the lowest point of the channel section from the free surface. The *depth of flow section*, d, is the depth of flow *normal* to the direction of flow. The *stage*, Z, is the elevation or vertical distance of free surface above a specified datum (Fig. 1-8). The *top width*, B, is the width of channel section at the free surface. The *flow area*, A, is the cross-sectional area of flow *normal* to the direction of flow. The *wetted perimeter*, P, is defined as the length of line of intersection of channel wetted surface with a cross-sectional plane normal to the flow direction. The *hydraulic radius*, R, and *hydraulic depth*, D, are defined as

$$R = \frac{A}{P}$$
$$D = \frac{A}{B} \tag{1-3}$$

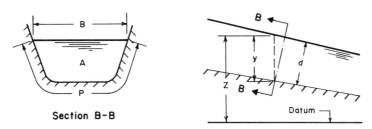

Figure 1-8 Definition sketch.

Expressions for A, P, R, B and D for typical channel cross sections are presented in Table 1-1.

1-5 Velocity distribution

The flow velocity in a channel section usually varies from one point to another. This is due to shear stress at the bottom and at the sides of the channel and due to the presence of free surface. Figure 1-9 shows typical velocity distributions in different channel cross sections.

The flow velocity may have components in all three Cartesian coordinate directions. Most of the time, however, the components of velocity in the vertical and transverse directions are small and may be neglected. Therefore, only the flow velocity in the direction of flow needs to be considered. This velocity component varies with depth from the free surface. A typical variation of velocity with depth is shown in Fig. 1-10.

Energy Coefficient

As discussed in the previous paragraphs, the flow velocity in a channel section usually varies from one point to another. Therefore, the mean velocity head in a channel section, $(V^2/2g)_m$, is not the same as the velocity head, $V_m^2/(2g)$, computed by using the mean flow velocity, V_m, in which the subscript m refers to the mean values. This difference may be taken into consideration by introducing an *energy coefficient*, α. This coefficient is also referred to as the *velocity-head*, or *Coriolis, coefficient*. An expression for the energy coefficient may be derived as follows.

Referring to Fig. 1-11, the mass of liquid flowing through area ΔA per unit time $= \rho V \Delta A$, in which $\rho =$ mass density of the liquid. Since the kinetic energy of mass m traveling at velocity V is $(1/2)mV^2$, we can write

$$\text{Kinetic energy transfer through area } \Delta A \text{ per unit time} = \frac{1}{2}\rho V \Delta A V^2 \qquad (1\text{-}4)$$

$$= \frac{1}{2}\rho V^3 \Delta A$$

Hence,

$$\text{Kinetic energy transfer through area } A \text{ per unit time} = \frac{1}{2}\rho \int V^3 \, dA \qquad (1\text{-}5)$$

Table 1-1 Properties of typical channel cross sections

Section	Area, A	Wetted perimeter, P	Hydraulic radius, R	Top width, B	Hydraulic depth, D	
Rectangular	$B_o y$	$B_o + 2y$	$\dfrac{B_o y}{B_o + 2y}$	B_o	y	
Trapezoidal	$(B_o + sy)y$	$B_o + 2y\sqrt{1+s^2}$	$\dfrac{(B_o + sy)y}{B_o + 2y\sqrt{1+s^2}}$	$B_o + 2sy$	$\dfrac{(B_o + sy)y}{B_o + 2sy}$	
Triangular	sy^2	$2y\sqrt{1+s^2}$	$\dfrac{sy}{2\sqrt{1+s^2}}$	$2sy$	$0.5y$	
Circular	$\frac{1}{8}(\theta - \sin\theta)D_o^2$	$\frac{1}{2}\theta D_o$	$\frac{1}{4}\left(1 - \dfrac{\sin\theta}{\theta}\right)D_o$	$D_o \sin\frac{1}{2}\theta$	$\left(\dfrac{\theta - \sin\theta}{\sin\frac{1}{2}\theta}\right)\dfrac{D_o}{8}$	

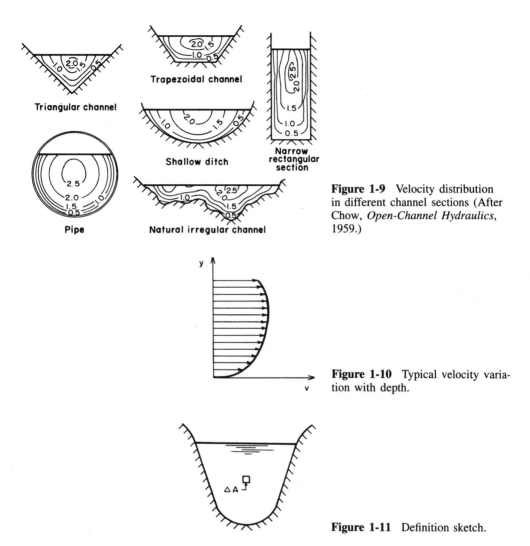

Figure 1-9 Velocity distribution in different channel sections (After Chow, *Open-Channel Hydraulics*, 1959.)

Figure 1-10 Typical velocity variation with depth.

Figure 1-11 Definition sketch.

It follows from Eq. 1-4 that the kinetic energy transfer through area ΔA per unit time may be written as $(\gamma V \Delta A) V^2/(2g) =$ weight of liquid passing through area ΔA per unit time × velocity head ($\gamma =$ specific weight of the liquid). Now, if V_m is the mean flow velocity for the channel section, then the weight of liquid passing through total area per unit time $= \gamma V_m \int dA$; and the velocity head for the channel section $= \alpha V_m^2/(2g)$, in which $\alpha =$ velocity-head coefficient. Therefore, we can write

$$\text{Kinetic energy transfer through area } A \text{ per unit time} = \rho \alpha V_m \frac{V_m^2}{2} \int dA \qquad (1\text{-}6)$$

Hence, it follows from Eqs. 1-5 and 1-6 that

$$\alpha = \frac{\int V^3 dA}{V_m^3 \int dA} \tag{1-7}$$

Figure 1-12 shows a typical cross section of a natural river, comprising the main river channel and the flood plain. The flow velocity in the flood plain is usually very low compared with that in the main section. In addition, the variation of flow velocity in each subsection is small. Therefore, each subsection may be assumed to have the same flow velocity throughout. In such a case, the integration of various terms of Eq. 1-7 may be replaced by summation, as follows:

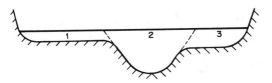

Figure 1-12 Typical river cross section.

$$\alpha = \frac{V_1^3 A_1 + V_2^3 A_2 + V_3^3 A_3}{V_m^3 (A_1 + A_2 + A_3)} \tag{1-8}$$

in which

$$V_m = \frac{V_1 A_1 + V_2 A_2 + V_3 A_3}{A_1 + A_2 + A_3} \tag{1-9}$$

By substituting Eq. 1-9 into Eq. 1-8 and simplifying, we obtain

$$\alpha = \frac{(V_1^3 A_1 + V_2^3 A_2 + V_3^3 A_3)(A_1 + A_2 + A_3)^2}{(V_1 A_1 + V_2 A_2 + V_3 A_3)^3} \tag{1-10}$$

Note that Eq. 1-10 is written for a section which may be divided into three subsections, each having uniform velocity distribution. For a general case in which total area A may be subdivided into N such subareas, each having uniform velocity, an equation similar to Eq. 1-10 may be written as

$$\alpha = \frac{\left(\sum_{i=1}^{N} V_i^3 A_i\right) \cdot \left(\sum_{i=1}^{N} A_i\right)^2}{\left(\sum_{i=1}^{N} V_i A_i\right)^3} \tag{1-11}$$

Momentum Coefficient

Similar to the energy coefficient, a coefficient for the momentum transfer through a channel section may be introduced to account for nonuniform velocity distribution. This coefficient, also called *Boussinesq coefficient*, is denoted by β. An expression for this may be obtained as follows.

The mass of liquid passing through area ΔA per unit time $= \rho V \Delta A$. Therefore, the momentum passing through area ΔA per unit time $= (\rho V \Delta A)V = \rho V^2 \Delta A$. By integrating this expression over the total area, we get

$$\text{Momentum transfer through area } A \text{ per unit time} = \rho \int V^2 \, dA \qquad (1\text{-}12)$$

By introducing the momentum coefficient, β, we can write the momentum transfer through area A in terms of the mean flow velocity, V_m, for the channel section, i.e.,

$$\text{Momentum transfer through area } A \text{ per unit time} = \beta \rho V_m^2 \int dA \qquad (1\text{-}13)$$

Hence, by equating Eqs. 1-12 and 1-13, we obtain

$$\beta = \frac{\int V^2 \, dA}{V_m^2 \int dA} \qquad (1\text{-}14)$$

The values of α and β for typical channel sections (Temple 1986, Watts et al. 1967; Chow 1959) are listed in Table 1-2. For turbulent flow in a straight channel having a rectangular, trapezoidal, or circular cross section, α is usually less than 1.15 (Henderson 1966). Therefore, it may not be included in the computations, since it is nearly equal to unity and its value is not precisely known.

Table 1-2 Values of α and β for typical sections*

Channel section	α	β
Regular channels	1.10–1.20	1.03–1.07
Natural channels	1.15–1.50	1.05–1.17
Rivers under ice cover	1.20–2.00	1.07–1.33
River valleys, overflooded	1.50–2.00	1.17–1.33

*Compiled from data given by Chow (1959).

Example 1-1

The velocity distribution in a channel section may be approximated by the equation $V = V_o(y/y_o)^n$, in which V is the flow velocity at depth y; V_o is the flow velocity at depth y_o, and n = a constant. Derive expressions for the energy and momentum coefficients.

Solution   Let us consider a unit width of the channel. Then, we can replace area A in the equations for the energy and momentum coefficients by the flow depth y. Now,

$$V_m = \frac{\int V \, dA}{\int dA}$$

For a unit width, this equation becomes

$$V_m = \frac{\int V \, dy}{\int dy}$$

By substituting the expression for V into this equation, we obtain

$$V_m = \frac{\int_0^{y_o} V_o(\frac{y}{y_o})^n \, dy}{\int_0^{y_o} dy}$$

$$= \frac{V_o}{y_o^n} \frac{y^{n+1}}{n+1} \bigg|_0^{y_o} \frac{1}{y_o}$$

$$= \frac{V_o}{n+1}$$

By substituting $V = V_o(y/y_o)^n$, $V_m = V_o/(n+1)$, and $dA = dy$ into Eq. 1-7, we obtain

$$\alpha = \frac{\int_0^{y_o} V_o^3 (y/yo)^{3n} dy}{[V_o/(n+1)]^3 \int_0^{y_o} dy}$$

$$= \frac{(V_o^3/y_o^{3n})[y_o^{3n+1}/(3n+1)]}{y_o[V_o/(n+1)]^3}$$

$$= \frac{(n+1)^3}{3n+1}$$

Substitution of $V = V_o(y/y_o)^n$ and $V_m = V_o/(n+1)$ into Eq. 1-14 yields

$$\beta = \frac{\int_0^{y_o} V_o^2 (y/y_o)^{2n} \, dy}{[V_o/(n+1)]^2 \int_0^{y_o} dy}$$

$$= \frac{(V_o^2 y_o)/(2n+1)}{[V_o/(n+1)]^2 y_o}$$

$$= \frac{(n+1)^2}{2n+1}$$

1-6 Pressure Distribution

The pressure distribution in a channel section depends upon the flow conditions. Let us consider several possible cases, starting with the simplest one and then proceeding progressively to more complex situations.

Static Conditions

Let us consider a column of liquid having cross-sectional area ΔA, as shown in Fig. 1-13. The horizontal and vertical components of the resultant force acting on the liquid column are zero, since the liquid is stationary. Let p = pressure intensity at the bottom of the liquid column. Then the force due to pressure at the bottom of the column acting vertically upwards = $p\Delta A$. The weight of the liquid column = $\rho g y \Delta A$, and the weight acts vertically downward. Since the vertical component of the resultant force is zero, we can write

$$p\Delta A = \rho g y \Delta A$$

or

$$p = \rho g y \qquad (1\text{-}15)$$

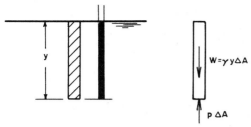

Figure 1-13 Pressure distribution in
stationary fluid.

Free-body diagram

In other words, the pressure intensity is directly proportional to the depth below the free surface; the relationship between the pressure intensity and depth plots as a straight line, and the liquid rises to the level of the free surface in a piezometer, as shown in Fig. 1-13. The linear relationship is based on the assumption that ρ is constant. This is true most of the time except at very large depths, where large pressures result in increased density.

Horizontal, Parallel Flow

Let us now consider the forces acting on a vertical column of liquid flowing in a horizontal, frictionless channel (Fig. 1-14). Let us assume that there is no acceleration in the direction of flow and the flow velocity is parallel to the channel bottom and is uniform over the channel section. Thus the streamlines are parallel to the channel bottom. Since there is no acceleration in the direction of flow, the component of the resultant force in this direction is zero. Referring to the free-body diagram shown in Fig. 1-14 and noting that the vertical component of the resultant force acting on the column of liquid is zero, we may write

$$\rho g y \Delta A = p \Delta A$$

or

$$p = \rho g y = \gamma y \tag{1-16}$$

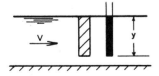

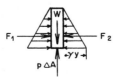

Free-body diagram   **Figure 1-14** Parallel horizontal flow.

in which $\gamma = \rho g = $ specific weight of the liquid. Note that this pressure distribution is the same as if the liquid were stationary; it is, therefore, referred to as the *hydrostatic pressure distribution*.

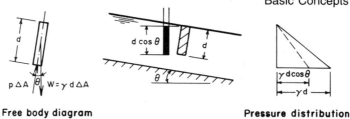

Free body diagram   Pressure distribution

Figure 1-15 Parallel flow in a sloping channel.

Parallel Flow in a Sloping Channel

Let us now consider the flow conditions in a sloping channel such that there is no acceleration in the flow direction and the flow velocity is uniform at a channel cross section and is parallel to the channel bottom; i.e., the streamlines are parallel to the channel bottom. Figure 1-15 shows the free-body diagram of a column of liquid normal to the channel bottom. The cross-sectional area of the column is ΔA. If θ = slope of the channel bottom, then the component of the weight of column acting along the column is $\rho g d \Delta A \cos \theta$ and the force acting at the bottom of the column is $p \Delta A$. There is no acceleration in a direction along the column length, since the flow velocity is parallel to the channel bottom. Hence, we can write $p \Delta A = \rho g d \Delta A \cos \theta$, or $p = \rho g d \cos \theta = \gamma d \cos \theta$. By substituting $d = y \cos \theta$ into this equation (y = flow depth measured vertically), we obtain

$$p = \gamma y \cos^2 \theta \qquad (1\text{-}17)$$

Note that in this case the pressure distribution is not hydrostatic in spite of the fact that we have parallel flow and there is no acceleration in the direction of flow. However, if the slope of the channel bottom is small, then $\cos \theta \simeq 1$ and $d \simeq y$. Hence,

$$p \simeq \rho g d \simeq \rho g y \qquad (1\text{-}18)$$

We will assume during several derivations in the subsequent chapters that the slope of the channel bottom is small. This assumption has two advantages: The pressure distribution may be assumed to be hydrostatic if the streamlines are parallel and straight; and the flow depths measured vertically or normal to the channel bottom are approximately the same.

Curvilinear Flow

In the previous three cases, the streamlines were straight and parallel to the channel bottom. However, in several real-life situations, the streamlines have pronounced curvature. To determine the pressure distribution in such flows, let us consider the forces acting in the vertical direction on a column of liquid with cross-sectional area ΔA, as shown in Fig. 1-16.

$$\text{Mass of the liquid column} = \rho y_s \Delta A \qquad (1\text{-}19)$$

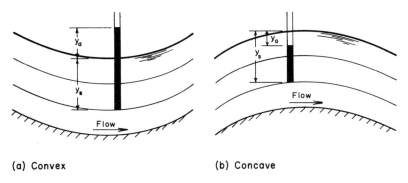

(a) Convex   (b) Concave

Figure 1-16 Curvilinear flow.

If r = radius of curvature of the streamline and V is the flow velocity at the point under consideration, then

$$\text{Centrifugal acceleration} = \frac{V^2}{r} \tag{1-20}$$

and

$$\text{Centrifugal force} = \rho y_s \Delta A \frac{V^2}{r} \tag{1-21}$$

Hence the pressure head acting at the bottom of the liquid column due to centrifugal acceleration is

$$y_a = \frac{1}{g} y_s \frac{V^2}{r} \tag{1-22}$$

The pressure due to centrifugal force is in the same direction as the weight of column if the curvature is convex, as shown in Fig. 1-16(a), and it is in a direction opposite to the weight if the curvature is concave (Fig. 1-16b). Therefore, the total pressure head acting at the bottom of the column is an algebraic sum of the pressure due to centrifugal action and the weight of the liquid column, i.e.,

$$\text{Total pressure head} = y_s \left(1 \pm \frac{1}{g} \frac{V^2}{r} \right) \tag{1-23}$$

A positive sign is used if the streamline is convex; and a negative sign is used if the streamline is concave. Note that the first term in Eq. 1-23 is the pressure head due to static conditions, while the second term is the pressure head due to centrifugal action. Thus, the liquid in a piezometer inserted into the flow will rise as shown in Fig. 1-16(a). In other words, pressure is increased due to centrifugal action in convex flows and it is decreased in concave flows (Fig. 1-16b).

1-7 Reynolds Transport Theorem

The *Reynolds transport theorem* relates the flow variables for a specified fluid mass to that of a specified flow region. We will utilize it in later chapters to derive the governing equations for steady and unsteady flow conditions. To simplify the presentation of its application, we include a brief description in this section; for details, see Roberson and Crowe (1990).

We call a specified fluid mass the *system* and a specified region the *control volume*. The boundaries of a system separate it from its *surroundings* and the boundaries of a control volume are referred to as the *control surface*. The three well-known conservation laws of mass, momentum, and energy describe the interaction between a system and its surroundings. However, in hydraulic engineering, we are usually interested in the flow in a region as compared to following the motion of a fluid particle or the motion of a quantity of mass. The Reynolds transport theorem relates the flow variables in a control volume to that of a system.

Let the *extensive* property of a system be B and the corresponding *intensive* property be β. The intensive property is defined as the amount of B per unit mass, m, of a system, i.e.,

$$\beta = \lim_{\Delta m \to 0} \frac{\Delta B}{\Delta m} \tag{1-24}$$

Thus, the total amount of B in a control volume is

$$B_{cv} = \int_{cv} \beta \rho d\mathcal{V} \tag{1-25}$$

in which ρ = mass density, $d\mathcal{V}$ = differential volume of the fluid, and the integration is over the control volume.

We will consider mainly one-dimensional flows in this book. The control volume will be fixed in space and will not change its shape (i.e., it will not stretch or contract) with respect to time. For such a control volume for one-dimensional flow, the following equation relates the system properties to those in the control volume:

$$\frac{dB_{sys}}{dt} = \frac{d}{dt} \int_{cv} \beta \rho d\mathcal{V} + (\beta \rho A V)_{out} - (\beta \rho A V)_{in} \tag{1-26}$$

in which the subscripts *in* and *out* refer to the quantities for the inflow and outflow from the control volume and V = flow velocity. The system is assumed to occupy the entire control volume—i.e., the system boundaries coincide with the control surface.

Let us now discuss the application of this equation to a control volume. As an example, the time rate of change of momentum of a system is equal to the sum of the forces exerted on the system by its surroundings (Newton's second law of motion). To apply this equation to describe the conservation of momentum of the water of mass m in a control volume, the extensive property B is the momentum of water = mV, and the corresponding intensive property, $\beta = \lim_{\Delta m \to 0} V(\Delta m / \Delta m) = V$. To describe the conservation of mass, B is the mass of water and the corresponding intensive property, $\beta = \lim_{\Delta m \to 0}(\Delta m / \Delta m) = 1$.

1-8 Computer Program

A computer program for computing the energy and momentum coefficients is presented in Appendix A. A trapeziodal rule is used for computing the integrals.

1-9 Summary

In this chapter, commonly used terms were defined, classification of flows using several different criteria was outlined, and the properties of a channel section were presented. The distribution of velocity and pressure in a channel section was discussed, and two coefficients were introduced to account for the nonuniform velocity distribution. The distribution of pressure was discussed, and a brief description of the Reynolds transport theorem was presented to facilitate its application in later chapters.

Problems

1-1. Derive expressions for the flow area, A, wetted perimeter, P, hydraulic radius, R, top water-surface width, B, and hydraulic depth, D, for the following channel cross sections:
  i. Rectangular (bottom width $= B_o$)
  ii. Trapezoidal (bottom width $= B_o$, side slopes $= 1V : sH$)
  iii. Triangular (side slopes $= 1V : sH$)
  iv. Partially full circular (diameter $= D$)
  v. Standard horseshoe (Fig. 1-17)

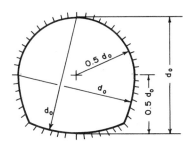

Figure 1-17 Horseshoe section.

1-2. The discharge in a given channel is proportional to $AR^{2/3}$ if the flow is uniform. For a circular conduit having an inside diameter D, prove that the discharge is maximum when the flow depth is $0.94D$.

1-3. Compute $(R/R_f)^{2/3}$ and $AR^{2/3}/(AR^{2/3})_f$ for different values of y/D for a circular conduit flowing partially full, in which $y =$ flow depth; $D =$ conduit diameter; and the subscript f refers to the values for the full section. At what values of the ratio y/D do the curves have maximum value?

1-4. Determine the energy and momentum coefficients for the velocity distribution, $V = 5.75V_o \log(30y/k)$, in which $V_o =$ flow velocity at the free surface; $y_o =$ flow depth, and $k =$ height of surface roughness. Assume the channel is very wide and rectangular.

1-5. The flow velocities measured at various flow depths in a wide rectangular flume are listed in the following table. Write a computer program to determine the values of α and β. Use Simpson's rule for the numerical integration.

y (m)	0.0	0.2	0.4	0.6	0.8	1.0	1.2
V (m/s)	0.0	3.87	4.27	4.53	4.72	4.87	5.0

1-6. At a bridge crossing, the mean flow velocities (in m/s) were measured at the midpoints of various subareas, as shown in Fig. 1-18. Compute the values of α and β for the cross section.

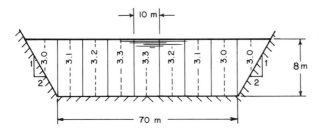

Figure 1-18 Velocities at bridge crossing.

1-7. Write a computer program to compute α and β for the flow in a channel having a general cross section. By using this program, compute α and β for the velocity distribution shown in Fig. 1-19.

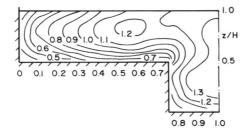

Figure 1-19 Dimensionless isovels (After Knight and Hamed, 1984).

1-8. Figure 1-20 shows the velocity distribution measured on the scale model of a canal. By using the computer program of Problem 1-7, compute the energy and momentum coefficients.

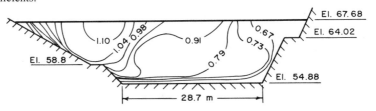

Figure 1-20 Velocity distribution (After Babb and Amorocho, 1965).

1-9. While computing the bending moment and the shear force acting on the side walls of the spillway chute of Fig. 1-21, a structural engineer assumed that the water pressure varies linearly from zero at the free surface to $\rho g y$ at the invert of the chute, where y = flow depth measured vertically. What are the computed values for the bending moment and the shear force at the invert level? Are the computed results correct? If not, compute the percentage error.

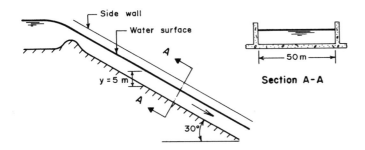

Figure 1-21 Spillway chute.

1-10. A spillway flip bucket has a radius of 20 m (Fig. 1-22). If the flow velocity at section B-B is 20 m/s and the flow depth is 5 m, compute the pressure intensity at point C.

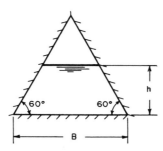

Figure 1-22 Flip bucket.

1-11. In a partially full channel having a triangular cross section (Fig. 1-23), the rate of discharge $Q = kAR^{2/3}$, in which k = a constant; A = flow area; and R = hydraulic radius. Determine the depth at which the discharge is maximum. For the triangular channel section shown, $A = [B - (h/\sqrt{3})]h$, and $P = B + (4h/\sqrt{3})$.

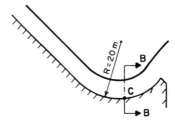

Figure 1-23 Triangular channel cross section.

1-12. In the following situations, is the flow uniform or nonuniform?
  i. Flow in a channel contraction or expansion
  ii. Flow at a channel entrance

     iii. Flow in the vicinity of a bridge pier

     iv. Flow at the end of a long prismatic channel

1-13. In the following situations, is the flow steady or unsteady?

     i. Flow in a storm sewer during a large storm

     ii. Flow in a power canal following the shutting down of turbines

     iii. Flow in a power canal when the turbines have been producing constant power

     iv. Flow in an estuary during a tide

1-14. In the following cases, is the flow laminar or turbulent?

     i. Flow in a wide rectangular channel at a flow velocity of 1 m/s at 1 m flow depth

     ii. Flow in a wide rectangular channel at a flow velocity of 0.1 m/s at 2 mm flow depth

1-15. Is it possible to have uniform flow in a frictionless sloping channel? Give reasons for your answer.

1-16. Is it possible to have uniform flow in a horizontal channel? Justify your answer.

1-17. If the angle between the flow surface and horizontal axis is θ and the angle between the channel bottom and horizontal is ϕ, prove that the pressure intensity at the channel bottom is

$$p = \frac{1}{1 + \tan\theta \tan\phi} \rho g y$$

in which y = flow depth measured vertically.

1-18. For an assumed velocity distribution of $V = 5.75 V_f \log\left(\frac{30y}{k}\right)$, prove that $\alpha = 1 + 3r^2 - 2r^3$ and $\beta = 1 + r^2$. In the preceding expression, $r = V_{max}/\overline{V} - 1$; $\overline{V}$ = mean velocity; and $V_{\max}$ = maximum velocity.

1-19. Show that the bending moment on the side walls of a steep channel with a bottom slope θ for a flow depth of y is $\frac{1}{6}\gamma y^3 \cos^4\theta$. Derive an expression for the shear force.

REFERENCES

Babb, A. F., and Amorocho, J. 1965. *Flow conveyance efficiency of transitions and check structures in a trapezoidal channel,* Dept. of Irrigation, University of California, Davis.

Chow, V. T. 1959. *Open-channel hydraulics,* McGraw-Hill, New York, NY.

French, R. H. 1985. *Open-channel hydraulics,* McGraw-Hill, New York, NY.

Henderson, F. M. 1966. *Open channel flow,* Macmillan, New York, NY.

Hulsing, H.; Smith, W.; and Cobb, E. D. 1966. Velocity head coefficient in open channels, *Water Supply Paper* 1869-C, U.S. Geological Survey, Washington, D.C.

Knight, D. W., and Hamed, M. E. 1984. Boundary shear in symmetrical compound channels. *Jour. Hydraulic Engineering,* Amer. Soc. Civil Engrs., 110, no. 10: 1412-1430.

Li, D., and Hager, W. 1991. Correction coefficient for uniform channel flow, *Canadian Jour. of Civil Engineering* (Feb.).

Roberson, J.A., and Crowe, C.T. 1990, *Engineering fluid mechanics,* Houghton Mifflin, Boston, MA.

Song, C. C. S. 1984. Modeling of mixed-transient flows, *Proc., Southeastern Conference on Theoretical and Applied Mechanics,* SECTAM XII, I(May): 431–435.

Temple, D. M. 1986. Velocity distribution coefficients for grass-lined channels, *Jour. of Hydraulic Engineering,* Amer. Soc. Civil Engrs., 112, no. 3: 193–205.

Watts, F. J.; Simons, D. B.; and Richardson, E. V. 1967. Variation of α and β values in lined open channels. *Jour. Hydraulics,* Division, Amer. Soc. Civil Engrs. 93, HY6: 217–234 (see also Discussions: vol. 94 (1968), HY3: 834–37; HY6: 1560–64; and vol. 95 (1969), HY3: 1059).

2 Conservation Laws

Hydraulic jump in 1:25 scale sectional hydraulic model of the spillway of Lower Granite Dam; discharge = 150,000 ft^3/sec; upstream reservoir level = El. 733 ft; downstream water level = El. 641 ft (Courtesy of U.S. Army Corps of Engineers, Walla Walla District).

2-1 Introduction

The following equations describe steady, free-surface flows: conservation of mass, conservation of momentum, and conservation of energy. In this chapter, these equations are derived and their application to the analysis of these flows is demonstrated.

For simplicity, only one-dimensional flows are considered in this chapter. In these flows, the flow velocity is only in the direction of flow and the components of flow velocity in the transverse and vertical directions are zero.

2-2 Continuity Equation

Civil engineers deal primarily with the flow of incompressible liquids, i.e., the mass density, ρ, of the liquid remains constant. Therefore, the law of conservation of mass between different channel cross sections implies that the volumetric flow rates at these sections are equal to each other provided there is no lateral inflow or outflow.

To derive the continuity equation, let us consider the flow of an incompressible liquid in a channel, as shown in Fig. 2-1. There is no inflow or outflow across the channel boundaries. Let the flow be steady. Let us denote the instantaneous flow velocity at a point by v, the flow depth by y, the mass density by ρ, the flow area by A, the top water-surface width by B, and use subscripts 1 and 2 to designate quantities for sections 1 and 2 respectively. Then, we may write

$$\text{Rate of mass inflow through area } dA_1 \text{ at section 1 } = \rho_1 v_1 dA_1 \qquad (2\text{-}1)$$

$$\text{Rate of mass outflow through area } dA_2 \text{ at section 2 } = \rho_2 v_2 dA_2 \qquad (2\text{-}2)$$

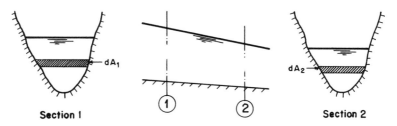

Figure 2-1 Notation for continuity equation.

According to the law of conservation of mass, the rate of mass inflow at section 1 must equal the rate of mass outflow at section 2, since the volume of liquid stored in the channel between sections 1 and 2 remains unchanged, i.e.,

$$\int \rho_1 v_1 \, dA_1 = \int \rho_2 v_2 \, dA_2 \tag{2-3}$$

Since the liquid is assumed incompressible, $\rho_1 = \rho_2$. Therefore,

$$\int v_1 \, dA_1 = \int v_2 \, dA_2 \tag{2-4}$$

If the flow velocity is assumed uniform at each section, then Eq. 2-4 may be written as

$$V_1 \int dA_1 = V_2 \int dA_2 \tag{2-5}$$

or

$$V_1 A_1 = V_2 A_2 \tag{2-6}$$

Note that we will obtain Eq. 2-6 even if the velocity distribution is nonuniform, provided V_1 and V_2 are the mean flow velocities at sections 1 and 2, respectively. In terms of volumetric flow rate, Q, this equation becomes

$$Q_1 = Q_2 \tag{2-7}$$

This equation in hydraulic engineering is usually referred to as the *continuity equation.*

2-3 Momentum Equation

To derive the momentum equation, let us consider the steady flow of an incompressible liquid in a channel, as shown in Fig. 2-2. The channel is prismatic and there is no lateral inflow or outflow. Referring to this figure and using subscripts 1 and 2 to designate quantities for sections 1 and 2,

$$\text{Time rate of mass inflow at section 1} = \frac{\gamma}{g} Q \tag{2-8}$$

in which γ = specific weight of the liquid. If V_1 is the mean flow velocity at section 1, then

$$\text{Time rate of momentum inflow at section 1} = \frac{\gamma}{g} \beta_1 Q V_1 \tag{2-9}$$

in which β_1 = the momentum coefficient introduced to account for the nonuniform velocity distribution. Similarly, we can write for section 2 that

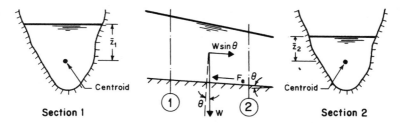

Figure 2-2   Notation for momentum equation.

$$\text{Time rate of momentum outflow} = \frac{\gamma}{g}\beta_2 Q V_2 \qquad (2\text{-}10)$$

Hence, it follows from Eqs. 2-9 and 2-10 that

Time rate of increase of momentum of the volume of liquid

$$\text{between sections 1 and 2} = \frac{\gamma}{g}Q(\beta_2 V_2 - \beta_1 V_1) \qquad (2\text{-}11)$$

The following forces are acting on the volume of liquid between sections 1 and 2.

$$\text{Pressure force at section 1, } P_1 = \gamma \bar{z}_1 A_1 \qquad (2\text{-}12)$$

$$\text{Pressure force at section 2, } P_2 = \gamma \bar{z}_2 A_2 \qquad (2\text{-}13)$$

in which $\bar{z}$ = depth of the centroid of flow area A.

Component of the weight of the liquid between sections 1
and 2 in the downstream direction $= W \sin \theta \qquad (2\text{-}14)$

in which W = weight of the volume of the liquid between sections 1 and 2; and $\theta =$ slope of the channel bottom. Let us neglect the shear stress at the free surface between air and liquid and let us designate the external force due to shearing force between the liquid and the channel bottom and sides by F_e. Then the resultant force, F_r, acting on the volume of liquid in the downstream direction is

$$F_r = \gamma A_1 \bar{z}_1 - \gamma A_2 \bar{z}_2 + W \sin \theta - F_e \qquad (2\text{-}15)$$

According to the Newton's second law of motion, the time rate of change of momentum of the liquid volume is equal to the resultant of the external forces acting on the liquid volume. Hence, noting that $\gamma = \rho g$, we obtain from Eqs. 2-11 and 2-15

$$\rho \beta_2 Q_2 V_2 - \rho \beta_1 Q_1 V_1 = \rho g A_1 \bar{z}_1 - \rho g A_2 \bar{z}_2 + W \sin \theta - F_e \qquad (2\text{-}16)$$

Note that F_e is the external shearing force acting on the volume of liquid and does not depend upon the losses occuring inside the liquid segment.

Equation 2-16 is a general application of the momentum principle. Let us discuss how this is simplified for a special case. For a prismatic channel with horizontal bottom, the component of the weight of the liquid in the downstream direction is zero. If we assume that the channel bottom and sides are smooth, then the shearing force is zero.

If the flow velocity is uniform at sections 1 and 2, then $\beta_1 = \beta_2 = 1$. With these simplifications, Eq. 2-16 for a smooth, horizontal channel becomes

$$Q_2 V_2 - Q_1 V_1 = g A_1 \bar{z}_1 - g A_2 \bar{z}_2 \tag{2-17}$$

From the continuity equation, $Q_1 = Q_2 = Q$ (say). Then Eq. 2-17 may be written as

$$\frac{Q^2}{g A_1} + \bar{z}_1 A_1 = \frac{Q^2}{g A_2} + \bar{z}_2 A_2 \tag{2-18}$$

Note that each side of this equation is similar except for the subscripts designating quantities for sections 1 and 2, respectively. Let us define

$$F_s = \frac{Q^2}{g A} + \bar{z} A \tag{2-19}$$

in which F_s is referred to as the *specific force*, or *momentum function*. Since each term on the right-hand side of Eq. 2-19 represents force per unit weight, we will refer to F_s as the specific force. The concept of specific force is very helpful in the application of the momentum equation; we will illustrate this in later sections.

2-4 Euler's Equation of Motion

Let us consider a rectangular fluid element along a streamline in a nonviscous fluid, as shown in Fig. 2-3. Let the length of the fluid element along the streamline be Δs, the length normal to the streamline be Δn, and the thickness of the fluid element perpendicular to the plane of paper be unity. Since the fluid is assumed nonviscous, there is no frictional force acting on the fluid element.

If p = pressure intensity at section 1, then the pressure at section 2 will be $p + (\partial p / \partial s) \Delta s$. Hence,

$$\text{Pressure force acting on the upstream face} = p \Delta n \tag{2-20}$$

and

$$\text{Pressure force acting on the downstream face} = \left(p + \frac{\partial p}{\partial s} \Delta s \right) \Delta n \tag{2-21}$$

If ρ = mass density of the fluid, then

$$\text{Weight of the fluid element} = \rho g \Delta s \Delta n \tag{2-22}$$

$$\text{Component of this weight in the } s\text{-direction} = \rho g \Delta s \Delta n \sin \theta \tag{2-23}$$

in which $\sin \theta = -(\partial z / \partial s)$ and z = height above the datum, measured positive in the upward direction. Now, the resultant force acting on the element in the downstream direction is

$$F_r = p \Delta n - \left(p + \frac{\partial p}{\partial s} \Delta s \right) \Delta n - \rho g \Delta s \Delta n \frac{\partial z}{\partial s} \tag{2-24}$$

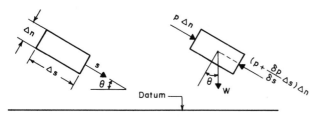

Figure 2-3 Forces acting on fluid element.

This equation, upon simplification, becomes

$$F_r = -\frac{\partial p}{\partial s}\Delta s\,\Delta n - \rho g\,\Delta s\,\Delta n\frac{\partial z}{\partial s} \tag{2-25}$$

According to the Newton's second law of motion, the resultant force is equal to the mass of the fluid element times acceleration of the fluid element, a_s; i.e.,

$$\rho\,\Delta s\,\Delta n a_s = -\frac{\partial p}{\partial s}\Delta s\,\Delta n - \rho g\,\Delta s\,\Delta n\frac{\partial z}{\partial s} \tag{2-26}$$

This equation may be simplified as

$$\rho a_s = -\frac{\partial}{\partial s}(p + \gamma z) \tag{2-27}$$

Since the flow velocity, $V = V(s, t)$, acceleration, a_s, in the s-direction may be written as

$$\begin{aligned} a_s &= \frac{dV_s}{dt} = \frac{\partial V_s}{\partial t} + \frac{\partial V_s}{\partial s}\frac{ds}{dt} \\ &= \frac{\partial V_s}{\partial t} + V_s\frac{\partial V_s}{\partial s} \end{aligned} \tag{2-28}$$

The first term on the right-hand side of this equation, $\partial V_s/\partial t$, represents the *local* acceleration; the second term, $V_s(\partial V_s/\partial s)$, represents the *convective* acceleration.

Substitution of Eq. 2-28 into Eq. 2-27 yields

$$\rho\left(\frac{\partial V_s}{\partial t} + V_s\frac{\partial V_s}{\partial s}\right) + \frac{\partial}{\partial s}(p + \gamma z) = 0 \tag{2-29}$$

This equation is called the Euler's equation of motion. Note that the only assumption we made is that the fluid is nonviscous; otherwise, the equation is valid along a streamline for unsteady, nonuniform flows.

Let us now discuss how this equation may be simplified for a number of special cases.

Steady Flow

The local acceleration in steady flow is zero, i.e., $(\partial V_s/\partial t) = 0$. Hence Eq. 2-29 becomes

$$\rho V_s\frac{dV_s}{ds} + \frac{d}{ds}(p + \gamma z) = 0 \tag{2-30}$$

Note that we have the total derivatives in Eq. 2-30 instead of the partial derivatives, since both p and V_s for steady flow are now functions of s only. By multiplying throughout by ds and integrating the resulting equation, we obtain

$$\frac{1}{2}\rho V_s^2 + p + \gamma z = \text{constant} \tag{2-31}$$

Dividing by γ, this equation becomes

$$z + \frac{p}{\gamma} + \frac{V_s^2}{2g} = H = \text{constant} \tag{2-32}$$

The constant of integration H is referred to as the *total head*, or *energy head*. Equation 2-32 is referred to as the *Bernoulli equation*. This equation is valid along a streamline. However, if the flow is irrotational, then it can be shown that this equation is valid throughout the flow field (Roberson and Crowe 1990). Recall that the Euler equation was valid for nonviscous fluid and we assumed in the derivation that ρ is constant. Therefore, we may say that the Bernoulli equation is valid for steady, irrotational, incompressible, and nonviscous flow.

Each term of Eq. 2-32 represents energy/unit weight and has the dimensions of length. In addition, note that the total head comprises three parts: datum head, z; pressure head, p/γ; and velocity head, $V_s^2/(2g)$. The datum head represents the potential energy, whereas the velocity head represents the kinetic energy.

Steady, Uniform Flow

Both the local and convective accelerations in steady, uniform flow are zero. Hence, Eq. 2-29 becomes

$$\frac{d}{ds}(p + \gamma z) = 0 \tag{2-33}$$

By integrating this equation, we obtain

$$\frac{p}{\gamma} + z = \text{constant} \tag{2-34}$$

The term $(p/\gamma + z)$ is referred to as the *piezometric head*. Note that this equation represents hydrostatic pressure distribution.

Unsteady, NonUniform Flow

In the case of unsteady, nonuniform flow, neither the local nor convective acceleration is zero. By multiplying Eq. 2-29 throughout by ds and by integrating the resulting equation, we obtain

$$\rho \int \frac{\partial V_s}{\partial t}ds + \rho \int V_s\, dV_s + (p + \gamma z) = \text{constant} \tag{2-35}$$

Dividing by ρg, this equation becomes

$$\frac{1}{g}\int \frac{\partial V_s}{\partial t}ds + \frac{V_s^2}{2g} + \frac{p}{\gamma} + z = \text{constant} \tag{2-36}$$

A comparison of Eqs. 2-32 and 2-36 shows that an additional term, $\frac{1}{g} \int \frac{\partial V_s}{\partial t} ds$, is introduced due to unsteadiness. To evaluate the integral of this term, we need to know the variation of V_s with respect to time. Such an expression is not usually known. Therefore, this equation is not useful for a general analysis.

The preceding derivation shows us that the Bernoulli equation is valid only for steady flows. We will discuss its applications in the following sections.

2-5 Specific Energy

The concept of specific energy was introduced by Bakhmeteff in 1912. As we shall see in the following sections, it is very useful for the application of the Bernoulli equation.

Let us consider each term of the Bernoulli equation (Eq. 2-32) in detail. The flow velocity at a channel cross section may vary from point to point, depending upon the velocity distribution. However, we may use the mean velocity at the section to calculate the velocity head by introducing the energy coefficient, α. The sum of the other two terms, $(z + p/\gamma)$, represents the piezometric head at a point. The piezometric head is constant at a section if the pressure distribution is hydrostatic. Assuming that the velocity distribution is uniform (i.e., $\alpha = 1$) and the pressure distribution is hydrostatic (i.e., $p = \gamma y$), Eq. 2-32 may be written as

$$z + y + \frac{V^2}{2g} = H \tag{2-37}$$

Now, let us use the channel bottom as the datum. Then $z = 0$, and Eq. 2-37 simplifies to

$$y + \frac{V^2}{2g} = E \tag{2-38}$$

in which E is referred to as the *specific energy*. Note that E is the total head above the channel bottom.

To facilitate understanding the concept of specific energy, let us first consider it for a rectangular cross section having uniform velocity distribution, i.e., $\alpha = 1$. Let the channel width be B and the channel discharge be Q. Then, the discharge per unit width, q (hereinafter called the *unit discharge*), is $q = Q/B$, and $V = q/y$. Eq. 2-38 may now be written as

$$E = y + \frac{q^2}{2gy^2} \tag{2-39}$$

or

$$(E - y)y^2 = \frac{q^2}{2g} \tag{2-40}$$

For a specified unit discharge, q, the right-hand side of Eq. 2-40 is a constant. Hence, we may write this equation as

$$Ey^2 - y^3 = \text{constant} \tag{2-41}$$

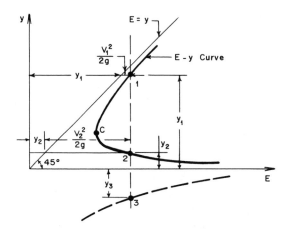

Figure 2-4 Specific energy curve for a given unit discharge.

This equation describes the relationship between E and y for a specified q. The E-y curve represented by this equation is plotted in Fig. 2-4. (The significance of the dotted curve is discussed later.) Mathematically, we can prove that the E-y curve has two asymptotes: $E - y = 0$ and $y = 0$. The first asymptote represents a straight line passing through the origin and inclined at 45° to the horizontal axis; and the second asymptote is the horizontal axis. From physical considerations, we may explain the existence of these asymptotes as follows. It follows from Eq. 2-38 that the specific energy, E, comprises two parts: the flow depth, y, and the velocity head, $V^2/(2g)$. The value of V decreases to pass the same amount of q as y increases, thereby decreasing the velocity head. Hence, referring to Fig. 2-4, the upper limb of the curve approaches the straight line, $E = y$, as the velocity head becomes very small for very large values of y. In a similar manner, the value of V increases to pass the specified q as the value of y decreases, thereby increasing velocity head. Hence, as y tends to zero, the velocity head tends to infinity, and the lower limb of the curve approaches the horizontal axis.

Equation 2-41 is a cubic equation in y for a given E. This equation may have three distinct roots. One of these roots is always negative. However, since it is not physically possible to have a negative depth, there can be only two different values of y for a given value of E. These two depths, say y_1 and y_2, are called the *alternate depths*. As a special case, it is possible that $y_1 = y_2$—i.e., at point C in Fig. 2-4. Such a depth is called the *critical depth, y_c,* and the corresponding flow is called the *critical flow*. This flow has a number of characteristics, which are discussed in the next chapter. A flow having depth greater than the critical depth is called the *subcritical flow* and a flow having depth less than the critical depth is called the *supercritical flow* (Chapter 1).

The E-y curve for a given value of q is presented in Fig. 2-4. To show the existence of three roots for a given value of E, the negative depths are plotted by a dotted curve.

The preceding discussion was for the specific energy curve for a given value of q. Let us now discuss how the curves for the other values of q will plot relative to that for q. Referring to Eq. 2-39, E increases as q increases for a given value of y. In other words, if we draw a line parallel to the E-axis for any given y, then the E-y curve for

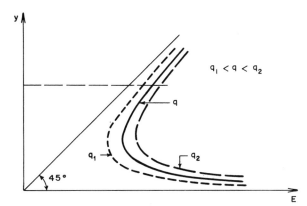

Figure 2-5 Specific energy curves for different unit discharges.

q_1 intersects it to the left of that for q if $q_1 < q$, and the E-y curve for q_2 intersects it to the right of that for q if $q_2 > q$. This is clear from the curves shown in Fig. 2-5.

In terms of discharge, Q, the Bernoulli equation for a general channel section may be written as

$$H = z + \frac{p}{\gamma} + \frac{\alpha Q^2}{2gA^2} \tag{2-42}$$

Now, let us consider channels having steep bottom slopes. As we discussed in Chapter 1, as a general case, $p = \gamma d \cos\theta$, in which d = depth of flow normal to the channel bottom and θ = angle between the channel bottom and the horizontal axis (Fig. 2-6). In addition, because of steep bottom slope, the flow depth d, measured normal to the channel bottom, is different from the flow depth y, measured vertically. Let us use the channel bottom as the datum. Then, noting that the total head above the channel bottom is referred to as the specific energy, we may write Eq. 2-42 as

$$E = d \cos\theta + \frac{\alpha Q^2}{2gA^2} \tag{2-43}$$

Since $d = y \cos\theta$, Eq. 2-43 becomes

$$E = y \cos^2\theta + \frac{\alpha Q^2}{2gA^2} \tag{2-44}$$

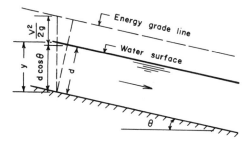

Figure 2-6 Definition sketch for steep channel.

Figure 2-7 shows the E-y curves for a channel having a steep slope for three rates of discharges, $Q_1 < Q < Q_2$. In this case, note that the angle between the horizontal axis and the straight line to which the upper limbs of the E-y curves are asymptotes is not 45°; this angle depends upon the slope of the channel bottom (see Prob. 2-1).

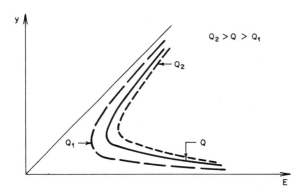

Figure 2-7 Specific-energy diagram for general-channel cross section

2-6 Application of Momentum and Energy Equations

The momentum and energy equations should yield the same results if properly applied to any flow problem. However, which of these equations should be preferred for a particular situation depends upon the problem under consideration. Although it is difficult to set rigid guidelines, the following discussion of the advantages and limitations of each equation should be helpful for such a selection.

The energy equation provides computational ease and conceptual simplicity, since energy is a scalar quantity, as compared to the momentum equation, in which different terms are vector quantities. Thus, in the latter case, the magnitude as well as the direction must be known. This may make the analysis more difficult and cumbersome.

The head losses to be included in the energy equation are the internal losses that occur in the volume of liquid. The losses to be considered in the momentum equation are those due to the external shear stress acting on the boundaries of the control volume. The local losses, such as those in a bend or in a hydraulic jump, usually occur in a short length of the channel. In such short lengths, the losses due to shear at the boundaries are very small and may be neglected. Therefore, the momentum equation is preferable in such situations, since the energy equation cannot be used directly because the amount of internal losses is not known. However, it is advantageous to use the energy equation first if there are some unknown external forces, e.g., forces on the sides in a channel expansion or contraction, provided the losses in the transition are negligible, and then to use the momentum equation to determine the magnitude of these external forces.

The energy and momentum equations may thus be used either alone or in sequence to solve a particular problem. The concepts of specific force (Sec. 2-3) and specific energy (Sec. 2-5) are very useful for such applications. The discussion of a channel transition, hydraulic jump, and hydraulic jump at a sluice gate outlet presented in the following sections should help in understanding the application of these concepts.

2-7 Channel Transition

A *channel transition* may be defined as a change in the channel cross section, e.g., change in the channel width and/or channel bottom slope. Such a geometric change may be accomplished over a long distance or it may be sudden. A channel transition is usually designed so that the losses at the transition are small. Thus, the energy losses in the transition may be neglected; and consequently the energy equation is more appropriate for their analysis.

For illustration purposes, we will consider a constant-width rectangular channel having a bottom step, as shown in Fig. 2-8. We want to determine whether the water surface rises or drops downstream of the transition for a specified flow depth and flow velocity upstream of the transition.

Since the channel width is constant, the unit discharge, q, is the same on both sides of the transition and the same specific energy curve is applicable to the upstream and downstream sides. Because the energy losses in the transition are assumed to be negligible, the total head H_1 is equal to H_2, in which the subscripts 1 and 2 refer to the quantities for the upstream and downstream sides of the transition, respectively. Referring to Fig. 2-8(a), $E_1 = H_1$ and $E_2 = H_2 - \Delta z$. Hence $E_2 = E_1 - \Delta z$.

On the specific energy diagram of Fig. 2-8, the point corresponding to flow conditions at section 1 is marked as 1. To determine the point corresponding to section 2, a vertical line is drawn such that $E = E_2$ as shown in Fig. 2-8(a). The flow depths corresponding to the points where this line intersects the specific-energy curve are the possible downstream depths. In this case, there are three such points, marked as 2, 2′, and 2″. Point 2″ corresponds to a negative depth, which is not physically possible. Hence we shall not consider this point any further in our discussion. Of the other two points, 2 and 2′, let us determine which one is actually possible. We see no particular problem in going from point 1 to 2 along the specific-energy curve (Fig. 2-8a). However, to go from 2 to 2′, two different paths (Henderson 1966) may be followed, as shown in Fig. 2-8(b) and 2-8(c). For the path along the vertical line 2-2′ (Fig. 2-8b), we have to move off the specified specific-energy curve and pass through the curves corresponding to higher unit discharges. Higher unit discharges are possible only if the channel width is reduced at the transition, as shown by a hypothetical channel in this figure. However, since there is no such contraction in this case, this path is not feasible. Similarly, a decrease in E is necessary to follow the second path, 2-C-2′, as shown in Fig. 2-8(c). E decreases only if the channel bottom rises, as shown by a hypothetical channel in this figure, i.e., the channel bottom rises until $E = E_c$ and then drops again until $E = E_2$. There is no such rise or drop in the bottom of the channel under consideration. Hence the second path, 2-C-2′, is not possible either. Therefore, only one depth is possible, which corresponds to point 2 of Fig. 2-8(a).

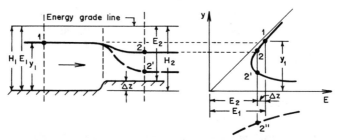

(a) Possible channel depths downstream of transition

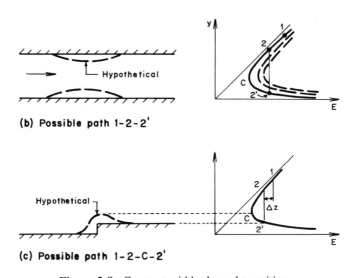

(b) Possible path 1-2-2'

(c) Possible path 1-2-C-2'

Figure 2-8   Constant-width channel transition.

Following a similar argument, we can show that if the flow upstream of the transition is supercritical, then the possible flow depth downstream of the transition is that corresponding to point 2 and not that corresponding to point 2' (Fig. 2-9).

From the preceding discussion, we conclude that for a step rise in the channel bottom, the flow depth decreases downstream of the step if the flow upstream of the transition is subcritical and the flow depth increases if the upstream flow is supercritical. These conclusions were drawn by considering all possible paths on a specific-energy diagram. Let us now derive them mathematically in a more rigorous manner.

If the pressure distribution is hydrostatic and $\alpha = 1$, then the total head, H, at a channel section may be written as

$$H = z + y + \frac{V^2}{2g} \tag{2-45}$$

or

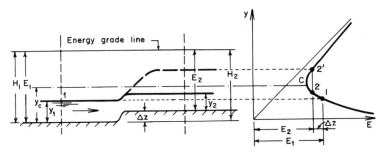

Figure 2-9 Supercritical flow upstream of transition.

$$H = z + y + \frac{Q^2}{2gA^2} \qquad (2\text{-}46)$$

We are interested in determining the sign of variation of y with a variation in the elevation of channel bottom, z. Assuming the downstream flow direction as positive for distance x measured along the channel bottom, flow depth increases if dy/dx is positive and it decreases if dy/dx is negative. By differentiating Eq. 2-46 with respect to x, we obtain

$$\frac{dH}{dx} = \frac{dz}{dx} + \frac{dy}{dx} + \frac{Q^2}{2g}\frac{d}{dx}\left(\frac{1}{A^2}\right) \qquad (2\text{-}47)$$

Now,

$$\frac{d}{dx}\left(\frac{1}{A^2}\right) = \frac{-2}{A^3}\frac{dA}{dx} \qquad (2\text{-}48)$$

and

$$\frac{dA}{dx} = \frac{dA}{dy}\frac{dy}{dx} \qquad (2\text{-}49)$$

For a small change in the flow depth, Δy, change in the flow area is $\Delta A \simeq B\Delta y$, in which B = top water-surface width. In the limit, as $\Delta y \to 0$, we may write $dA = Bdy$. Hence, Eq. 2-49 becomes

$$\frac{dA}{dx} = B\frac{dy}{dx} \qquad (2\text{-}50)$$

In addition, we define the Froude number as

$$\begin{aligned}\mathbf{F}_r^2 &= \frac{V^2}{gA/B}\\ &= \frac{BQ^2}{gA^3}\end{aligned} \qquad (2\text{-}51)$$

On the basis of Eqs. 2-50 and 2-51, Eq. 2-47 may be written as

$$\frac{dH}{dx} = \frac{dz}{dx} + (1 - \mathbf{F}_r^2)\frac{dy}{dx} \qquad (2\text{-}52)$$

Note that this equation is valid only if the pressure distribution is hydrostatic. If there are no losses, then $dH/dx = 0$, and Eq. 2-52 becomes

$$\frac{dz}{dx} = (\mathbf{F}_r^2 - 1)\frac{dy}{dx} \tag{2-53}$$

This equation describes the variation of the flow depth for any variation in the bottom elevation. If there is a step rise in the channel bottom, then $dz/dx > 0$. For the right-hand side of Eq. 2-53 to be positive, there are two possible situations: $(\mathbf{F}_r^2 - 1)$ and dy/dx are both positive or both negative. The first condition implies that if $\mathbf{F}_r > 1$ (i.e., flow is supercritical), then $dy/dx > 0$; i.e., flow depth increases at the step. Similarly, the second condition implies that if $\mathbf{F}_r < 1$ (i.e., flow is subcritical), then $dy/dx < 0$, i.e., flow depth decreases at the step. Following a similar argument, it can be shown from Eq. 2-53 that for a drop in the channel bottom, the flow depth decreases if the flow upstream of the step is supercritical, and it increases if the upstream flow is subcritical.

Example 2-1

A 4-m-wide rectangular channel is carrying 10 m^3/s at a depth of 2.5m. There is a step rise of 0.2m in the channel bottom. Assuming there are no losses at the transition, determine the flow depth downstream of the bottom step. Does the water surface rise or fall at the step?

Given:

Q = 10 m³/s
B = 4 m
y_1 = 2.5 m
Δz = 0.2 m
No head losses in the transition.

Determine:

y_2 = ?
Change in water-surface level = ?

Solution:

$$\begin{aligned}
V_1 &= \frac{Q}{A_1} \\
&= \frac{10}{4 \times 2.5} \\
&= 1 \text{ m/s}
\end{aligned}$$

$$\begin{aligned}
E_1 &= y_1 + \frac{V_1^2}{2g} \\
&= 2.5 + \frac{1}{2 \times 9.81} \\
&= 2.55 \text{ m}
\end{aligned}$$

Since there are no losses, referring to Fig. 2-10,

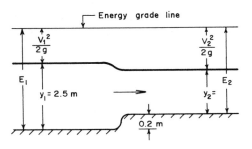

Figure 2-10 Definition sketch for bottom step.

$$E_2 = E_1 - 0.2$$
$$= 2.55 - 0.2$$
$$= 2.35 \text{ m}$$

Now

$$E_2 = y_2 + \frac{Q^2}{2gA_2^2}$$

By substituting values of E_2, Q, and $A_2 = 4y_2$ into this equation and simplifying, we have

$$y_2^3 - 2.35y_2^2 + 0.32 = 0$$

Solution of this equation by trial and error yields three roots: 2.29, 0.405, and −0.345 m. The third root is physically impossible because of the negative depth. In addition, only the first root is possible, since the upstream flow is subcritical, i.e., $\mathbf{F}_r < 1$; the second root requires that the flow has to pass through the critical depth at the step. Hence, the only possible downstream depth, $y_2 = 2.29$ m. Using the channel bottom upstream of the transition as the datum,

$$\text{Water level downstream of the transition} = 0.2 + 2.29$$
$$= 2.49 \text{ m}$$

Thus, the water-surface level drops by $2.5 - 2.49 = 0.01$ m.

2-8 Hydraulic Jump

A hydraulic jump is formed whenever supercritical flow changes to subcritical flow. At the jump location, there is a sharp discontinuity in the water surface and considerable amount of energy is dissipated due to turbulence. A detailed description of the hydraulic jump is presented in Chapter 7; at present, we are interested only in developing a relationship between the flow depths and the flow velocities upstream and downstream of the jump. The flow depths upstream and downstream of the jump are called *sequent depths*, or *conjugate depths*.

To simplify the derivation, we will consider a rectangular, horizontal channel. Since the amount of energy loss in the jump is not known a priori, we cannot apply the energy equation directly. However, since the length of the jump is usually short, the losses due to shear at the channel bottom and sides are small as compared to the pressure forces and may be neglected. In addition, since the channel is horizontal, the component of the weight of water in the downstream direction is zero. Therefore, referring to Fig. 2-11, specific force, F_s, at section 1 is equal to that at section 2; i.e.,

$$\frac{Q^2}{gA_1} + \bar{z}_1 A_1 = \frac{Q^2}{gA_2} + \bar{z}_2 A_2 \tag{2-54}$$

Rearranging the terms of this equation, we obtain

$$\frac{Q^2}{g}\frac{A_2 - A_1}{A_1 A_2} = \bar{z}_2 A_2 - \bar{z}_1 A_1 \tag{2-55}$$

For a rectangular channel, $A = By$ and $\bar{z} = \frac{1}{2}y$. By substituting these relationships into Eq. 2-55 and simplifying, we obtain

$$\frac{Q^2}{g}(y_2 - y_1) = \frac{1}{2}B^2 y_1 y_2 (y_2^2 - y_1^2) \tag{2-56}$$

or

$$\frac{Q^2}{g}(y_2 - y_1) = \frac{1}{2}B^2 y_1 y_2 (y_2 + y_1)(y_2 - y_1) \tag{2-57}$$

Substituting $Q = By_1 V_1$ and simplifying Eq. 2-57 becomes

$$\frac{y_1 V_1^2}{g} = \frac{1}{2}y_2(y_2 + y_1) \tag{2-58}$$

Dividing throughout by y_1^2 yields

$$\frac{2V_1^2}{gy_1} = \frac{y_2}{y_1}\left(\frac{y_2}{y_1} + 1\right) \tag{2-59}$$

Now, the Froude number, $\mathbf{F}_{r_1} = V_1/\sqrt{gy_1}$. Hence, Eq. 2-59 may be written as

$$\left(\frac{y_2}{y_1}\right)^2 + \frac{y_2}{y_1} - 2\mathbf{F}_{r_1}^2 = 0 \tag{2-60}$$

Solution of this equation yields

$$\frac{y_2}{y_1} = \frac{1}{2}\left(-1 + \sqrt{1 + 8\mathbf{F}_{r_1}^2}\right) \tag{2-61}$$

Note that the negative sign with the radical term is neglected because it gives a negative ratio, which is physically impossible. This equation specifies a relationship between the depths upstream and downstream of the jump in terms of $\mathbf{F}_{r_1}$. Proceeding similarly, we can derive the following equation in terms of $\mathbf{F}_{r_2}$

$$\frac{y_1}{y_2} = \frac{1}{2}\left(-1 + \sqrt{1 + 8\mathbf{F}_{r_2}^2}\right) \tag{2-62}$$

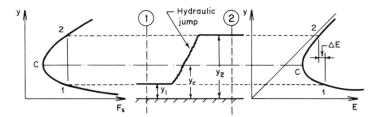

Figure 2-11 Hydraulic jump.

Thus if the flow depth and flow velocity on one side of the jump are known, then their values on the other side can be determined by using Eq. 2-62 or 2-63 and the continuity equation. The energy losses can then be computed from the energy equation.

It is easier to visualize the relationships between E, F_s, flow depths on the upstream and downstream sides, and the energy loss in the jump by plotting the specific energy and specific force diagrams side by side, as shown in Fig. 2-11.

Example 2-2

The reservoir level upstream of a 30-m-wide spillway for a flow of 800 m³/s is at El. 200 m. The downstream river level for this flow is at El. 100 m. Determine the invert level of a stilling basin having the same width as the spillway so that a hydraulic jump is formed in the basin. Assume the losses in the spillway are negligible.

Given:

$$Q = 800 \text{ m}^3/\text{s}$$
$$B = 30 \text{ m}$$
Upstream water level = El. 200 m
Downstream water level = El. 100 m

Determine:
The elevation of the stilling basin = ?

Solution: Let the invert elevation be z. Referring to Fig. 2-12, $y_2 = 100 - z$. Since the losses on the spillway face are negligible,

$$V_1 = \sqrt{2g(200 - z)}$$

Now, $Q = BV_1y_1$. Hence

$$
\begin{aligned}
y_1 &= \frac{800}{30 \times \sqrt{19.62(200 - z)}} \\
&= \frac{6.02}{\sqrt{200 - z}}
\end{aligned}
$$

Substituting expressions for y_1 and V_1

$$
\begin{aligned}
\mathbf{F}_{r_1}^2 &= \frac{V_1^2}{gy_1} \\
&= \frac{19.62(200 - z)}{9.81 \times 6.02/\sqrt{200 - z}} \\
&= 0.332(200 - z)^{1.5}
\end{aligned}
$$

41

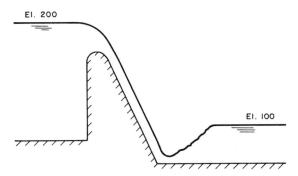

Figure 2-12 Sketch for Example 2-2.

Substitution of expressions for y_1, y_2, and $\mathbf{F}_{r_1}^2$ into Eq. 2-61 yields

$$\frac{100 - z}{6.02/\sqrt{200 - z}} = \frac{1}{2}\left(-1 + \sqrt{1 + 8 \times 0.332(200 - z)^{1.5}}\right)$$

Simplifying this equation, we obtain

$$(100 - z)\sqrt{200 - z} = -3.01 + 3.01\sqrt{1 + 2.656(200 - z)^{1.5}}$$

Solving this equation by trial and error

$$z = 84.18 \text{ m}$$

Thus, the stilling basin should be set at El. 84.18 to form the jump.

2-9 Hydraulic Jump at Sluice Gate Outlet

Figure 2-13 shows the outflow conditions downstream of a sluice gate used to control outflow from the reservoir. A hydraulic jump is formed in this case just downstream of the gate. A combined use of the specific-energy and specific-force diagrams, as shown in this figure, illustrates the usefulness of these concepts.

Assuming that there are no losses at the gate, $E_1 = E_2$. However, because of an additional external force between sections 1 and 2 (i.e., thrust on the gate, F_g), F_{s1} is not equal to F_{s2}. There is loss of energy in the hydraulic jump between sections 2 and 3. Hence, E_2 is not equal to E_3. Since the losses due to shear at the channel bottom and sides between sections 2 and 3 are small and can be neglected, $F_{s3} = F_{s2}$ provided the channel bottom slope is either zero or can be assumed as zero. Referring to Fig. 2-13, the thrust on the gate, $F_g = \gamma(F_{s_1} - F_{s_2})$; and the energy losses in the jump $= E_2 - E_3$.

The following example illustrates the application of the specific-energy and specific-force diagrams.

Example 2-3

A hydraulic jump is formed in a 5-m wide outlet at a short distance downstream of a control gate (Fig. 2-13). If the flow depths just upstream and downstream of the gate are 10 m and 2 m, respectively, and the outlet discharge is 150 m³/s, determine:

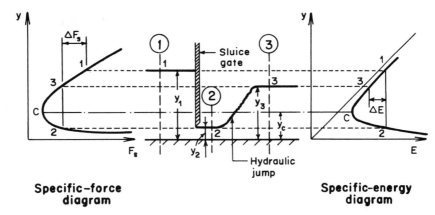

Figure 2-13   Hydraulic jump at sluice gate outlet.

i. *Flow depth downstream of the jump;*
ii. *Thrust on the gate; and*
iii. *Head losses in the jump.*

Assume there are no losses in the flow through the gate.

Given:

$$Q = 150 \text{ m}^3/\text{s}$$
$$B = 5 \text{ m}$$
$$y_1 = 10 \text{ m}$$
$$y_2 = 2 \text{ m}$$

Determine:

$$y_3 = ?$$
Thrust on the gate = ?
Head losses in the jump = ?

Solution:

$$q = 150/5 = 30 \text{ m}^3/\text{s}$$

Referring to Fig. 2-13,

$$
\begin{aligned}
V_2 &= \frac{q}{y_2} \\
    &= \frac{30}{2} = 15 \text{ m/s} \\
\mathbf{F}_{r2}^2 &= \frac{V_2^2}{g y_2} \\
    &= \frac{(15)^2}{9.81 \times 2} \\
    &= 11.47
\end{aligned}
$$

i. *Depth downstream of jump*

Note that in this example, section 2 is upstream of the jump and section 3 is downstream of the jump. Hence, substituting this value of $\mathbf{F}_{r2}^2$ into Eq. 2-61, we obtain

$$y_3 = \frac{1}{2}y_2(\sqrt{1 + 8\mathbf{F}_{r2}^2} - 1)$$

$$= \frac{1}{2}2(\sqrt{1 + 8 \times 11.47} - 1)$$

$$= 8.63 \text{ m}$$

ii. *Head loss in the jump*

$$E_2 = E_3 + H_l$$

$$H_l = E_2 - E_3$$

Hence

$$H_l = \left(y_2 + \frac{q^2}{2gy_2^2}\right) - \left(y_3 + \frac{q^2}{2gy_3^2}\right)$$

Substituting the values for y_2, y_3, and q gives

$$H_l = 2 + \frac{(30)^2}{2 \times 9.81(2)^2} - 8.63 - \frac{(30)^2}{2 \times 9.81(8.63)^2}$$

$$= 4.22 \text{ m}$$

iii. *Thrust on the gate*

Referring to Fig. 2-13 and considering unit width

$$\frac{P_f}{\gamma} = F_{s_1} - F_{s_2}$$

$$= \left(\frac{y_1^2}{2} + \frac{q^2}{gy_1}\right) - \left(\frac{y_2^2}{2} + \frac{q^2}{gy_2}\right)$$

$$= \left[\frac{(10)^2}{2} + \frac{(30)^2 \cdot}{9.81 \times 10}\right] - \left[\frac{(2)^2}{2} + \frac{(30)^2}{9.81 \times 2}\right]$$

$$= 11.3$$

Therefore,

$$\text{Thrust on the gate} = \gamma P_f \times \text{gate width}$$

$$= 9.81 \text{ (kN/m}^3) \times 11.3 \text{ (m}^2) \times 5 \text{ (m)}$$

$$= 554.4 \text{ kN}$$

2-10 Summary

In this chapter, the continuity and momentum equations describing the steady, free-surface flows were derived. The concepts of specific energy and specific force were introduced and their applications to the analysis of different problems were illustrated.

Problems

2-1. Determine an expression for the slope of the straight line to which the upper limb of the specific energy curve is an asymptote for a channel having a bottom slope of θ.

2-2. Plot the specific energy versus depth curves for $Q = 400$, 600, and 800 m³/s in a trapezoidal channel having a bottom width of 20 m and side slopes of 2H:1V. Assume the bottom slope is small. From these curves, determine the critical depth for each discharge.

2-3. Derive an expression for the specific force for a rectangular channel.

2-4. The flow depth and the flow velocity upstream of a 0.2-m sudden step rise in the bottom of a 5-m-wide rectangular channel are 5 m and 4 m/s, respectively. Assuming there are no losses at the transition, determine:
  i. The flow depth downstream of the step and the change in the water level
  ii. The flow depth and the water level downstream of the step if the channel bottom has a 0.2-m drop instead of the rise, as in (i).

2-5. A 10-m-wide rectangular channel is carrying a discharge of 200 m³/s at a flow depth of 5 m.
  i. If the channel bottom has a sudden rise of 0.3 m, determine the depth of flow downstream of the step. Does the water surface rise or drop at the step?
  ii. Compute the flow depth and the water-surface level downstream of the bottom step if the step is a drop of 0.2 m.

  In both cases, assume that there are no losses at the bottom step.

2-6. If the discharge in the channel of Prob. 2-5 is 400 m³/s, determine the flow depth downstream of the step if the step is: (i) a rise, and (ii) a drop. Does the water surface rise or drop downstream of the step in each case?

2-7. An 8-m-wide rectangular channel carries a flow of 96 m³/s at a flow depth of 4 m. The channel width is constricted to 6 m in a length of 5 m. Assuming the channel transition has straight and vertical sides and there are no losses, plot the water-surface profile in the transition.

2-8. A hydraulic jump is formed in a 4-m-wide outlet just downstream of the control gate, which is located at the upstream end of the outlet. The flow depth upstream of the gate is 20 m. If the outlet discharge is 40 m³/s, determine
  i. Flow depth downstream of the jump;
  ii. Thrust on the gate; and
  iii. Energy losses in the jump. Assume there are no losses in the flow through the gate.

2-9. On the specific-energy diagram for a rectangular channel, prove that the slope of the straight line joining the critical depth for different unit discharges is $\frac{2}{3}$.

2-10. An 8-m-wide rectangular channel has a flow velocity and flow depth of 4 m/s and 4 m, respectively. The channel bottom is at El. 700. Assuming no losses, design a transition so that the water level downstream of the transition is at El. 703.54 if
  i. The channel width remains constant; and
  ii. The channel bottom level downstream of the transition is at El. 700.2 m.

2-11. Fig. 2-14 shows a step rise in the channel bottom. If the channel width is constant and there are no losses, determine:
  i. Flow depth downstream of the transition; and
  ii. Maximum step height so that upstream water levels are not affected.

2-12. If there is no step in the channel bottom of Fig. 2-14 (i.e., channel bottom is horizontal), determine the minimum channel width such that the upstream water level remains unchanged. Assume there are no losses at the channel transition.

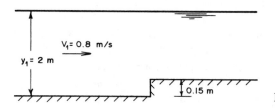

Figure 2-14 Channel of Prob. 2-12.

2-13. A 5-ft-wide rectangular channel carries a flow of 80 ft³/s at a depth of 8 ft. The channel width is reduced to 4 ft in a length of 10 ft. By assuming that there are no losses in the contraction,

   i. Compute the water surface profile for horizontal channel bottom; and
   ii. Determine the variation of bottom elevation so that the water level remains constant.

2-14. The discharge in a 20-m-wide, rectangular, horizontal channel is 80 m³/s at a flow depth of 0.5 m upstream of a hydraulic jump. What are the flow depth downstream of the jump and the head losses in the jump.

2-15. An 8-m-wide rectangular channel is carrying a flow of 54 m³/s at a flow depth of 0.6 m. A sluice gate located at the downstream end of the channel controls the flow depth y_2. Determine y_2 so that a hydraulic jump is formed upstream of the gate.

2-16. The channel bottom at the junction of two channels is raised by 0.1 m. The upstream channel is 10 m wide and the downstream channel is 8 m wide. If the channel discharge is 10 m³/s and the depth in the upstream channel is 1.5 m, determine the flow depth downstream of the junction assuming the losses at the junction to be $0.2V_2^2/(2g)$, where $V_2 =$ flow velocity in the downstream channel.

2-17. The flow depth in an 8-m-wide rectangular channel upstream of a 0.15-m sudden drop is 2 m and the flow velocity is 3 m/s. Assuming no head losses, compute the flow depth downstream of the drop.

2-18. The flow velocity in a rectangular channel is 3 m/s at a depth of 3 m. If the channel bottom has a step rise of 0.3 m, determine the flow depth downstream of the step assuming no losses at the step.

2-19. Determine the required depth in the river downstream of a 20-m wide spillway to form a hydraulic jump at its toe for the following data. The upstream reservoir level is at El. 160 m, the spillway discharge is 1200 m³/s and the river bottom level is at El. 120. Assume the losses on the spillway face are negligible and the stilling basin walls are vertical. Compute the energy loss in the jump.

2-20. A 6-m wide rectangular channel is carrying a flow of 18 m³/s. Plot a diagram between the flow depth and the specific energy for these conditions. What are the alternate and sequent depths corresponding to $y = 0.3$ m? Determine the head loss in the jump.

2-21. The flow depth and velocity in a 6-ft wide rectangular channel are 5 ft and 3 ft/sec, respectively. The channel width is reduced to 5 ft in a distance of 15 ft.

   i. Compute and plot the flow depth in the contraction assuming the bottom is horizontal and the head losses are negligible.

ii. If we want to keep the water surface in the contraction as horizontal, determine the variation of the bottom elevation.

2-22. The width of a rectangular channel is reduced from 5 ft to 4 ft along with a smooth raising of the channel bottom. If the flow depths upstream and downstream of the contraction are 3 ft and 2.5 ft, respectively, for a flow of 50 ft^3/sec, determine the height of the step. Assume the losses at the transition are negligible.

2-23. The flow depth upstream of the hydraulic jump in a rectangular channel is 1.5 ft and the Froude number is 5. Compute the flow depth downstream of the jump and the head losses.

2-24. A hydraulic jump is formed just downstream of a sluice gate located at the entrance of a channel. There is a constant-level lake upstream of the sluice gate. The flow depth and velocity in the channel downstream of the jump are 5.2 ft and 4.3 ft/sec, respectively. What is the lake-water level? Assume the losses for flow through the gate are negligible.

2-25. Plot a family of curves between E and y for a trapezoidal channel with bottom width of 8 m and side slopes of 2H : 1V for $Q = 10, 20, 40$, and 50 m^3/s. Draw the locus of the critical depths on these curves.

2-26. Prove that the Froude number, $\mathbf{F}_r$, for a channel having steep bottom slope is $\mathbf{F}_r = V/\sqrt{gD\cos\theta/\alpha}$, where D = hydraulic depth and α = velocity-head coefficient.

2-27. Two different flow depths, y_1 and y_2 (called alternate depths), are possible in a channel for a specified discharge Q and specific energy, E. If y_1 is known for a given value of Q in a trapezoidal channel, determine y_2 by using the bisection method. Write the governing equation, computational steps you will use, and a computer program based on this procedure.

2-28. The reservoir level upstream of an overflow spillway is at El. 400 ft. The downstream water level for the design flow of 80,000 ft^3/sec is at El. 220 ft. If the spillway width at the entrance to the stilling basin is 200 ft, determine the invert level of the basin so that a hydraulic jump is formed in the basin at design flow. No baffle piers, chute blocks, or end sill are to be provided.

REFERENCES

Bakhmeteff, B. A.   1932. *Hydraulics of open channels*, McGraw-Hill, New York, NY.

Chow, V. T.   1959. *Open-channel hydraulics*, McGraw-Hill, New York, NY.

Henderson, F. M.   1966. *Open channel flow*, Macmillan, New York, NY.

Roberson, J.A. and Crowe, C.T.   1990. *Engineering fluid mechanics,* Houghton Mifflin, Boston, MA.

3 Critical Flow

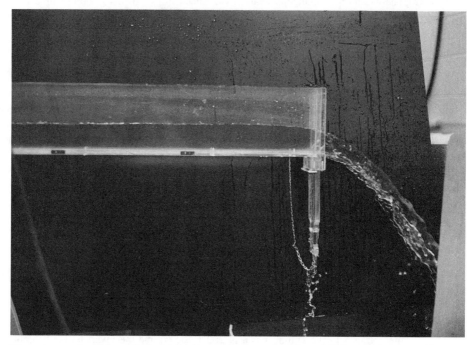

Flow passes through critical depth just upstream of a free fall in a laboratory flume (Courtesy of Professor M. C. Quick, University of British Columbia, Vancouver, B.C., Canada).

3-1 Introduction

In Chapter 2, we called the depth at which the specific energy was minimum for a given discharge as the *critical depth* and the corresponding flow as the *critical flow*. In this chapter, we show that there may be more than one critical depth for a specified discharge in a compound channel.* Critical flow has a number of special properties. We discuss these properties and show how they can be used for engineering applications. Then, a procedure for computing the critical depth in a channel is presented. The chapter concludes by discussing the possible number of critical depths in a compound channel and how to determine their values.

3-2 Properties of Critical Flow

For simplicity, we will first consider channels having a rectangular cross section and then channels having a non-rectangular cross section.

Rectangular Channel

In this section, we will discuss different properties of critical flow in a channel having a rectangular cross section.

Specific Energy.   As we discussed in Chapter 2, the specific energy for a rectangular channel having hydrostatic pressure distribution and uniform velocity may be written as

$$E = y + \frac{V^2}{2g} \tag{3-1}$$

or

$$E = y + \frac{q^2}{2gy^2} \tag{3-2}$$

---

*A channel having a cross section comprising the main flow section and one or two overbank flow sections (e.g., a stream with flood plains) is called a compound channel in this text.

in which y = flow depth and q = discharge per unit width. We know from calculus that $dE/dy = 0$ for E to be minimum or maximum for a given q. Hence, differentiating Eq. 3-2 with respect to y and equating the resulting expression to zero, we obtain

$$\frac{dE}{dy} = 1 - \frac{q^2}{gy^3} = 0 \tag{3-3}$$

According to the preceding definition, the depth at which E is minimum is called the critical depth, y_c. In Section 3-6 we discuss that this definition is not sufficient to determine the critical depths in a compound channel.

It follows from Eq. 3-3 that

$$y_c = \sqrt[3]{\frac{q^2}{g}} \tag{3-4}$$

When $dE/dy = 0$, E may be minimum or maximum. For E to be minimum, d^2E/dy^2 is positive at that depth. Let us prove that this is the case when $y = y_c$. By differentiating Eq. 3-3 with respect to y, we obtain

$$\frac{d^2E}{dy^2} = \frac{3q^2}{gy^4} \tag{3-5}$$

On the basis of Eq. 3-4, this equation becomes

$$\frac{d^2E}{dy^2} = \frac{3}{y_c} \tag{3-6}$$

The right-hand side of Eq. 3-6 is always positive. Hence, E is minimum at $y = y_c$ for a given value of q.

We will derive in the following paragraphs three important properties of critical flows from Eq. 3-4:

1. It follows from Eq. 3-4 that

$$q^2 = gy_c^3 \tag{3-7}$$

By designating V_c as the flow velocity at critical flow, Eq. 3-7 may be written as

$$\frac{V_c^2}{2g} = \frac{1}{2}y_c \tag{3-8}$$

Hence, the velocity head in critical flow is *one-half* of the critical depth.

2. By substituting Eq. 3-8 into Eq. 3-1, we obtain

$$E = y_c + \frac{1}{2}y_c$$

or

$$y_c = \frac{2}{3}E \tag{3-9}$$

i.e., the critical depth is equal to *two-thirds* of the specific energy.

3. It follows from Eq. 3-8 that

$$\frac{V_c^2}{gy_c} = 1$$

or the Froude number is

$$\mathbf{F}_r = \frac{V_c}{\sqrt{gy_c}} = 1 \tag{3-10}$$

This equation shows that the Froude number, $\mathbf{F}_r = 1$ at critical flow.

Unit discharge.   To determine the variation of unit discharge q with y for a specified value of E, let us rewrite Eq. 3-2 as

$$q^2 = 2gEy^2 - 2gy^3 \tag{3-11}$$

It is clear from this equation that $q = 0$ when $y = 0$, and also when $y = E$. Thus we have two points on the q-y curve for a given E. To study the shape of this curve, let us determine the locations of the maxima and minima of this curve and the value of q at these points. For q to be maximum or minimum, $dq/dy = 0$. Hence, differentiating Eq. 3-11 with respect to y and simplifying, we obtain

$$q\frac{dq}{dy} = gy(2E - 3y) \tag{3-12}$$

Equating the derivative to zero and simplifying, we obtain

$$y(2E - 3y) = 0 \tag{3-13}$$

Equation 3-13 has two roots: $y = 0$ and $y = \frac{2}{3}E$. We showed previously that $q = 0$ when $y = 0$. Hence, we will not gain any more information by studying this root further. The second root yields the same depth as the critical depth (see Eq. 3-9). To verify whether the flow is maximum or minimum at this depth, we have to determine the sign of d^2q/dy^2. Differentiating Eq. 3-12 with respect to y, we get

$$q\frac{d^2q}{dy^2} + \left(\frac{dq}{dy}\right)^2 = 2gE - 6gy \tag{3-14}$$

Substitution of $dq/dy = 0$ and $y = \frac{2}{3}E$ into this equation yields

$$\frac{d^2q}{dy^2} = -\frac{2gE}{q} \tag{3-15}$$

It is clear from this equation that the second derivative of q with respect to y is always negative. Hence, for a given E, the unit discharge, q, is maximum at critical depth, y_c. An expression for the maximum discharge may be obtained by substituting $y = \frac{2}{3}E$ into Eq. 3-11 and then simplifying the resulting expression. This procedure yields

$$q_{max}^2 = \frac{8}{27}gE^3 \tag{3-16}$$

Based on the preceding information, a typical q-y curve for a specified E may be plotted as shown in Fig. 3-1. The q-y curves for two other values of specific energy, such that $E_1 < E < E_2$, are also shown in this figure.

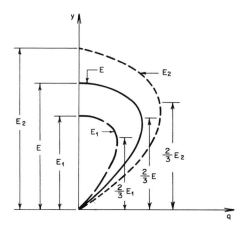

Figure 3-1   Variation of unit discharge.

Specific force.   As we discussed in Chapter 2, the expression for specific force, F_s, for a rectangular channel is

$$F_s = \frac{q^2}{gy} + \frac{1}{2}y^2 \tag{3-17}$$

The maxima and minima for the F_s-y curve may be determined as follows:

By differentiating Eq. 3-17 with respect to y and equating the resulting expression to zero, we get

$$\frac{dF_s}{dy} = -\frac{q^2}{gy^2} + y = 0 \tag{3-18}$$

Noting that $V = q/y$, this equation may be written as

$$\frac{V^2}{2g} = \frac{1}{2}y \tag{3-19}$$

This equation is same as Eq. 3-8, which is valid when the flow is critical. To determine whether F_s is maximum or minimum at critical depth, let us differentiate Eq. 3-18 with respect to y again and substitute $y = y_c$. This procedure yields

$$\frac{d^2F_s}{dy^2} = 1 + \frac{2q^2}{gy_c^3} \tag{3-20}$$

Since the right-hand side of this equation is always positive, the specific force is *minimum* at critical depth.

Wave celerity.   So far our discussion of critical flows has been in terms of the specific energy and the specific force. Another parameter of great importance in free-surface flows is the celerity of a small wave. The *celerity* is defined as the wave velocity with respect to the velocity of the medium in which the wave is traveling.

To derive an expression for the wave celerity, let us consider a small wave in a horizontal, frictionless channel, as shown in Fig. 3-2(a). The wave is considered to be small if $|\delta y| << y$. Let us assume that this wave is traveling in the downstream direction with absolute wave velocity V_w and that, as a result of the wave motion, the flow velocity is changed from V to $V + \delta V$. By superimposing a constant velocity V_w in the upstream direction, we may transform the unsteady flow of Fig. 3-2(a) to the steady flow of Fig. 3-2(b).

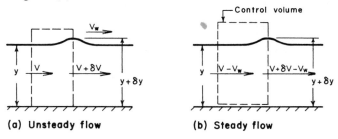

(a) **Unsteady flow**   (b) **Steady flow**

Figure 3-2 Definition sketch for wave propagation.

Let the thickness of the control volume perpendicular to the paper be unity. Since the channel is horizontal, the component of the weight of water in the control volume in the downstream direction is zero. Similarly, there is no shear force acting on the channel boundary, since the channel is assumed to be frictionless. Thus, the pressure force F_1, acting on the upstream side of the control volume, and the pressure force F_2, acting on the downstream side, are the only forces acting on the control volume in the flow direction. Expressions for these forces are

$$F_1 = \frac{1}{2}\rho g y^2 \tag{3-21}$$

$$F_2 = \frac{1}{2}\rho g (y + \delta y)^2 \tag{3-22}$$

in which ρ = mass density of water. Hence, the resultant force, F_r, acting on the control volume is

$$F_r = F_1 - F_2$$
$$= \frac{1}{2}\rho g \left[y^2 - (y + \delta y)^2 \right] \tag{3-23}$$

Now,

Time rate of change of momentum =
$$\rho y (V - V_w)[(V + \delta V - V_w) - (V - V_w)] \tag{3-24}$$

By equating the resultant force to the time rate of change of momentum and dividing throughout by ρg, it follows from Eqs. 3-23 and 3-24 that

$$\frac{y}{g}(V - V_w)\left[(V + \delta V - V_w) - (V - V_w)\right] = \frac{1}{2}\left[y^2 - (y + \delta y)^2\right] \tag{3-25}$$

By neglecting the higher-order terms, this equation may be simplified as

$$(V - V_w)\delta V = -g\delta y \tag{3-26}$$

The continuity equation for Fig. 3-2(b) may be written as

$$y(V - V_w) = (y + \delta y)(V + \delta V - V_w) \tag{3-27}$$

Neglecting the higher-order terms and simplifying, this equation becomes

$$y\delta V = -(V - V_w)\delta y \tag{3-28}$$

Combining Eqs. 3-26 and 3-28, we obtain

$$(V - V_w)^2 = gy$$

or

$$V_w = V \pm \sqrt{gy} \tag{3-29}$$

By definition, celerity, c, is the wave velocity relative to the medium in which the wave is traveling,— i.e., $V_w = V \pm c$. Hence, it follows from Eq. 3-29 that

$$c = \sqrt{gy} \tag{3-30}$$

We proved in a previous section that the Froude number $\mathbf{F}_r = 1$ when the flow is critical. By substititing the expression for $\mathbf{F}_r$ and using subscript c to denote various quantities for critical flow, we obtain

$$\frac{V_c}{\sqrt{gy_c}} = 1$$

or

$$V_c = \sqrt{gy_c} \tag{3-31}$$

Hence, on the basis of Eqs. 3-30 and 3-31, we may write

$$V_c = c \tag{3-32}$$

Thus, the celerity of a small wave is equal to the flow velocity when the flow is critical. Since $V < V_c$ in subcritical flows, it follows that $V < c$ in these flows. Similarly, we may prove that $V > c$ in supercritical flows.

Three different flow situations for the propagation of a disturbance are possible, depending upon the relative magnitudes of V and c, i.e., whether the flow is subcritical, critical, or supercritical. These three cases are shown in Fig. 3-3. In subcritical flow, the wave travels in the upstream and downstream directions at velocities $(V - c)$ and $(V + c)$, respectively, since $V < c$, as shown in Fig. 3-3(a). In critical flow, since $c = V$, the upper end of the wave remains stationary, and only the downstream end travels in the downstream direction at velocity $V + c$ (Fig. 3-3b). In supercritical flow, since $V > c$, the upstream and the downstream ends travel in the downstream direction at

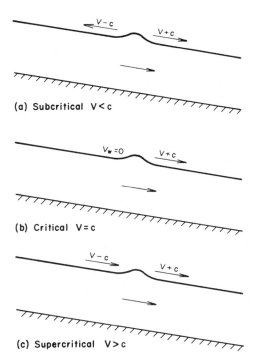

(a) Subcritical $V < c$

(b) Critical $V = c$

(c) Supercritical $V > c$ **Figure 3-3** Wave propagation.

velocities $(V - c)$ and $(V + c)$ respectively (Fig. 3-3c). In other words, supercritical flow carries the wave downstream and the wave does not travel in the upstream direction.

This property, which defines the direction of travel of a disturbance, may be utilized in the field to determine the type of flow as follows. Generate a small-amplitude disturbance on the flow surface by throwing a small object and note the directions in which this disturbance travels. If the disturbance travels both in the upstream and in the downstream directions, then the flow is subcritical. However, if the flow carries the disturbance in the downstream direction, then the flow is supercritical.

Whether the disturbance in a flow travels in the upstream direction or not has some practical significance. For example, since a disturbance does not travel in the upstream direction in supercritical flow, it means that the flow upstream of a specified location does not "know" what is happening on the downstream side of that location. In other words, to change the flow conditions at a section, flow conditions must be changed at an upstream location. In hydraulic-engineering jargon, this condition is referred to as the supercritical flows have upstream *control*. Following a similar argument, we can show that subcritical flows have downstream control.

The question might be asked, Can we keep on changing flow conditions at a downstream location in supercritical flow without affecting the flow upstream of that location? The answer to this question depends upon what happens to the flow upstream of that location: If the flow changes to subcritical flow, then the upstream flow is affected. However, if the flow conditions remain supercritical, then the flow conditions do not change upstream of that location.

Regular Nonrectangular Cross Section

Let us now develop relationships for critical flow in a prismatic channel having regular, nonrectangular cross sections (e.g., trapezoidal, triangular, circular, parabolic, etc.). We call a cross section regular if the top water-surface width, B, is a continuous function of y and there are no flood plains or overbanks. Channels having flood plains are considered in Section 3-6.

Specific energy. To simplify the derivation, let us assume that the pressure distribution is hydrostatic and the velocity is uniform. Then, the specific energy,

$$E = y + \frac{Q^2}{2gA^2} \tag{3-33}$$

For E to be minimum, $dE/dy = 0$. Hence, differentiating Eq. 3-33 with respect to y and equating it to zero, we obtain

$$\frac{dE}{dy} = 1 - \frac{Q^2}{gA^3}\frac{dA}{dy} = 0 \tag{3-34}$$

Since $dA/dy = B$ for a regular cross section, Eq. 3-34 may be written as

$$1 - \frac{BQ^2}{gA^3} = 0$$

or

$$\frac{V^2}{2g} = \frac{D}{2} \tag{3-35}$$

in which $D = A/B$ is defined as the *hydraulic depth*. If we differentiate Eq. 3-34 with respect to y again, we can show that d^2E/dy^2 is positive provided $3B^2/A > dB/dy$, a condition which is usually satisfied. Therefore, E is minimum at a depth when $dE/dy = 0$. We refer to this depth as the *critical depth*. It is clear from Eq. 3-35 that the velocity head is one-half the hydraulic depth when the flow is critical. Recall that the velocity head was one-half the flow depth in a rectangular cross section.

For critical flow, $\mathbf{F}_r = 1$. Hence, we can derive the following expression for $\mathbf{F}_r$ from Eq. 3-35:

$$\mathbf{F}_r = \frac{V}{\sqrt{gD}} \tag{3-36}$$

For flows in a steep channel having nonuniform velocity, the following expression for $\mathbf{F}_r$ may be derived by introducing the velocity-head coefficient, α, and the slope of the channel bottom:

$$\mathbf{F}_r = \frac{V}{\sqrt{gD\cos\theta/\alpha}} \tag{3-37}$$

Specific force. Now, let us prove that the specific force, F_s, is minimum when the flow is critical. As we discussed in Chapter 2,

$$F_s = \frac{Q^2}{gA} + \bar{z}A \tag{3-38}$$

For F_s to be minimum, $dF_s/dy = 0$ and $d^2F_s/dy^2 > 0$. By differentiating Eq. 3-38 with respect to y, we obtain

$$\frac{dF_s}{dy} = \frac{d}{dy}\left(\frac{Q^2}{gA}\right) + \frac{d}{dy}(\bar{z}A) = 0 \tag{3-39}$$

Let us now consider the terms of this equation one by one. Since Q is constant,

$$\begin{aligned}\frac{d}{dy}\left(\frac{Q^2}{gA}\right) &= -\frac{Q^2}{gA^2}\frac{dA}{dy} \\ &= -\frac{BQ^2}{gA^2}\end{aligned} \tag{3-40}$$

The derivative of the second term may be evaluated as follows. The moment of area A about the top water surface is $\bar{z}A$. Referring to Fig. 3-4, the change in the moment of area A due to a small change in the flow depth, Δy, about the top water surface is

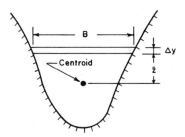

Figure 3-4 Definition sketch for $\Delta(\bar{z}A)$.

$$\Delta(\bar{z}A) = [A(\bar{z} + \Delta y) + (B\Delta y)\frac{1}{2}\Delta y] - \bar{z}A \tag{3-41}$$

By neglecting terms of higher order, this equation becomes

$$\Delta(\bar{z}A) = A\Delta y \tag{3-42}$$

In the limit, as $\Delta y \to 0$, this equation may be written as

$$d(\bar{z}A) = A\, dy \tag{3-43}$$

By substituting Eqs. 3-40 and 3-43 into Eq. 3-39 and simplifying the resulting equation, we obtain

$$\frac{V^2}{2g} = \frac{D}{2} \tag{3-44}$$

This condition is satisfied when the flow is critical. Hence, the specific force is minimum at critical depth. Note that we have not proved the necessary condition for F_s to be minimum, i.e., that $d^2 F_s/dy^2$ is positive when $y = y_c$. This is left as an exercise for the reader.

3-3 Application of Critical Flow

It is clear from the preceding properties that the discharge and flow depth have a unique relationship when the flow is critical (Eqs. 3-4 and 3-35). Utilizing this unique relationship, several flow-measuring devices have been developed. These devices are called *critical-flow meters*.* As we discussed in Section 3-2, the discharge is maximum when the flow is critical for a given specific energy. Therefore, the length of bridges and other structures on a channel may be reduced by producing critical flow at that section (see Example 3-1).

Critical flow may be produced in a channel by raising the channel bottom, by reducing the channel width, or by a combination of these measures. Let us now discuss how to determine the size of bottom step or the magnitude of channel contraction to produce critical flow.

Constant-width Channel with Bottom Step

Figure 3-5 shows the variation of water surface due to a step rise in the channel bottom. We showed in Chapter 2 that the water level at the step rises if the flow upstream of the step is supercritical, and it drops if the flow is subcritical. We may ask the question, Is there any upper limit on the size of the step such that the upstream water levels are not affected? A casual look at the specific-energy diagram indicates that there is a limit. As we raise the channel bottom, the point on the specific-energy curve moves toward point C, which corresponds to the critical flow. Thus, if we have subcritical flow at section 1, then the maximum height of this step, $(\Delta z)_{max}$, is equal to $E_1 - E_c$, as shown in Fig. 3-5. In this expression, the subscripts 1 and c refer to section 1 and critical flow, respectively. Raising the bottom further requires a further reduction in the specific energy. However, that is not possible, since E is minimum when the flow is critical. Therefore, if we raise the bottom level more than this maximum amount, either the unit discharge is reduced if the upstream water level is constant or the upstream water level is raised to produce an increase in the specific energy to yield the specified discharge. Similarly, if the flow is supercritical upstream of the step, then there is an upper limit on the height of the step that will not affect the upstream water level or the channel discharge. Referring to Fig. 3-5, this limiting height is again $(\Delta z)_{max}$ if the flow depth at section 1 is y_2.

Horizontal, Variable-width Channel

Figure 3-6 shows the variation of water level produced by a reduction in the channel width with the channel bottom remaining horizontal. The water depth decreases when

---

*Critical-flow meters are discussed in Chapters 7 and 10.

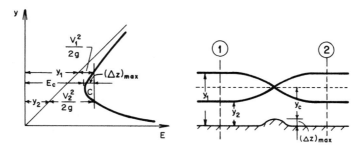

Figure 3-5 Water-surface variation for a bottom step.

the width decreases if the upstream flow is subcritical and it increases if the flow upstream of the constriction is supercritical. Similar to the step rise discussed previously, there is an upper limit by which the channel width may be contracted. We may reduce the channel width until critical flow is produced at the constriction. A further reduction in the channel width either reduces the unit discharge or raises the upstream water level.

The constriction in the channel width, a step rise in the channel bottom, or a combination of these two, such that the upstream water level for a specified discharge is influenced, is called a *choke*.

Example 3-1

A bridge is planned on a 50-m-wide rectangular channel carrying a flow of 200 m³/s at a flow depth of 4.0 m. For reducing the length of the bridge, what is the minimum channel width such that the upstream water level is not influenced for this discharge?

Given:
$$Q = 200 \text{ m}^3/\text{s}$$
$$B = 50 \text{ m}$$
$$y = 4.0 \text{ m}$$

Determine:
Minimum channel width at the bridge site = ?

Solution:

$$V = \frac{200}{50 \times 4}$$
$$= 1.0 \text{ m/s}$$

$$E = 4. + \frac{(1)^2}{2 \times 9.81}$$
$$= 4.05 \text{ m}$$

For the discharge to be maximum at the bridge site for a given upstream specific energy of 4.05 m, the flow should be critical. Hence, it follows from Eq. 3-9 that

$$y_c = \frac{2}{3} E$$
$$= \frac{2}{3} \times 4.05$$
$$= 2.7 \text{ m}$$

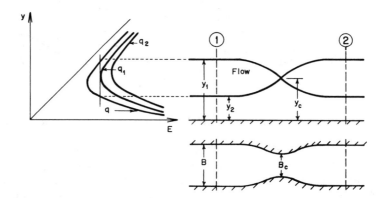

Figure 3-6 Water-level variation in a variable-width, horizontal channel.

The unit discharge q corresponding to this critical depth may be computed from Eq. 3-4; i.e.,

$$q = \sqrt{gy_c^3}$$
$$= \sqrt{9.81 \times (2.7)^3}$$
$$= 13.9 \, \text{m}^3/\text{s/m}$$

The width that gives this unit discharge is

$$B_c = \frac{200}{13.9}$$
$$= 14.4 \, \text{m}$$

Therefore, the channel width may be reduced from 50 to 14.4 m without affecting the upstream level for a flow of 200 m³/s.

3-4 Location of Critical Flow

In the previous section, we showed that critical flow may be produced in a channel by raising the channel bottom and/or by decreasing the channel width.* However, we did not discuss where the critical depth will occur. For this purpose, we will consider a rectangular channel: first, a constant-width channel having a variable bottom level, followed by a horizontal channel having a variable width.

By neglecting the losses in a transition, we derived the following equation in Chapter 2 for a *rectangular channel* having constant width (Eq. 2-53) and variable bottom level:

$$\frac{dz}{dx} = \left(\mathbf{F}_r^2 - 1 \right) \frac{dy}{dx} \qquad (3\text{-}45)$$

---

*The occurrence of critical flow at a change in the bottom slope and at a free overfall is discussed in Chapter 5.

It is clear from this equation that the right-hand side is equal to zero when the flow is critical, i.e., when $\mathbf{F}_r = 1$ or when $dy/dx = 0$. Hence, the flow is critical at a point where $dz/dx = 0$, i.e., at the highest point of the step. The proof that it is not at the lowest point of the step is left as an exercise for the reader (*Hint:* The second derivative of z with respect to x must be negative at the highest point).

For a horizontal rectangular channel having a variable width, the total head,

$$H = z + y + \frac{1}{2g}\left(\frac{Q}{By}\right)^2 \tag{3-46}$$

By differentiating this equation with respect to x, assuming there are no losses (i.e., $dH/dx = 0$), and there is no lateral inflow or outflow, and noting that $B = B(x)$, we obtain

$$\frac{dy}{dx} + \frac{Q^2}{2g}\frac{d}{dx}\left(\frac{1}{(By)^2}\right) = 0 \tag{3-47}$$

Upon expansion of the second term, this equation becomes

$$\frac{dy}{dx} - \frac{Q^2}{gB^2y^3}\frac{dy}{dx} - \frac{Q^2}{gy^2B^3}\frac{dB}{dx} = 0 \tag{3-48}$$

By definition $\mathbf{F}_r^2 = Q^2/(gB^2y^3)$. Hence we may write this equation as

$$\left(1 - \mathbf{F}_r^2\right)\frac{dy}{dx} - \mathbf{F}_r^2\frac{y}{B}\frac{dB}{dx} = 0 \tag{3-49}$$

For critical flow, $\mathbf{F}_r = 1$. Hence, it follows from Eq. 3-49 that $dB/dx = 0$. In other words, critical flow occurs at the point of minimum channel width. The reader may prove that it does not occur where the channel width is maximum.

3-5 Computation of Critical Depth

For the analysis and design of open channels, it is necessary to know the critical depth. Procedures for computing the critical depth in a channel having a regular cross section are discussed in this section, and a procedure for determining the critical depths in a compound channel is presented in Section 3-6.

The critical depth for a specified discharge may be computed from the equation $\mathbf{F}_r = 1$. The effects of nonuniform velocity distribution may be considered by including the velocity-head coefficient, α. The channel bottom slope may be large. Then, based on Eq. 3-37, this equation becomes

$$\frac{V}{\sqrt{gD\cos\theta/\alpha}} = 1 \tag{3-50}$$

Since $Q = VA$, Eq. 3-50 becomes

$$\frac{Q/A}{\sqrt{gD\cos\theta/\alpha}} = 1 \tag{3-51}$$

Let us define the section factor as $Z = A\sqrt{D}$. Then this equation may be written as

$$Z = A\sqrt{D} = \frac{Q/\sqrt{\cos\theta}}{\sqrt{g/\alpha}} \tag{3-52}$$

The left-hand side of this equation is a function of the properties of the channel cross section and the value of y_c. Thus there is only one critical depth for a specified discharge in a given channel if $A\sqrt{D}$ for the channel cross section increases monotonically with y. Or, critical flow at a given value of y_c in a channel is possible only for one value of discharge. In Section 3-6 we discuss how multiple critical depths are possible for a specified discharge in a compound channel.

An explicit relationship may be derived (see Prob. 3-11) to determine the critical depth in a rectangular, triangular, or parabolic channel. For general applications, however, the critical depth may be determined by solving Eq. 3-52 using design curves (Chow 1959), by a trial-and-error procedure, or by using numerical methods for the solution of a nonlinear algebraic equation. These procedures are discussed in the following paragraphs.

Design Curves

Design curves are presented in Fig. 3-7. Let $Z_c = A\sqrt{D}$, where Z_c = section factor for the critical depth. If we want to determine the critical depth for a specified discharge, then we know the values of Q, θ, and α. Therefore, we can compute the left-hand side of Eq. 3-52. Let us divide this computed value by $B_o^{2.5}$ for a trapeziodal cross section and by $D_o^{2.5}$ for a circular section (B_o = channel bottom width and D_o = conduit diameter). The resulting value is then equal to $Z_c/B_o^{2.5}$ or $Z_c/D_o^{2.5}$, depending upon the cross section. Now, y_c/B_o or y_c/D_o may be read directly from Fig. 3-7, corresponding to this value of $Z_c/B_o^{2.5}$ or $Z_c/D_o^{2.5}$.

Trial-and-Error Procedure

In the trial-and-error procedure, we substitute expressions for flow area, A, and hydraulic depth, D, for the channel cross section into Eq. 3-52 and then solve the resulting equation by trial-and-error.

Numerical Methods

Several numerical methods are available for the solution of a nonlinear algebraic equation (McCracken and Dorn 1964)— e.g., bisection method, Newton method, secent method, and method of successive approximations. Of these methods, we will present only the application of the Newton method (For the solution of a system of equations, this method is called the Newton-Raphson method).

To determine the roots of an algebraic equation by the Newton method, we write the equation as

$$F(y) = 0 \tag{3-53}$$

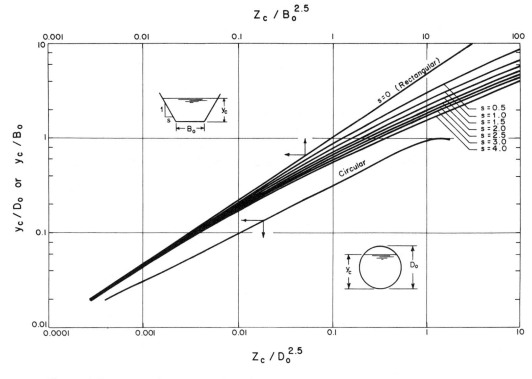

$Z_c / B_o^{2.5}$

$Z_c / D_o^{2.5}$

Figure 3-7 Curves for computation of critical depth (After Chow, *Open-Channel Hydraulics*, 1959).

Let us substitute $D = A/B$ into Eq. 3-52 and rewrite it as

$$F(y) = A^{3/2} B^{-1/2} - \frac{Q/\sqrt{\cos \theta}}{\sqrt{g/\alpha}} = 0 \tag{3-54}$$

To solve Eq. 3-54 by the Newton method, we need the expression for dF/dy. An expression for this derivative may be obtained by differentiating Eq. 3-54 with respect to y and noting that $dA/dy = B$; i.e.,

$$\frac{dF}{dy} = \frac{3}{2} A^{1/2} B B^{-1/2} - \frac{1}{2} A^{3/2} B^{-3/2} \frac{dB}{dy} \tag{3-55}$$

This equation may be simplified as

$$\frac{dF}{dy} = \frac{3}{2} A^{1/2} B^{1/2} - \frac{1}{2} \left(\frac{A}{B} \right)^{3/2} \frac{dB}{dy} \tag{3-56}$$

For a trapezoidal section having side slopes of 1 vertical to s horizontal, $dB/dy = 2s$. For any other channel section, an expression for dB/dy may be obtained similarly.

To start the iterative procedure, we need an initial estimate for y_c. The number of iterations are reduced considerably if this estimate is close to the actual value of y_c. For such an initial estimate, Eq. 3-52 may be approximately solved by assuming the channel as rectangular. For example, an approximate value for the initial estimate for a trapezoidal channel is

$$y_c = \left(\frac{Q^2}{g B_o^2} \right)^{1/3} \tag{3-57}$$

in which B_o = bottom width.

The following example illustrates these procedures. First, the design curves are used; then a trial-and-error procedure is employed. The computer programs of Appendices B-1 and B-2 illustrate the application of the Newton and the bisection methods.

Example 3-2

Compute the critical depth in a trapezoidal channel for a flow of 30 m^3/s. The channel bottom width is 10.0 m, side slopes are 2H : 1V, the bottom slope is negligible, and $\alpha = 1$.

Given:

B_o = 10.0 m
s = 2
θ = 0.0
Q = 30 m^3/s
α = 1

Determine:

y_c = ?

Solution:

Design Curves.   Substituting the values of Q, θ, g, and α into the left-hand side of Eq. 3-52, we obtain

$$
\begin{aligned}
Z_c &= \frac{Q/\sqrt{\cos\theta}}{\sqrt{g/\alpha}} \\
&= \frac{30/\sqrt{\cos 0}}{\sqrt{9.81/1}} \\
&= 9.58
\end{aligned}
$$

Now,

$$\frac{Z_c}{B_o^{2.5}} = \frac{9.58}{(10)^{2.5}} = 0.030$$

Corresponding to $Z_c/B_o^{2.5} = 0.030$ on the abcissa and for $s = 2$, we read the ordinate of Fig. 3-7 as

$$\frac{y_c}{B_o} = 0.09$$

Therefore,

$$y_c = 0.09 \times 10 = 0.9 \text{ m}$$

Trial-and-Error Procedure.   For critical flow,

$$\frac{Q/\sqrt{\cos\theta}}{\sqrt{g/\alpha}} = A\sqrt{D} \qquad (3\text{-}52)$$

By substituting the values of Q, g, θ, and α into this equation, we obtain

$$A\sqrt{D} = \frac{120}{\sqrt{9.81}} = 9.58$$

Now, we have to determine the flow depth y_c for which $A\sqrt{D}$ for the specified channel cross section is 9.58. By substituting the specified values into the expressions for the channel properties of a trapezoidal section, we obtain

$$\begin{aligned}
A &= \frac{1}{2}(10.0 + 10.0 + 4.0y_c)y_c \\
&= (10.0 + 2.0y_c)y_c \\
B &= 10.0 + 4.0y_c \\
D &= \frac{A}{B} \\
&= \frac{(10.0 + 2.0y_c)y_c}{10.0 + 4.0y_c}
\end{aligned}$$

Substituting these expressions for A and D into Eq. 3-52 and simplifying, we obtain

$$A\sqrt{D} = (10.0 + 2.0y_c)y_c\sqrt{\frac{(10.0 + 2.0y_c)y_c}{10.0 + 4.0y_c}} = 9.58$$

Upon simplification, this equation becomes

$$\left[(10.0 + 2.0y_c)y_c\right]^{3/2} - 9.58\sqrt{10.0 + 4.0y_c} = 0$$

By solving this equation by trial and error, we obtain

$$y_c = 0.91\,\text{m}$$

Numerical Methods.   Computer programs of Appendices B-1 and B-2 for computing the critical depth in a trapeziodal channel use the Newton method and the bisection method, respectively. The initial estimate of YI = 5.0 m for the Newton method was determined from Eq. 3-57. The initial estimated values for YR and YL for the bisection method are arbitrarily assigned as 0.1 and 10 m. The critical depth computed by these methods is 0.91 m.

3-6 Critical Depths in Compound Channels

It is possible to have critical flow at more than one depth in a channel with a compound cross section. In this section, we will discuss how to determine the total number of critical depths for a given discharge in a compound channel and how to compute their values one by one.

General Remarks

We proved in Section 3-2 that the specific energy at critical depth is minimum for a given discharge in channels having regular cross sections (rectangular, trapezoidal, circular, etc.). This characteristic of critical flow has been widely used to determine the critical depth in a channel. For regular channel cross sections, the specific energy for a specified discharge is minimum only at one depth; hence these procedures give only one critical depth. However, several investigators (Blalock and Sturm 1981, 1983; Konemann 1982; Petryk and Grant 1978) have analytically and experimentally shown that there may be more than one critical depth for channels with overbank or floodplain flow (Fig. 3-8). Schoellhamer, Peters, and Larock (1985) studied the problem using separate Froude numbers for the main channel and for the floodplains. It was shown that the Froude number for the main channel for their experimental section was equal to 1 at two different depths, thereby indicating that there is more than one critical depth. Because of the possibility of multiple critical depths, it is necessary to determine their values in order to compute correctly the water-surface flow profiles. Blalock and Sturm (1981) outlined difficulties associated with several available methods for critical-depth computations presented by Petryk and Grant (1978), Soil Conservation Service (1976), U.S. Army Corps of Engineers (1982), and U.S. Geological Survey (1976). They defined a compound-channel Froude number for the entire section. The flow depths at which specific energy was minimum were called critical depths. Their Froude number correctly locates the minima on the specific energy diagram. However, the flow may be critical even when the specific energy is not minimum, as shown by Chaudhry and Bhallamudi (1988).

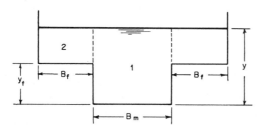

Figure 3-8 Compound-channel cross section.

Only one critical depth is determined in a compound channel section using the algorithms presented by the U.S. Army Corps of Engineers (1982), the Soil Conservation Service (1976), and Davidian (1984). Blalock and Sturm (1981) defined a Froude number that correctly locates the points of minimum specific energy. We present an algorithm that first determines the possible number of critical depths in a cross section for a given discharge and then computes their values one by one in an efficient manner.

In this section, we present the necessary expressions and outline a computational procedure to determine the critical depths in a compound channel. These expressions are derived for a symmetric cross section by assuming small bottom slope and hydrostatic pressure distribution even near the critical depth (For the detailed derivation of these expressions, the reader should see Chaudhry and Bhallamudi, 1988).

We first determine the characteristic directions of the governing equations (continuity and momentum) and write an expression for the compound channel Froude number, $\mathbf{F}_{cr}$. For critical flow, $\mathbf{F}_{cr} = 1$. By substituting expressions for the momentum coefficient in terms of the parameters for the main channel and for the flood plains into this equation and introducing the nondimensional parameters

$$
\begin{aligned}
y_r &= \frac{y}{y_f} \\
b_r &= \frac{B_f}{B_m} \\
b_f &= \frac{B_f}{y_f} \\
n_r &= \frac{n_m}{n_f}
\end{aligned}
\tag{3-58}
$$

we obtain

$$
\frac{g B_m^2 y_f^3}{Q^2} = C
\tag{3-59}
$$

in which

$$
\begin{aligned}
C &= \frac{1}{y_r + 2b_r(y_r - 1)} \left[\left(\frac{m}{y_r} \right)^2 + \left(\frac{1-m}{y_r - 1} \right)^2 \frac{1}{2b_r} \right] \\
&+ \frac{2m(1-m)}{3[y_r + 2b_r(y_r - 1)]} \left(\frac{5}{y_r(y_r - 1)} - \frac{2}{b_f + y_r - 1} \right) \\
&\times \left[\left(\frac{m}{y_r} \right) - \left(\frac{1-m}{y_r - 1} \right) \frac{1}{2b_r} \right]
\end{aligned}
\tag{3-60}
$$

and

$$
m = \frac{1}{1 + 2n_r \left(A_2 / A_1 \right)^{\frac{5}{3}} \left(P_1 / P_2 \right)^{\frac{2}{3}}}
\tag{3-61}
$$

It is clear from Eq. 3-60 that C is a function of y_r, n_r, b_r, and b_f. For any channel cross section, the C-y_r relationship may be plotted as shown in Fig. 3-9. Those values of y_r at which $C = k = (g B_m^2 y_f^3)/Q^2$ are the critical depths for flows over the flood plain; i.e., the abscissa of the intersection point of this C-y_r curve with the horizontal line $C = k$ gives the value of y_r corresponding to the critical depth.

A procedure for solving Eq. 3-59 in order to determine the critical depths is illustrated by the following example.

Example 3-3

Determine the critical depths for $Q = 1.7$, 2.5, and 3.5 m^3/s in a channel having the following dimensions: $B_m = 1.0$ m; $B_f = 3.0$ m; $y_f = 1.0$ m; and $n_r = 0.9$.

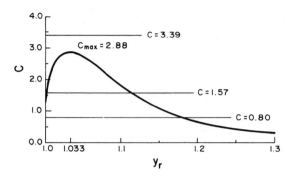

Figure 3-9 C-y_r curve for $b_m = 1$, $n_r = 0.9$, and $b_f = 3$.

Given:

$$B_m = 1 \text{ m}$$
$$B_f = 3 \text{ m}$$
$$y_f = 1 \text{ m}$$
$$n_r = 0.9$$

Determine:

Critical depths for three rates of discharge = ?

Solution: The C-y_r curve for this cross section is plotted in Fig. 3-9 for $b_m = B_m/y_f = 1.0$, $b_f = 3.0$ and $n_r = 0.9$. When $y_r \to 1$, $m \to 1$ and $C \to 1$. When $y_1 \to \infty$, $m \to 0$ and $C \to 0$. These properties of the C-y_r curve are clear from Fig. 3-9. In addition, this curve has a maximum value, C_{max}, equal to 2.88, which occurs at $y_r = 1.033$. Depending upon the value of discharge (and hence k), three typical cases are possible. These are discussed in the following paragraphs.

$Q = 1.7 \ m^3/s$. For this discharge, $k = 3.39$. The horizontal line corresponding to $k = 3.39$ does not intersect the C-y_r curve (Fig. 3-9); therefore, there is no solution for Eq. 3-59—i.e., critical flow cannot occur for this discharge if the flow depth $y > y_f$. However, there is a critical depth for flow in the main channel only, i.e., for $y < y_f$. This critical depth, y_c, may be determined from the equation

$$y_c = \left(\frac{Q^2}{g B_m^2} \right)^{1/3} \tag{3-62}$$

For $Q = 1.7 \text{m}^3/\text{s}$ and for this channel cross section, $y_c = 0.67$ m. This corresponds to the minimum specific energy, i.e., point C on the curve for $Q = 1.7\text{m}^3/\text{s}$ of Fig. 3-10.

$Q = 2.5 \ m^3/s$. For this discharge, $k = 1.57$. Thus k is greater than 1.0 but less than C_{max} (= 2.88). The horizontal line corresponding to $k = 1.57$ intersects the C-y_r curve twice (Fig. 3-10), giving two critical depths : $y_{c_2} = 1.002$ m and $y_{c_3} = 1.12$ m. Since $k > 1$, critical flow also occurs when the flow is only in the main channel, at depth $y_{c_1} = [Q^2/(g B_m^2)]^{\frac{1}{3}} = 0.86$ m. Hence for this discharge there are three critical depths, $y_{c_1} = 0.86$ m, $y_{c_2} = 1.002$ m, and $y_{c_3} = 1.12$ m. Specific energy is locally minimum at two of these depths, i.e., at points C_1 and C_3, and locally maximum at the third, i.e., at point C_2, as shown by the curve for $Q = 2.5 \text{ m}^3/\text{s}$ in Fig. 3-10. In other words, one of the three possible critical depths would not have been computed if only the minimum specific

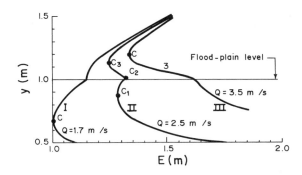

Figure 3-10 Specific energy vs depth curve for cross section of Fig. 3-8.

energy had been used as the criteria for determining the critical depths. The importance of this conclusion will become apparent while plotting the water-surface profiles in Chapter 5 (Examples 5-4 and 5-5).

$Q = 3.5 \, m^3/s$.   For this discharge, $k = 0.8$. Since $k < 1$, critical depth can not occur when the flow is only in the main channel. However, the horizontal line corresponding to $k = 0.8$ intersects the C-y_r curve at $y_r = 1.18$ (Fig. 3-10), giving critical depth $y_c = 1.18$ m. This is illustrated by the single minimum specific energy point C of the curve for $Q = 3.5 \text{m}^3/s$ in Fig. 3-10 occurring when the flow is over the banks.

Algorithm for Computing the Critical Depths

The following conclusions may be drawn from the typical cases of the preceding example:

1. For $k > C_{\max}$, there is only one critical depth, and it is less than the depth of the floodplains, y_f.
2. For $1 \leq k < C_{\max}$, three critical depths are possible. Two of these critical depths are for the flow over the banks and the third occurs when the flow is only in the main channel.
3. For $k < 1$, there is only one critical depth, and it is greater than the depth of the floodplains, y_f.

These conclusions may be utilized as follows to formulate an algorithm for computing the critical depths.

To solve Eq. 3-59 for y_r by an iterative procedure, we may write this equation as

$$y^* = \frac{2b_r}{2b_r + 1} + \frac{1}{C(2b_r + 1)}\left[\left(\frac{m}{y_r}\right)^2 + \left(\frac{1 - m}{y_r - 1}\right)^2 \frac{1}{2b_r}\right]$$
$$+ \frac{2m(1 - m)}{3C(2b_r + 1)}\left[\frac{5}{y_r(y_r - 1)} - \frac{2}{b_f + y_r - 1}\right]\left[\frac{m}{y_r} - \left(\frac{1 - m}{y_r - 1}\right)\frac{1}{2b_r}\right]$$

$$(3\text{-}63)$$

First, we estimate a value of y_r and then compute y^* from Eq. 3-63. If $|y^* - y_r|$ is less than a specified tolerance, then $y*$ is the correct value of y_r; otherwise y_r is set equal to $y*$ and the iterations are repeated until a solution is obtained.

The procedure for computing the critical depths for a specified discharge in a given channel (i.e., Q, B_m, B_f, y_f, n_m, and n_f are known) may be summarized as follows:

1. Calculate k.
2. If k is less than 1, solve Eq. 3-63 for y_r by using an iterative procedure, and then compute $y_c = y_r y_f$.
3. If k is greater than or equal to 1, follow steps 4 through 9.
4. Compute C_{max}. To do this, C is computed from Eq. 3-60 for different values of y_r, starting with an initial y_r value of approximately 1.0 and continuing until a maximum value of C is reached.
5. If k is greater than C_{max}, then $y_c = [Q^2/(gB_m^2)]^{1/3}$.
6. For k less than C_{max}, follow steps 7 through 9.
7. Compute y_{c_1} from the equation $y_{c_1} = [Q^2/(gB_m^2)]^{1/3}$.
8. Solve Eq. 3-63 for y_r using an iterative procedure. Then, $y_{c_3} = y_r y_f$.
9. Solve for y_{c_2}. To do this, C is computed for different values of y_r from Eq. 3-60, starting with an initial value of y_r close to 1.0 and continuing until the computed C value is equal to k. The value of y_r at which C is equal to k gives y_{c_2}/y_f. Then, $y_{c_2} = y_r y_f$.

3-7 Summary

In this chapter, several properties of critical flow were discussed and a number of expressions were derived for a rectangular cross section. Then, regular nonrectangular cross sections were considered. A number of engineering applications of critical flows were discussed. Mathematically, it was shown where critical depth occurs in a channel of varying width or bottom elevation. Three procedures for computing the critical depth in a regular cross section were presented. The chapter concluded with a discussion of the possibility of multiple critical depths in a compound channel and how to determine their values.

Problems

3-1. Derive an expression for the Froude number, $\mathbf{F}_r$ by assuming that the velocity-head coefficient, α, is a function of the flow depth.

3-2. Write a general-purpose computer program for computing the critical depth in a regular prismatic channel using: (i) the bisection method; (ii) the Newton method. Of these two methods, which method do you prefer and why?

Use this program to compute the critical depth in a circular conduit having a diameter of 4 m and carrying a discharge of 16.0 m³/s. Assume $\alpha = 1$.

3-3. A trapezoidal channel having a bottom width of 20 m and side slopes of 2H:1V is carrying 60 m³/s. Assuming $\alpha = 1.1$, determine the critical depth.

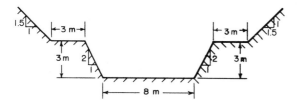

Figure 3-11 Cross section for Prob. 3-4.

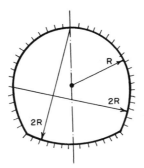

Figure 3-12 Horseshoe section.

3-4. For the channel section shown in Fig. 3-11, determine the critical depth for a flow of 80 m³/s.

3-5. For a discharge of 850 m³/s, compute the critical depth in a tunnel having a standard horseshoe section (Fig. 3-12). The flow area, A, top water-surface width, B, and hydraulic radius, R, at different flow depths, y, are listed in the following table:

y (m)	A (m²)	R (m)	B (m)
0	0.0	0.00	0.0
2	23.4	1.34	16.6
4	58.3	2.68	18.2
6	95.7	3.70	19.2
8	134.8	4.50	19.8
10	174.6	5.15	20.0
12	214.4	5.65	19.6
14	252.5	5.99	18.3
16	287.0	6.13	16.0
18	315.4	6.01	12.0
20	331.7	5.08	0.0

3-6. If y_1 and y_2 are the alternate depths in a rectangular channel, prove that

$$y_c = \sqrt[3]{\frac{2(y_1 y_2)^2}{y_1 + y_2}}$$

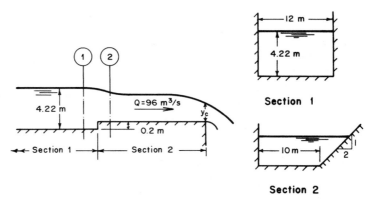

Figure 3-13   Drainage canal of Prob. 3-9

3-7. Derive expressions for the critical depth in a prismatic channel having the following cross sections and assuming in each case that the slope of the channel bottom is small:

   i. Trapezoidal
   ii. Triangular
   iii. Circular

3-8. A 50-m-wide rectangular channel is carrying a flow of 250 m^3/s at a flow depth of 5 m. To produce critical flow in this channel, determine

   i. The height of the step in the channel bottom if the width remains constant;
   ii. The reduction in the channel width if the channel-bottom level remains unchanged;
   iii. A combination of the width reduction and the bottom step.

3-9. The drainage canal shown in Fig. 3-13 has a flow of 96 m^3/s. If the flow depth at section 1 is 4.22 m, what is the depth at section 2? Assume there are no losses in the transition.

   Determine the flow depth at the downstream end if the canal ends in a free overfall. Assume that critical depth occurs at the overfall.

3-10. Write a general-purpose computer program to determine the critical depth in a channel for a specified discharge and for the following channel cross sections:

   i. Circular
   ii. Trapezoidal
   iii. Triangular
   iv. Horseshoe

3-11. Show that the critical depth in a channel having a triangular, rectangular, or parabolic cross section may be determined from the following explicit equation:

$$y_c = \left[\frac{Q^2}{g} \frac{(m+1)^3}{4k^2} \right]^{1/(2m+3)}$$

in which the x and y coordinates of the sides of the half-section may be defined as $x = ky^m$. For a triangular cross section, $k = s$ and $m = 1$; for a rectangular cross section, $k = \frac{1}{2}B_o$ and $m = 0$; and for the parabolic cross section, $k = (1/a)^n$ and $m = 1/n$, with the equation for the parabola being $y = ax^n$.

3-12. A mountain creek has a parabolic cross section with the top water-surface width of 9 ft at a depth of 3 ft. Determine the critical depth for a flow of 50 ft^3/sec.

3-13. An 8-ft-diameter concrete-lined sewer is laid at a bottom slope of 1 ft/mi. Compute the critical depth for a discharge of 100 ft^3/sec.

3-14. A trapezoidal irrigation channel is 10 ft wide at the bottom and has side slopes of 1V : 2H. For a flow of 100 ft³/sec, determine the critical depth.

3-15. A 5-ft-diameter circular culvert carries a flow of 15 ft³/sec. Determine the critical depth.

3-16. For the horseshoe tunnel shown in Fig. 1-17 ($d_o = 30$ ft), determine the critical depth for a discharge of 300 ft³/sec.

3-17. A 4-ft-diameter culvert barrel carries a flow of 10 ft³/sec and discharges freely into a lake. What is the depth just upstream of the free fall?

3-18. A high-level rectangular outlet at a dam is 10 ft wide, 15 ft high, and 500 ft long. The invert level at the entrance is at El. 122 and the botom slope is 0.001. A sluice gate at the entrance controls the flow through the outlet. If the coefficient of discharge, C_d, is 0.7 ($Q = C_d A \sqrt{2gH}$), determine the thrust on the gate for the reservoir level of El 182. The water level in the river into which the outlet discharges is at El. 131 for this flow.

3-19. In order to reduce the flow velocity at a section, a fisheries biologist tied a 6-in diameter tree log at the bottom of a stream. The flow velocity and the flow depth prior to the installation of the log were 2 ft/sec and 4 ft, respectively. Determine the change in the flow velocity and flow depth just downstream of the log.

3-20. For the channel of Problem 3-8, what is the minimum channel width without affecting the upstream water level.

3-21. Write a computer program to determine the critical depth in a circular cross section. Use the bisection and the Newton methods.

3-22. The flow velocity and flow depth in a 5-m-wide rectangular channel are 1.5 m/s and 4 m, respectively. Design a converging transition so that the flow is critical in the transition. Assume the channel bottom to be horizontal and losses in the transition to be negligible.

REFERENCES

Black, R. G. 1982. Discussion of Blalock and Sturm (1981). *Jour. Hydraulics Division*, Amer. Soc. Civil Engrs., 108, no. 6:798–800.

Blalock, M. E., and Sturm, T. W. 1981. Minimum specific energy in compound channel, *Jour. Hydraulics Division*, Amer. Soc. Civil Engrs. 107, no. 6:699–717. (See also closure, vol. 109 (1983):483–86.)

Chaudhry, M. H., and Bhallamudi, S. M. 1988. Computation of critical depth in compound channels, *Jour. Hydraulic Research*, Inter. Assoc. of Hydraulic Research, 26, no. 4:377–95.

Chow, V. T. 1959. *Open-channel hydraulics*. McGraw-Hill, New York, NY.

Davidian, J. 1984. Computation of water-surface profiles in channels. In *Techniques of Water-Resources Investigations*. U.S. Geological Survey, Book 3, Chapter A15.

French, R. H. 1985. *Open-channel hydraulics*, McGraw-Hill, New York, NY.

Henderson, F. M. 1966. *Open channel flow*, Macmillan, London.

Knight, D. W.; Demetriou, J. D.; and Hamed, M. E. 1984. Stage discharge relationships for compound channels. In *Channels and channel control structures,* edited by K. V. H. Smith, 4.21–4.35. Springer-Verlag, New York, NY.

Konemann, N. 1982. Discussion of Blalock and Sturm (1981). *Jour. Hydraulics Division*, Amer. Soc. Civil Engrs. 108, no. 3:462–64.

McCracken, D.D., and Dorn, W.S.   1964. *Numerical methods and FORTRAN programming,* John Wiley & Sons, New York, NY.

Myers, B. C., and Elsawy, E.M.   1975. Boundary shear in channel with flood plains, *Jour. Hydraulics Division,* Amer. Soc. Civil Engrs. 101, no. 7:933–46.

Petryk, S., and Grant, E. U.   1978. Critical flow in rivers with floodplains, *Jour. Hydraulics Division,* Amer. Soc. Civil Engrs. 104, no. 7:583–94.

Rajaratnam, N., and Ahmadi, R. M.   1979. Interaction between main channel and floodplain flows, *Jour. Hydraulics Division,* Amer. Soc. Civil Engrs. 105, no. 5:573–88.

Schoellhamer, D. H.; Peters, J. C.; and Larock, B. E.   1985. Subdivision Froude number. *Jour. Hydraulics Division,* Amer. Soc. Civil Engrs. 111, no. 7:1099–1104.

Soil Conservation Service.   1976. WSP-2 computer program. *Technical Release No. 61.*

U.S. Army Corps of Engineers   1982. HEC-2; water surface profile. *User's Manual.* Hydrologic Engineering Center, Davis, CA.

U.S. Geological Survey.   1976. Computer applications for step-backwater and floodway analysis. *Open File Report 76-499.*

Wright, R. R. and Carstens, M. R.   1970. Linear momentum flux to overbank sections. *Jour. Hydraulics Division,* Amer. Soc. Civil Engrs. 96, no. 9:1781–93.

4 Uniform Flow

Part of 242-mile long Colorado River Aqueduct. The trapezoidal canal is 6-m wide at the bottom and 16.8 m wide at the top and is 3.4 m deep. The maximum discharge is 51 m³/s. (Courtesy of the Metropolitan Water District of Southern California).

4-1 Introduction

In free-surface flow, the component of the weight of water in the downstream direction causes acceleration of flow (it causes deceleration if the bottom slope is negative), whereas the shear stress at the channel bottom and sides offers resistance to flow. Depending upon the relative magnitude of these accelerating and decelerating forces, the flow may accelerate or decelerate. For example, if the resistive force is more than the weight component, then the flow velocity decreases and, to satisfy the continuity equation, the flow depth increases. The converse is true if the weight component is more than the resistive force. However, if the channel is long and prismatic (i.e., channel cross section and bottom slope do not change with distance), then the flow accelerates or decelerates for some distance until the accelerating and resistive forces are equal. From that point on, the flow velocity and flow depth remain constant (Fig. 4-1). Such a flow, in which the flow depth does not change with distance, is called *uniform flow*, and the corresponding flow depth is called the *normal depth*.

Uniform flow is discussed in this chapter. An equation relating the bottom shear stress to different flow variables is first derived. Various empirical resistance formulas used for the free-surface flows are then presented. A procedure for computing the normal depth for a specified discharge in a channel of known properties is outlined.

4-2 Flow Resistance

Leonardo da Vinci described the resistance offered by the channel bottom and sides to free-surface flows and its effects on the velocity distribution in an excellent manner as follows (Rouse and Ince 1963):

> The water of straight rivers is the swifter the farther away it is from the walls, because of resistance.
>
> Water has higher speed on the surface than at the bottom. This happens because water on the surface borders on air which is of little resistance, because lighter than water, and the water at the bottom is touching the earth which is of higher resistance, because heavier than water and not moving. From this follows that the part which is more distant from the bottom has less resistance than that below.

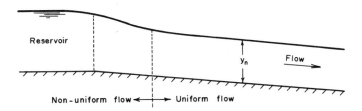

Figure 4-1 Uniform and nonuniform flows.

Because of the variation in resistance along the wetted perimeter and because of the shape of the channel cross section, secondary currents are usually set up in free-surface flows even if the channel is straight. In addition, the shear resistance offered to flow at the channel boundaries is not uniform. However, to simplify the analysis, we will assume that the flow is one-dimensional—i.e., there are no secondary currents in the flow and the shear resistance to flow at the boundaries is uniform.

4-3 Equations of Flow Resistance

We present several equations in this section relating various flow variables and the channel resistance. To keep the derivation general, we first derive an equation for nonuniform flow and then show that uniform flow is a special case of nonuniform flow.

Chezy Equation

To derive Chezy equation, we make the following assumptions:

1. The flow is steady.
2. The slope of the channel bottom is small.
3. The channel is prismatic.

Let us consider a control volume of length Δx, as shown in Fig. 4-2. On the upstream side of this control volume, let the distance be x, flow velocity be V, and the flow depth be y. Then the values of these variables at the downstream side will be $x + \Delta x$, $V + (dV/dx)\Delta x$, and $y + (dy/dx)\Delta x$.

The following forces are acting on the control volume: pressure force on the upstream side, F_1; pressure forces on the downstream side, F_2 and F_3; a component of the weight of water in the control volume in the downstream direction, W_x; and the shear force, F_f, acting on the channel bottom and the sides. Referring to Fig. 4-2, the expression for these forces may be written as follows:

$$\text{Presssure force } F_1 = \gamma A \bar{z} \qquad (4\text{-}1a)$$

in which $\bar{z}$ = depth of the centroid of flow area A below the water surface and γ = specific weight of water.

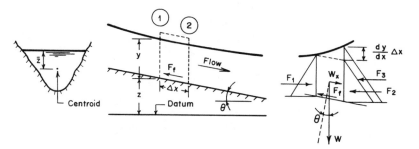

Figure 4-2 Definition sketch.

The component of the weight of water in the downstream direction is

$$W_x = \gamma A \Delta x \sin \theta \tag{4-1b}$$

in which θ = angle between the channel bottom and the horizontal axis. Since the channel-bottom slope is assumed to be small, $\sin \theta \simeq \tan \theta \simeq -dz/dx$. Note that the negative sign is due to the fact that z decreases as x increases. Hence, we may write Eq. 4-1(b) as

$$W_x = -\gamma A \frac{dz}{dx} \Delta x \tag{4-1c}$$

The pressure force acting on the downstream side of the control volume may be divided into two parts, as shown in Fig. 4-2. F_2 is the pressure force due to flow depth, y, and F_3 is the pressure force for the increase in depth in distance, Δx. Expressions for F_2 and F_3 are

$$F_2 = \gamma A \bar{z}$$

and

$$F_3 = \gamma A \frac{dy}{dx} \Delta x \tag{4-1d}$$

Note that in the expression for F_3, we have neglected the higher-order term, which corresponds to the small triangle at the top.

If the average shear stress acting on the channel bottom and sides is τ_o, then the shearing force is

$$F_f = \tau_o P \Delta x \tag{4-1e}$$

in which P = wetted perimeter.

Referring to Fig. 4-2, the resultant force, F_r, acting on the control volume in the downstream direction is

$$F_r = \Sigma F = F_1 - (F_2 + F_3) + W_x - F_f \tag{4-2}$$

Substituting Eqs. 4-1(a) through 4-1(e) into Eq. 4-2 and simplifying, we obtain

$$F_r = -\gamma A \Delta x \left(\frac{dy}{dx} + \frac{dz}{dx} + \frac{P\tau_o}{\gamma A} \right) \tag{4-3}$$

To apply the Reynolds transport theorem,[*] the intensive property, $\beta = V$. Therefore,

$$\sum F = \frac{\partial}{\partial t} \int_x^{x+\Delta x} \rho V A \, dx + (\rho A V^2)_2 - (\rho A V^2)_1 \tag{4-4}$$

Since the flow is assumed to be steady, the first term on the right-hand side of this equation is zero. By substituting for $\sum F_r$ from Eq. 4-3 into Eq. 4-4, dividing both sides by $\gamma A \Delta x$, and applying the mean value theorem to the right-hand side, we obtain

$$\frac{V}{g} \frac{dV}{dx} = - \left(\frac{dy}{dx} + \frac{dz}{dx} + \frac{P}{A} \frac{\tau_o}{\gamma} \right) \tag{4-5}$$

It follows from this equation that

$$\tau_o = -\gamma R \left(\frac{dy}{dx} + \frac{dz}{dx} + \frac{V}{g} \frac{dV}{dx} \right) \tag{4-6}$$

in which $R = A/P$ = hydraulic radius. This equation may be simplified as

$$\begin{aligned}
\tau_o &= -\gamma R \frac{d}{dx} \left(y + z + \frac{V^2}{2g} \right) \\
&= -\gamma R \frac{dH}{dx} \\
&= \gamma R S_f
\end{aligned} \tag{4-7}$$

in which S_f = slope of the energy grade line = $-dH/dx$. Note that we are using a negative sign, since H decreases as x increases.

If the flow is steady and uniform, then by definition $dV/dx = 0$ and $dy/dx = 0$. In addition, since $S_o = -dz/dx$, Eq. 4-7 may be written as

$$\tau_o = \gamma R S_o \tag{4-8}$$

Based on a dimensional analysis, we may write

$$\tau_o = k\rho V^2 \tag{4-9}$$

in which k is a dimensionless constant that depends upon the Reynolds number, roughness of the channel bottom and sides, etc. It follows from Eqs. 4-7 and 4-9 that

$$V = \sqrt{\frac{g}{k} R S_f} \tag{4-10}$$

This equation may be written as

$$V = C \sqrt{R S_f} \tag{4-11}$$

[*]We may derive the same equation by applying Newton's second law to the liquid in the control volume; i.e., the resultant force is equal to the time rate of change of momentum. The time rate of change of momentum = $\rho A V[(V + dV/dx \Delta x) - V]$. By equating this to the expression for F_r from Eq. 4-3 and simplifying, we obtain Eq. 4-5.

in which C = Chezy constant. This equation was introduced by the French engineer Antoine Chezy in 1768 while designing a canal for the water-supply system of Paris (Henderson 1966; Chow 1959).

Note that Eq. 4-11 is valid for nonuniform, steady flow. For uniform flow we will use Eqs. 4-8 and 4-9 instead of Eqs. 4-7 and 4-9. As a result we get the following equation instead of Eq. 4-11:

$$V = C\sqrt{RS_o} \qquad (4\text{-}12)$$

Note that Eq. 4-11 for nonuniform flow and Eq. 4-12 for uniform flow are similar except that we use the slope of the energy-grade line, S_f, for nonuniform flow, but we use the slope of the channel bottom, S_o (which has the same value as the slope of the energy grade line or the slope of the water surface), for the uniform flow.

It is clear from Eq. 4-11 or 4-12 that C has dimensions of $\sqrt{\text{length}}/\text{time}$. This coefficient, like the Darcy Weisbach friction factor f, depends upon the channel roughness and the Reynolds number, $\mathbf{R}_e$. In addition, it may depend upon the channel cross-sectional shape as well, although this dependence appears to be small (American Society of Civil Engineers 1963) and may be neglected. Because the channel roughness may vary over a wide range, its effect on C has not been as thoroughly investigated as that on f.

Let us now compare the Chezy equation, Eq. 4-11, for open channels with the Darcy-Weisbach friction formula for pipes,

$$h_f = \frac{fL}{D}\frac{V^2}{2g} \qquad (4\text{-}13)$$

in which h_f = head loss in a pipe of diameter D and length L. The slope of the energy-grade line, $S = h_f/L$. Therefore, we may write this equation as

$$V = \sqrt{\frac{2gDS}{f}} \qquad (4\text{-}14)$$

Noting that the hydraulic radius, R, for a pipe is equal to $D/4$, Eq. 4-11 becomes

$$V = C\sqrt{\frac{DS}{4}} \qquad (4\text{-}15)$$

It follows from the preceding two equations that $C = \sqrt{8g/f}$.

Figure 4-3 shows the Moody diagram plotted with C as the ordinate instead of f (Henderson 1966). This diagram is divided into three regions: hydraulically smooth, transition, and fully rough. A flow may be considered *hydraulically smooth* even though the channel surface is rough provided the projections of the surface roughness are covered by the laminar sublayer. As the Reynolds number increases, the thickness of this layer decreases and the effect of roughness projections on flow becomes important. Then, the flow is in the *transition* region. However, when the roughness projections are not covered by the viscous sublayer and dominate the flow because losses are due to form drag, flow may be classified as *fully rough*. These flow types may be classified using the value of a dimensionless number, $R_s = kV^*/\nu$. In this expression, ν is the

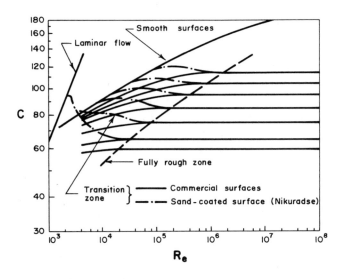

Figure 4-3 Modified Moody diagram (After F. M. Henderson *Open Channel Flow*. Reprinted with the permission of Macmillan Publishing Company. Copyright © 1966 by Macmillan Publishing Company).

kinematic viscosity of the liquid; k is a characteristic length parameter for the size of the channel-surface roughness; and V^*, known as the *shear velocity*, is defined as

$$V^* = \sqrt{\frac{\tau_o}{\rho}} = \sqrt{gRS_f} \tag{4-16}$$

The flow is considered smooth if $R_s < 4$; in transition region if $4 < R_s < 100$; and fully rough if $R_s > 100$.

Henderson (1966) lists the following expressions for C for smooth and rough flows:

Smooth flows

$$C = 28.6\mathbf{R}_e^{1/8} \qquad \text{if } \mathbf{R}_e < 10^5 \tag{4-17}$$

and

$$C = 4\sqrt{2g}\,\log_{10}\left(\frac{\mathbf{R}_e\sqrt{8g}}{2.51C}\right) \qquad \text{if } \mathbf{R}_e > 10^5 \tag{4-18}$$

Rough flows

$$C = -2\sqrt{8g}\,\log_{10}\left(\frac{k}{12R} + \frac{2.5}{\mathbf{R}_e\sqrt{f}}\right) \tag{4-19}$$

The preceding equations are derived from the data obtained from experiments conducted on flow through pipes. Therefore, they may be used only for small channels with fairly smooth surfaces. Empirical relationships and field observations should be employed for large channels with rough flow surfaces.

Manning Equation

Since the derivation of the Chezy equation in 1768, several researchers have tried to develop a rational procedure for estimating the value of the Chezy constant C. However, unlike the case of the Darcy-Weisbach friction factor for the closed conduits, these attempts have not been very successful, because C depends upon several parameters in addition to the channel roughness. Based on the field observations, Ganguillet and Kutter (Chow 1959) proposed a complex formula for C. Later, Gauckler and Hagen independently showed that

$$C \propto R^{1/6} \tag{4-20}$$

According to Henderson (1966), a French engineer named Flament incorrectly attributed this equation to an Irishman, R. Manning, and expressed it in the following form in 1891:

$$V = \frac{1}{n} R^{2/3} S_f^{1/2} \tag{4-21}$$

in which n = Manning coefficient. This is the Manning equation, which has been very widely used in English-speaking countries.

Again note that n is not a dimensionless constant and has dimensions of $T\,L^{-1/3}$. Therefore, we convert this equation so that the value of n is the same in both SI and English units.

Equation 4-21 is valid for SI units, i.e., V is in m/s, and R is in m. In foot-pound-second units, this equation becomes

$$V = \frac{1.49}{n} R^{2/3} S_f^{1/2} \tag{4-22}$$

in which V is in ft/sec and R is in ft. As a result of this conversion, the value of n is the same in both system of units.

The value of n depends mainly upon the surface roughness, amount of vegetation, and channel irregularity and—to a lesser degree—upon stage, scour and deposition, and channel alignment. Table 4-1 lists the average values of n for different flow surfaces. In addition, photographs published by the United States Geological Survey (Barnes 1967) and the Department of Agriculture (Ramser 1929; Scobey 1939) are excellent sources of reference for the selection of a value for n. A number of typical photographs are presented in Fig. 4-4. The following is the information for these photographs:

   a. $n = 0.024$ (Columbia River at Vernita, Washington): The channel bottom consists of slime-covered cobbles and gravel, the steep left bank is composed of cemented cobbles and gravel, and the right bank consists of cobbles set in gravel.

   b. $n = 0.030$ (Salt Creek at Roca, Nebraska): The bottom consists of sand and clay; the banks are smooth and free of vegetation.

   c. $n = 0.032$ (Salt River below Stewart Mountain Dam, Arizona): The bottom and banks consist of smooth 0.15-m-diameter cobbles, with a few 0.45-m-diameter boulders.

Table 4-1 Average values* of Manning n

Material	n
Metals	
Steel	0.012
Cast iron	0.013
Corrugated metal	0.025
Nonmetals	
Lucite	0.009
Glass	0.010
Cement	0.011
Concrete	0.013
Wood	0.012
Clay	0.013
Brickwork	0.013
Gunite	0.019
Masonary	0.025
Rock cuts	0.035
Natural streams	
Clean and straight	0.030
Bottom: gravel, cobbles and boulders	0.040
Bottom: cobbles with large boulders	0.050

*Compiled from Chow (1959).

d. $n = 0.036$ (West Fork Bitterroot River near Conner, Montana): The bottom is gravel and boulders; $d_{50} = 1.72$ m; left bank has overhanging bushes and the right bank has trees.

e. $n = 0.041$ (Middle Fork Flathead River near Essex, Montana): The bottom consists of boulders; $d_{50} = 1.4$ m; banks are composed of gravel and boulders and have trees and brushes.

f. $n = 0.049$ (Deep River at Ramseur, North Carolina): The bottom is mostly coarse sand and contains some gravel; the banks are fairly steep and have underbrush and trees.

g. $n = 0.050$ (Clear Creek near Golden, Colorado): The bottom and banks are composed of 0.7-m-diameter angular boulders.

h. $n = 0.060$ (Rock Creek Canal near Darby, Montana): The bottom and banks consists of boulders; $d_{50} = 2.1$ m.

i. $n = 0.070$ (Pond Creek near Louisville, Kentucky): The bottom is fine sand and silt; the banks are irregular with heavy growth of trees.

j. $n = 0.075$ (Rock Creek near Darby, Montana): The bottom consists of boulders; $d_{50} = 2.2$ m; the banks are composed of boulders and have brush and trees.

Figure 4-4 Photographs for typical Manning n (After Barnes 1967).

(f)

(g)

(h)

(i)

(j)

Figure 4-4 Continued

Christensen (1984b) investigated the range of validity of the Manning formula assuming that the Nikuradse equations (Nikuradse 1932) for the friction factors of closed conduits are valid for the free-surface flows. By substituting the approximation

$$\frac{1}{\sqrt{f}} = 2.916 \left(\frac{R}{k}\right)^{1/6}$$ (4-23)

for rough turbulent flows in circular conduits into Eq. 4-14 and noting that for closed conduits $R = D/4$, we obtain

$$V = 8.25 \frac{\sqrt{g}}{k^{1/6}} S^{1/2} R^{2/3}$$ (4-24)

A comparison of Eqs. 4-21 and 4-24 shows that they are identical if we replace $1/n$ of Eq. 4-21 by $8.25\sqrt{g}/k^{1/6}$ of Eq. 4-24. This relationship between n and k is almost identical to the Strickler's formula reported by Forchheimer in 1930.

Equation 4-24 has the following advantages over Eq. 4-21: Manning n is difficult to estimate, since it does not have any physical meaning. On the other hand, k is physically based and is directly related to the size of surface roughness, which can be measured. In addition, since k is raised to the one-sixth power, an error in estimating its value has a considerably less effect on the computed value of V as compared to that introduced by a similar error in the estimation of n.

Other Resistance Formulas

In Europe, the following resistance formula has been widely used (Jaeger 1961):

$$V = k_s R^{2/3} S_f^{1/2}$$ (4-25)

This is called Strickler's formula, and k_s is called Strickler constant. In SI units, k_s may be computed from

$$k_s = \frac{21.1}{k^{1/6}}$$ (4-26)

in which k = the mean size of the wall roughness. Typical mean values of k for various materials, taken from Jaeger (1961), are listed in Table 4-2.

Table 4-2 Roughness sizes*

Material	k (mm)
Cast iron, new	0.5–1.0
Cast iron, somewhat rusty	1.0–1.5
Cement mortar, smoothed	0.3–0.8
Cement mortar, left rough	1.0–2.0
Rough wooden boards	1.0–2.5
Rough masonary (blockwork)	8.0–15

*Compiled from Jaeger (1961).

A comparison of Eqs. 4-21 and 4-25 shows that the Manning and Strickler formulas are similar and that

$$k_s = \frac{1}{n} \qquad (4\text{-}27)$$

In Russia, the following formula for Chezy constant, C, has been widely used:

$$C = \frac{1}{n} R^a \qquad (4\text{-}28)$$

in which $a = 1.3\sqrt{n}$ if $R > 1$ m and $a = 1.5\sqrt{n}$ if $R < 1$ m. This formula was proposed by Pavlovskii (Chow 1959) in 1925.

4-4 Computation of Normal Depth

To analyze open-channel flow, it is usually necessary to know the normal depth, y_n. A number of procedures for computing the normal depth in a given channel for a specified discharge are discussed in this section. We will consider only the Manning formula in our discussions, since it is very widely used in English-speaking countries. These discussions are valid for Strickler formula as well if we replace n by $1/k_s$.

The Manning formula for uniform flow in terms of discharge may be written as

$$Q = VA = \frac{C_o}{n} A R^{2/3} S_o^{1/2} \qquad (4\text{-}29)$$

in which $C_o = 1.49$ in customary English units and $C_o = 1$ in SI units.

In this equation, A and R are function of the flow depth, y, and of the channel cross section, whereas n is a function of the flow surface and other factors discussed in the previous section. Thus, for a given channel section and specified bottom slope, only one discharge is possible for a given normal depth. However, if the value of this depth is known, then we can determine the corresponding discharge directly from Eq. 4-29.

We may write Eq. 4-29 as

$$Q = K S_o^{1/2} \qquad (4\text{-}30)$$

in which the conveyance factor, K, for the channel section is defined as

$$K = \frac{C_o}{n} A R^{2/3} \qquad (4\text{-}31)$$

Note that K is a function of the normal depth, properties of the channel section, and Manning n.

Equation 4-29 may be written as

$$A R^{2/3} = \frac{nQ}{C_o S_o^{1/2}} \qquad (4\text{-}32)$$

in which the left-hand side is referred to as the *section factor*. Thus, for the specified values of n, Q, and S_o, we solve this equation to determine the normal depth in a given channel. This may be done by using design charts presented by Chow (1959), by a trial-and-error procedure, or by using numerical methods for the solution of a nonlinear algebraic equation. These procedures are discussed in the following paragraphs.

Design Curves

Design curves are presented in Fig. 4-5 for a trapezoidal and for a circular channel section. If we want to determine the normal depth for a specified discharge in a given channel section, then we know Q, n, and S_o. Therefore, we can compute the right-hand side of Eq. 4-32. Let us divide this computed value by $B_o^{8/3}$ if the channel section is trapezoidal and by $D_o^{8/3}$ if the channel cross section is circular. The resulting value is then equal to $AR^{2/3}/B_o^{8/3}$ for a trapezoidal section and equal to $AR^{2/3}/D_o^{8/3}$ for a circular cross section. Now, y_n/B_o or y_n/D_o corresponding to the value of $AR^{2/3}/B_o^{8/3}$ or $AR^{2/3}/D_o^{8/3}$ may be read directly from Fig. 4-5.

Trial-and-Error Procedure

Substitute expressions for the flow area, A, and hydraulic radius, R, and the values of n, Q, and S_o into Eq. 4-32 and then solve the resulting equation by a trial-and-error procedure.

Numerical Methods

Several methods, such as the bisection method, method of successive approximations, and Newton method, are available (McCracken and Dorn 1964) for solving Eq. 4-32. We discuss only the Newton method.

To determine y_n by this method, we write Eq. 4-32 as

$$F(y_n) = AR^{2/3} - \frac{nQ}{C_o S_o^{1/2}} = 0 \tag{4-33}$$

For the Newton method, we need the first derivative of function F. An expression for this derivative may be obtained as follows:

$$\begin{aligned}
\frac{dF}{dy_n} &= \frac{d}{dy_n}\left(A^{5/3}P^{-2/3} - \frac{nQ}{C_o S_o^{1/2}}\right) \\
&= \frac{5}{3}P^{-2/3}A^{2/3}\frac{dA}{dy_n} - \frac{2}{3}A^{5/3}P^{-5/3}\frac{dP}{dy_n} \\
&= \frac{5}{3}BR^{2/3} - \frac{2}{3}R^{5/3}\frac{dP}{dy_n}
\end{aligned} \tag{4-34}$$

since $dA/dy_n = B$. For a trapezoidal section having side slopes of s horizontal to 1 vertical, $dP/dy_n = 2\sqrt{1+s^2}$. Similar expressions for other channel sections may be obtained.

The following example will help in understanding these procedures for determining the normal depth. The computer programs included in Appendix C illustrate the application of the Newton and bisection methods.

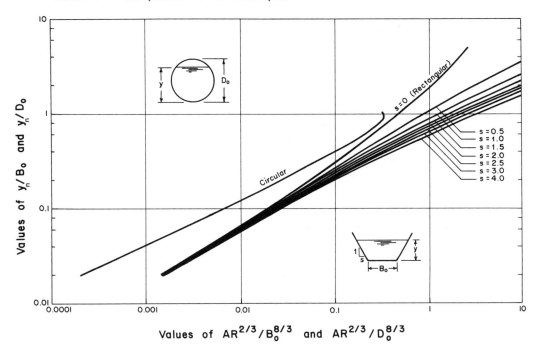

Figure 4-5 Curves for computation of normal depth (After Chow, *Open-Channel Hydraulics*, 1959).

Example 4-1

Compute the normal depth in a trapezoidal channel having a bottom width of 10 m, and side slopes of 2H to 1V and carrying a flow of 30 m^3/s. The slope of the channel bottom is 0.001 and n = 0.013.

Given:

$$Q = 30 \text{ m}^3/\text{s}$$
$$n = 0.013$$
$$B_o = 10 \text{ m}$$
$$s = 2$$
$$S_o = 0.001$$
$$C_o = 1.0$$

Determine:

$$y_n = ?$$

Solution:

We use the given procedures one by one to determine y_n.

Design curves. By substituting the values of n, Q, and S_o into the right-hand side of Eq. 4-32, we obtain

$$\frac{nQ}{C_o S_o^{1/2}} = \frac{0.013 \times 30}{1 \times (0.001)^{1/2}}$$

$$= 12.33$$

Hence, it follows from Eq. 4-32 that

$$AR^{2/3} = 12.33$$

Thus

$$\frac{AR^{2/3}}{B_o^{8/3}} = \frac{12.33}{(10)^{8/3}}$$

$$= 0.026$$

For $s = 2$ and $AR^{2/3}/B_o^{8/3} = 0.026$, we read from Fig. 4-5 that $y_n/B_o = 0.11$. Hence,

$$y_n = 1.1 \text{ m}$$

Trial-and-Error Procedure.   We earlier computed $AR^{2/3} = 12.33$ for the design-curve procedure. By using the data for the channel, we obtain the following expressions for A and R:

$$A = \frac{1}{2}y_n(10 + 10 + 2sy_n)$$

$$= y_n(10 + 2y_n)$$

$$P = B + 2\sqrt{s^2 + 1}\,y_n$$

$$= 10 + 4.47y_n$$

$$R = \frac{y_n(10 + 2y_n)}{10 + 4.47y_n}$$

Now, substituting these expressions for A and R into $AR^{2/3} = 12.33$ and simplifying the resulting equation, we obtain

$$\left[y_n(10 + 2y_n)\right]^{5/3} - 12.33(10 + 4.47y_n)^{2/3} = 0$$

A trial-and-error solution of this equation yields

$$y_n = 1.09 \text{ m}$$

Numerical Method.   To compute the normal depth in a trapezoidal channel, the computer program of Appendix C-1 uses the Newton method and the computer program of Appendix C-2 uses the bisection method. An initial estimate for the flow depth in the Newton method is used as 0.5 m; YL = 0.5 m and YR = 10 m are used as initial estimates for the bisection method.

4-5 Equivalent Manning Constant

So far we have assumed that the flow surface at a channel cross section has the same roughness along the entire wetted perimeter. However, this is not always true. For example, if the channel bottom and sides are made from different materials, then the Manning n for the bottom and sides may have different values. To simplify the computations, it becomes necessary to determine a value of n, designated by n_e that may be used for the entire section. This value of n_e is referred to as the equivalent n for the entire cross section.

Let us consider a channel section that may be subdivided into N subareas having wetted perimeter P_i and Manning constant, n_i, $(i = 1, 2, \ldots, N)$. By assuming that the mean flow velocity in each of the subareas is equal to the mean flow velocity (Horton 1933; Einstein 1934) in the entire section, the following equation may be derived:

$$n_e = \left(\frac{\sum P_i n_i^{3/2}}{\sum P_i} \right)^{2/3} \tag{4-35}$$

in which subscript i refers to values for the ith subarea.

Similarly, the following expression for the equivalent Manning constant n_e may be derived by assuming that the total force resisting the flow is equal to the sum of forces resisting the flow in each subarea (Muhlhofer 1933; Einstein and Banks 1951):

$$n_e = \frac{\left(\sum P_i n_i^2 \right)^{1/2}}{\left(\sum P_i \right)^{1/2}} \tag{4-36}$$

By utilizing the fact that the total discharge is equal to the sum of the discharges in each subarea, Lotter (1933) obtained the following equation for the equivalent Manning constant:

$$n_e = \frac{P R^{5/3}}{\sum \frac{P_i R_i^{5/3}}{n_i}} \tag{4-37}$$

Krishnamurthy and Christensen (1972) derived the following equation by assuming a logarithmic velocity distribution:

$$\ln n_e = \frac{\sum_{i=1}^{N} P_i y_i^{3/2} \ln n_i}{\sum P_i y_i^{3/2}} \tag{4-38}$$

in which $y_i = $ flow depth.

By utilizing the data for 36 natural-channel cross sections obtained by U.S. Gelogical Survey, Motayed and Krishnamurthy (1980) showed that the equivalent roughness computed by using Eq. 4-35 gives the least error of the above four equations listed for computing n_e.

4-6 Compound Cross Section

A compound cross section may be defined as a section in which various subareas have different flow properties, e.g., surface roughness, etc. A natural stream having overbank flow during a flood (Fig. 4-6) is a typical example of a compound section. The roughness of the overbanks is usually higher than that of the main channel; and, therefore, the flow velocity in the main channel is higher than that in the overbank flow.

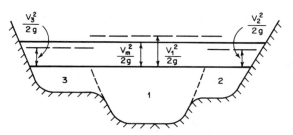

Figure 4-6 Compound channel section.

The analysis of flow in a compound section becomes complex if the flow in each subarea is considered separately. This requires the use of a two- or three-dimensional model or to apply a one-dimensional model separately to each subarea by considering the flow in each subarea as parallel flow and allowing for the exchange of mass and momentum between the adjacent subareas.

In a straight channel, the water surface should be level over the entire cross section, since the pressure along any horizontal line must be constant although the flow velocity may vary from one subarea to the next. Due to different flow velocity, the level of the energy grade line is different in each subarea. Thus, there is no common level for the energy grade line for the entire section. To avoid this complexity, we derive in this section expressions for the energy coefficient, α, and for S_f in terms of the conveyance factor, K, of the subareas. With these expressions, the flow in a compound section may be computed without knowing the individual flows in each subarea.

Let us subdivide the compound section into N subareas. We want to derive an expression for the energy coefficient, α, such that the velocity head for the entire section is $\alpha(V_m^2/2g)$, in which V_m = mean flow velocity in the compound section. In Chapter 1, we derived the following expression

$$\alpha = \frac{\sum_i^N V_i^3 A_i}{V_m^3 \sum_i^N A_i} \tag{4-39}$$

in which N = number of subareas. By substituting

$$V_m = \frac{\sum V_i A_i}{\sum A_i} \tag{4-40}$$

and $V_i = Q_i/A_i$ into Eq. 4-39 and simplifying the resulting equation, we obtain

$$\alpha = \frac{\left(\sum Q_i^3/A_i^2\right) \cdot \left(\sum A_i\right)^2}{\left(\sum Q_i\right)^3} \tag{4-41}$$

Now, the flow in subarea i may be written as

$$Q_i = K_i S_{fi}^{1/2} \tag{4-42}$$

or

$$S_{fi}^{1/2} = \frac{Q_i}{K_i} \tag{4-43}$$

Let us assume that S_f has the same value for all subareas, i.e., $S_{fi} = S_f$ $(i = 1, 2, \ldots, N)$. Then, on the basis of Eq. 4-43, we may write the following for each subarea:

$$\frac{Q_1}{K_1} = \frac{Q_N}{K_N}$$

$$\frac{Q_2}{K_2} = \frac{Q_N}{K_N}$$

$$\vdots$$

$$\frac{Q_i}{K_i} = \frac{Q_N}{K_N}$$

$$\frac{Q_N}{K_N} = \frac{Q_N}{K_N}$$

(4-44)

It follows from this equation that

$$Q_1 = K_1 \frac{Q_N}{K_N}$$

$$Q_2 = K_2 \frac{Q_N}{K_N}$$

$$\vdots$$

$$Q_N = K_N \frac{Q_N}{K_N}$$

(4-45)

The additon of the preceding equations yields

$$Q = \sum Q_i = \frac{Q_N}{K_N} \sum K_i$$

(4-46)

By substituting this expression for $\sum Q_i$, and $Q_i = K_i(Q_N/K_N)$ into Eq. 4-41 and simplifying the resulting equation, we obtain

$$\alpha = \frac{\left(\sum \frac{K_i^3}{A_i^2}\right)\left(\sum A_i\right)^2}{\left(\sum K_i\right)^3}$$

(4-47)

The elimination of Q_N/K_N from Eqs. 4-44 and 4-46 and squaring both sides give

$$S_f = \left(\frac{\sum Q_i}{\sum K_i}\right)^2$$

$$= \frac{Q^2}{\left(\sum K_i\right)^2}$$

(4-48)

Now we have expressions for both α and S_f such that we do not have to know the flow in each subarea, Q_i $(i = 1, 2, \ldots, N)$, to compute α and S_f.

4-7 Summary

In this chapter an expression relating the bottom shear stress to other flow variables was derived. Different empirical formulas for computing the friction losses were presented. Several procedures for computing the normal depth in a channel section for a specified discharge were outlined. Based on different simplifying assumptions, equations for computing the equivalent roughness coefficient were presented. The chapter was concluded by deriving expressions for the energy coefficient and for the slope of the energy grade line in a compound channel section.

Problems

4-1. A 5-m-wide rectangular channel is carrying a flow of 5 m^3/s. If the Manning $n = 0.013$ and the bottom slope, $S_o = 0.001$, determine the normal depth.

4-2. What will be the normal depth if the discharge in the channel of Example 4-1 is 50 m^3/s?

4-3. Develop a general-purpose computer program to compute the normal depth in a channel having any general cross section. Write the program such that A and R will be supplied by the user through a subroutine. Use this program to compute the normal depths in the channels of Probs. 4-1 and 4-4.

4-4. A channel with a cross section shown in Fig. 4-7 has a flow of 150 m^3/s. The slope of the channel bottom is two per thousand, and the Manning n for the flow surfaces is 0.03. Compute the normal and critical depths in the channel.

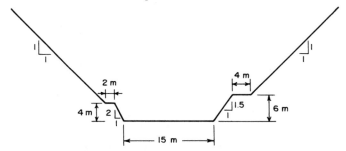

Figure 4-7 Channel cross section for Prob. 4-4.

4-5. A concrete-lined, trapezoidal irrigation canal has a bottom width of 10 m, side slopes of 1H : 1V, and longitudinal bottom slope of 0.0005. If the canal is several kilometers long, what is the flow depth near the downstream end for a flow of 60 m^3/s?

4-6. Analytically determine the depths at which the discharge and flow velocities are maximum in a circular conduit flowing partially full. Assume the flow is uniform and the Manning n does not vary with depth.

4-7. Prove that the most efficient cross sections (i.e., discharge is maximum) for a given flow area are as follows:
   i. Triangular section: vertex angle $= 90°$
   ii. Trapezoidal section: half-hexagon

4-8. Prove that a semicircle with its center at the middle of the water surface is the most efficient circular cross section.

4-9. For flow in a pipe flowing partially full, analytically prove that

$$\frac{Q_p}{Q_f} = \frac{\left(\theta - \frac{1}{2}\sin 2\theta\right)^{5/3}}{\pi \theta^{2/3}}$$

In this equation, the subscripts p and f refer to the pipe flowing partially full and flowing full, respectively; D_o = pipe diameter, and y_n = normal depth. S_o = pipe bottom slope if the pipe is flowing partially full, and it is equal to the slope of the energy-grade line for full pipe flow. For a partially full pipe,

$$\theta = \cos^{-1}\left(1 - \frac{2y_n}{D_o}\right)$$

Experimental data for flow in partially full pipes may be approximated by the following equation (Christensen 1984a):

$$\frac{Q_p}{Q_f} = 0.46 - 0.5\cos\left(\frac{\pi y_n}{D_o}\right) + 0.04\cos\left(\frac{2\pi y_n}{D_o}\right)$$

For different values of y_n/D_o, compute Q_p/Q_f using the preceding expression derived analytically and from that based on the experimental results. Plot and compare these results.

Several authors state that the difference between these results is due to the variation of n with the flow depth. Do you agree with this explanation? If so, give your reasons.

Assuming that n varies with depth compute and plot n_p/n_f with respect to y/D_o.

4-10. Compute the normal depth in an unlined tunnel having a standard horseshoe section (Fig. 4-8), with d_o = 25 m, carrying a flow of 800 m³/s, and having a bottom slope of 0.0005. Assume n for the flow surface is 0.03.

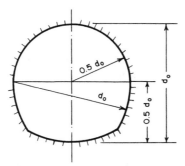

Figure 4-8 Channel cross section for Prob. 4-10.

4-11. Figure 4-9 shows the longitudinal profile of the Roman Aqueduct of Nimes (Hauck and Novak 1987). Assuming Manning n of 0.0125 and the channel width of 1.2 m for each segment of the aqueduct, compute the normal depth in different segments for flows of 210, 350, and 450 l/s.

4-12. The cross section of a drainage channel may be approximated as a trapezoidal section with the bottom width of 15 ft and side slope of 1.5H : 1V. If the channel drops 2.5 ft/mi, compute the flow depth for a flow of 150 ft³/sec. Assume n = 0.024.

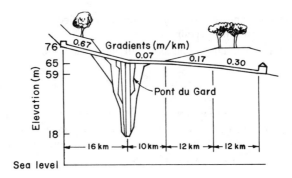

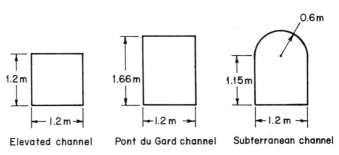

Figure 4-9   Roman Aqueduct of Nimes (After Hauck and Novak, 1987).

4-13. For the creek of Prob. 3-12, determine the flow depth if the bottom slope is 10 ft/mi and $n = 0.045$

4-14. An 8-ft-diameter concrete-lined sewer is laid at a bottom slope of 1 ft/mi. Find the flow depth for a flow of 30 ft^3/sec.

4-15. Determine the normal depth for the sewer of Problem 3-15 for a bottom slope of 2 ft/mi and $n = 0.014$

4-16. The flow depth in a long trapezoidal channel (bottom width = 8 m, side slopes 1 : 1) for a flow of 28 m^3/s is 3 m. The channel bottom slope is 0.0001. Determine the flow depth if the rate of discharge is doubled.

4-17. The flow depth at a section in a long rectangular channel changes from 4 ft to 5 ft. Determine the percent change in the rate of discharge.

4-18. A boulder-lined drainage channel overflowed its banks during a spring runoff for flows exceeding 20 ft^3/sec. The channel is 15 ft wide, is rectangular in shape, and drops 10 ft/mi. If you are the design engineer, what will be your options to prevent flooding for this flow?

4-19. Is the flow subcritical or supercritical in a 4-m-wide rectangular channel for a discharge of 9 m^3/s? The bottom slope is 0.005 and $n = 0.014$.

4-20. Compute the critical and normal depths in a trapezoidal channel (bottom width = 20 ft, side slopes = 1.5H : 1V) for a flow of 220 ft^3/sec. The bottom slope is 0.00032 and $n = 0.022$. Is the flow subcritical or supercritical?

4-21. The flow depth for a discharge of 15 m^3/s in a long canal having a trapezoidal cross section (bottom width = 10 m; side slopes = 1V : 2H) is 2 m. If the discharge is increased to 20 m³/s, what will be the flow depth?

4-22. For the symmetrical compound section shown in Fig. 4-10, determine

    i. Equivalent n_e;

    ii. Velocity-head coefficient, α;

    iii. Slope of the energy grade line, S_f.

Assume the flood plains and the main channel have the same bottom slope of 0.001, and Manning n for the main channel and for the floodplains are 0.021 and 0.039 respectively.

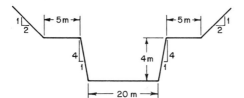

Figure 4-10 Compound section.

4-23. Compute the discharge in a 12-ft wide rock channel ($n = 0.035$) having a bottom slope of 0.001 and flow depth of 3 ft. What is the critical depth at this flow? Is the flow critical, subcritical, or supercritical?

REFERENCES

American Society of Civil Engineers. 1963. Friction factors in open channels, *Task Force Report, Jour. Hydraulics Division,* Amer. Soc. of Civil Engrs, 89, no. HY2 (March):97–143.

Barnes, H. H. 1967. *Roughness characteristics of natural channels,* U.S. Geological Survey, Water Supply Paper No. 1849, U. S. Government Press.

Berlamont, J. E., and Vanderstappen, N. 1981, Unstable turbulent flow in open channels, *Jour. Hydraulics Div.,* Amer. Soc. Civ. Engrs., 107, no. 4: 427–49.

Chow, V. T. 1959, *Open-channel hydraulics,* McGraw-Hill. New York, NY.

Christensen, B. A. 1984a. Discussion of "Flow velocities in pipelines," *Jour. Hydraulic Engineering,* Amer. Soc. of Civil Engrs., 110, no. 10:1510–12.

Christensen, B. A. 1984b. Analysis of partially filled circular storm sewers, *Proc. Hydraulics Div. Conference,* Amer. Soc. of Civil Engrs (August):163–67.

Einstein, H. A. 1934. Der hydraulische oder profil-radius, *Schweizerische Bauzeitung,* Zurich, 103, no. 8 (February 24):89–91.

Einstein, H. A., and Banks, R. B. 1951. Fluid resistance of composite roughness, *Trans.,* Amer. Geophysical Union 31, no. 4:603–10.

Forchheimer, P. 1930, *Hydraulik,* Teubner Verlag, Berlin, Germany, p. 146.

Hauck, G. F., and Novak, R. A. 1987. Interaction of flow and incrustation in the Roman Aqueduct of Nimes, *Jour. Hydraulic Engineering,* Amer. Soc. Civil Engrs., 113, no. 2:141–57.

Henderson, F. M. 1966. *Open channel flow,* Macmillan, New York, NY.

Hollinrake, P.G. 1987, 1988, 1989. The structure of flow in open channels, vols. 1–3, Research Reports SR96, 153, 209, Hydraulic Research Ltd., Wallingford, England.

Horton, R. A.   1933. Separate roughness coefficients for channel bottom and sides, *Engineering News Record* 111, no. 22 (November 30):652–53.

Jaeger, C.   1961. *Engineering fluid mechanics*, Blackie and Sons, London, England.

Krishnamurthy, M., and Christensen, B. A.   1972. Equivalent roughness for shallow channels, *Jour. Hydraulics Div.*, Amer. Soc. Civil Engrs. 98, no. 12:2257–63.

Lotter, G. K.   1933. Considerations on hydraulic design of channels with different roughness of walls, *Trans. All Union Scientific Research*, Institute of Hydraulic Engineering 9, Leningrad, 238–41.

Motayed, A. K., and Krishnamurthy, M.   1980. Composite roughness of natural channels, *Jour. Hydraulics Div.*, Amer. Soc. Civil Engrs., 106, no. 6:1111–16.

Muhlhofer, L.   1933. Rauhigkeitsuntersuchungen in einem stollen mit betonierter sohle und unverkleideten wanden, *Wasserkraft und Wasserwirtschaft,* 28, no. 8:85–88.

Myers, W. R. C.   1978. Momentum transfer in a compound channel, *Jour. Hydraulic Research,* Inter. Assoc. Hydraulic Research 16, no. 2:139–50.

Nikuradse, J.   1932. Gesetzmassigkeit der turbulenten stromung in glatten rohren, *Forschung Arb. Ing-Wesen*, I, Heft 356, Berlin, Germany.

Rajaratnam, N., and Ahmadi, R. M.   1979. Interaction between main channel and flood-plain flows, *Jour. Hydraulics Div.*, Amer. Soc. Civil Engrs., 105, no. 5:573–88.

Ramser, C. E.   1929. Flow of water in drainage channels, *Technical Bulletin No. 129,* U.S. Department of Agriculture (November).

Rouse, H., and Ince, S.   1963. *History of hydraulics*, Dover Publications, New York, NY.

Scobey, F. C.   1939. Flow of water in irrigation and similar canals, *Technical Bulletin No. 652,* U.S. Department of Agriculture (February).

Sturm, T. W., and King, D. A.   1988. Shape effects on flow resistance in horseshoe conduits, *Jour. Hydraulic Engineering,* Amer. Soc. Civil Engrs., 114, no. 11:1416–29.

Thompson, G. T. and Roberson, J. A.   1976. Theory of flow resistance for vegetated channels, *Trans.*, Amer. Soc. of Agric. Engrs. (gen. ed.) 19, no. 2:288–93.

Vanoni, V. A.   1941. Velocity distribution in open channels, *Civil Engineering*, Amer. Soc. Civil Engrs., 11:356–57.

5 Gradually Varied Flow

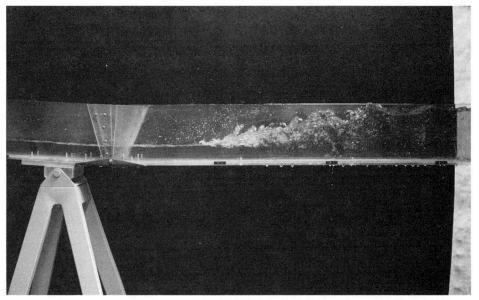

Gradually and rapidly varied flows in a change in bottom slope in a laboratory flume. (Courtesy of Prof. M. B. Quick, University of British Columbia, Vancouver, B. C., Canada).

5-1 Introduction

In Chapter 4, we considered uniform flow in which the flow depth does not vary with distance. Such flows occur in long and prismatic channels (i.e., the channel cross section and bottom slope do not change with distance). In reality, however, channel cross sections and bottom slopes are varied for economic reasons and to suit the existing topographical conditions. In addition, hydraulic structures are provided for flow control. Nonuniform flows are produced by changes in the channel geometry, while changing from one uniform-flow condition to another, and in the accelerating and decelerating flows. As we discussed in Chapter 1, such flows are called *gradually varied flows* if the rate of variation of depth with respect to distance is small and *rapidly varied flows* if the rate of variation is large. In other words, significant depth changes occur gradually over a long distance in gradually varied flows and in a short distance in rapidly varied flows. Since the analysis of gradually varied flows usually involves long channel lengths, the inclusion of friction losses due to boundary shear is necessary. These losses may, however, be neglected in the analysis of rapidly varied flows because the distances involved are short. In addition, the pressure distribution in gradually varied flow may be assumed hydrostatic because the streamlines are more or less straight and parallel. However, this is not the case in rapidly varied flows, where significant acceleration normal to streamlines may be produced by sharp curvatures in the streamlines.

A general discussion of the steady, gradually varied flow is presented in this chapter and of the rapidly varied flow, in Chapter 7. Equations describing the gradually varied flow are first derived. The classification of various water surface profiles is then presented. This is followed by a presentation of procedures for qualitatively sketching the surface profiles and for determining the discharge from a reservoir. The water-surface profiles in compound channels are then discussed.

5-2 Governing Equations

Equations describing the gradually varied flow in a prismatic channel having no lateral inflow or outflow are derived in this section. The following simplifying *assumptions* are made:

   1. The slope of the channel bottom is small.*

---

*An exaggerated vertical scale is used in the illustrations for clarity of presentation. Even though the slope of the channel bottom may appear to be large in these illustrations, it is actually small.

2. The channel is prismatic and there is no lateral inflow or outflow from the channel.

3. The pressure distribution at a channel section is hydrostatic.

4. The head losses in gradually varied flow may be determined by using equations for head losses in uniform flows.

The slope of channel bottom may be assumed small if it is less than 5 percent. In such a case, $\sin \theta \simeq \tan \theta \simeq \theta$, in which $\theta =$ angle of the channel bottom with horizontal and the flow depths measured vertically or normal to the bottom are approximately the same. Because the curvature of streamlines in the gradually varied flows is small, the assumption of hydrostatic pressure distribution is usually valid. The water-surface profiles measured during hydraulic-model investigations and during field observations compare very well with those computed by using the head-loss equations for steady-uniform flow.

By referring to Fig. 5-1, the total head at a channel section may be written as

$$H = z + y + \frac{\alpha V^2}{2g} \tag{5-1}$$

in which $H =$ elevation of the energy line above the datum; $z =$ elevation of the channel bottom above the datum; $y =$ flow depth; $V =$ mean flow velocity, and $\alpha =$ velocity-head coefficient.

Let us consider distance, x, as positive in the downstream flow direction. By differentiating both sides of Eq. 5-1 with respect to x and expressing V in terms of discharge, Q, we obtain

$$\frac{dH}{dx} = \frac{dz}{dx} + \frac{dy}{dx} + \frac{\alpha Q^2}{2g} \frac{d}{dx} \left(\frac{1}{A^2} \right) \tag{5-2}$$

Now, by definition

$$\frac{dH}{dx} = -S_f$$
$$\frac{dz}{dx} = -S_o \tag{5-3}$$

in which $S_f =$ slope of the energy-grade line and $S_o =$ slope of the channel bottom. The negative sign with S_f and S_o indicates that both H and z decrease as x increases. An expression for $(d/dx)(1/A^2)$ may be derived as follows (see also Sec. 2-9):

$$\begin{aligned} \frac{d}{dx} \left(\frac{1}{A^2} \right) &= \frac{d}{dA} \left(\frac{1}{A^2} \right) \frac{dA}{dx} \\ &= \frac{d}{dA} \left(\frac{1}{A^2} \right) \frac{dA}{dy} \frac{dy}{dx} \\ &= -\frac{2B}{A^3} \frac{dy}{dx} \end{aligned} \tag{5-4}$$

since $dA/dy = B$.

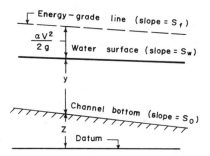

Figure 5-1 Definition sketch.

By substituting Eqs. 5-3 and 5-4 into Eq. 5-2 and rearranging the resulting equation, we obtain

$$\frac{dy}{dx} = \frac{S_o - S_f}{1 - (\alpha B Q^2)/(g A^3)} \tag{5-5}$$

This equation describes the variation of y with x.

Note that if the channel is not prismatic, then

$$\frac{dA}{dx} = \frac{\partial A}{\partial x} + \frac{\partial A}{\partial y}\frac{dy}{dx}$$

Thus Eqs. 5-4 and 5-5 will be modified accordingly (see Problem 5-9). On the basis of the expression for Froude number, $\mathbf{F}_r$, derived in Chap. 3, the second term in the denominator of Eq. 5-5 may be written as

$$\frac{\alpha B Q^2}{g A^3} = \frac{(Q/A)^2}{(g A)/(\alpha B)} \tag{5-6}$$
$$= \mathbf{F}_r^2$$

Hence, Eq. 5-5 takes the form

$$\frac{dy}{dx} = \frac{S_o - S_f}{1 - \mathbf{F}_r^2} \tag{5-7}$$

We will use this equation in the following sections to draw qualitative conclusions about the water-surface profiles.

5-3 Classification of Water-Surface Profiles

We designate water-surface profiles as follows: A letter is used to indicate the type of bottom slope and a numeral is used to indicate the relative position of the profile with respect to the critical-depth line (hereinafter referred to as CDL) and the normal-depth line (hereinafter called NDL). The critical depth and the normal depth are denoted by y_c and y_n, respectively.

The channel-bottom slopes may be classified into the following five categories: mild, steep, critical, horizontal (zero slope), and adverse (negative slope). The first letter of these names is used to indicate the type, i.e., M for mild, S for steep, C for critical, H for horizontal, and A for adverse slope.

The bottom slope is designated as *mild* slope if the uniform flow for the given discharge and Manning n is subcritical (i.e., $y_n > y_c$); it is *critical* slope if the uniform flow is critical (i.e., $y_n = y_c$); and it is *steep* slope if the uniform flow is supercritical (i.e., $y_n < y_c$). The normal depth is infinite if the bottom slope is horizontal and it is nonexistent if the bottom slope is negative. To summarize, the bottom slope is called

- Mild if $y_n > y_c$;
- Steep if $y_n < y_c$; and
- Critical if $y_n = y_c$.

Now, the relative position of the surface profile may be designated as follows. The normal-depth and critical-depth lines divide the space above the channel bottom into three regions, as shown in Fig. 5-2. However, there are only two regions in those cases when the normal depth does not exist, is infinite, or is the same as the critical depth. The region above both lines is designated as *Zone 1*; that between the upper and lower lines is designated as *Zone 2*; and the one between the lower line and the channel bottom is designated as *Zone 3*. Note that the upper line is the normal-depth line if the channel bottom slope is mild, and the upper line is the critical-depth line if the bottom slope is steep.

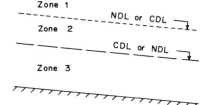

Figure 5-2 Zones for classification of surface profiles.

Thus, we have 13 different types of surface profiles: three for the mild slope, three for the steep slope, two for the critical slope (zone 2 does not exist, since $y_n = y_c$); two for the horizontal slope (zone 1 does not exist since $y_n = \infty$), and two for the adverse slope (there is no zone 1, since y_n does not exist).

Figure 5-3 shows different zones and profiles for all five types of bottom slopes.

5-4 General Remarks

By considering the signs of the numerator and the denominator of Eq. 5-7, we can make qualitative observations about various types of water-surface profiles. These observations allow us to sketch the profile without doing any calculations. For example, they indicate whether the depth increases or decreases with distance, how the profile ends at the upstream and at the downstream limits, etc. First, let us discuss some general points and then consider specific cases for illustration purposes.

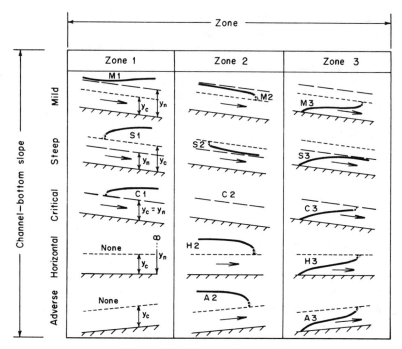

Figure 5-3 Water surface profiles.

The flow depth, y, increases with distance if dy/dx is positive and it decreases if dy/dx is negative. Thus, by determining the signs of the numerator and denominator of Eq. 5-7, we can say whether the flow depth for a particular profile increases or decreases with distance.

We discussed in Chapter 4 that the energy-grade line, water surface, and channel bottom are parallel to each other for uniform flow; i.e., $S_f = S_w = S_o$ when $y = y_n$. Therefore, it is clear from the Manning or Chezy formulas that for a given discharge, Q,

$$S_f > S_o \text{ if } y < y_n \tag{5-8}$$

and

$$S_f < S_o \text{ if } y > y_n \tag{5-9}$$

By using these two inequalities, we can determine the sign of the numerator of Eq. 5-7. Similarly, whether the flow is subcritical ($\mathbf{F}_r < 1$) or supercritical ($\mathbf{F}_r > 1$), we can determine the sign of the denominator of Eq. 5-7.

Now, let us discuss how the surface profiles approach the normal and critical depths and the channel bottom.

As $y \to y_n$ (reads as y tends to y_n), $S_f \to S_o$. Therefore, it follows from Eq. 5-7 that $dy/dx \to 0$ provided $\mathbf{F}_r \neq 1$ (i.e., flow is not critical). In other words, the surface profile approaches the normal-depth line asymptotically.

As $y \to y_c$, $\mathbf{F}_r \to 1$ and the denominator of Eq. 5-7 tends to zero. Therefore, dy/dx tends to ∞. Thus, the water-surface profile approaches the critical-depth line vertically. But, since it is physically impossible to have a vertical water surface, we assume that the surface profile approaches the critical-depth line at a very steep slope. The question arises as to why this conclusion about the vertical water surface derived theoretically is not realized in the real world. The reason for this discrepency is that as soon as the water surface has a sharp curvature, the pressure distribution is not hydrostatic. Therefore, Eq. 5-7 is not valid, and any conclusions we draw from this equation become questionable. As we discussed in the previous chapters, a hydraulic jump is formed when the flow changes from supercritical to subcritical.

As $y \to \infty$, $V \to 0$, and consequently both $\mathbf{F}_r$ and S_f tend to zero. Hence, it follows from Eq. 5-7 that $dy/dx \to S_o$ for very large values of y. Since we are assuming that S_o is small, we may say that the water surface profile almost becomes horizontal as y becomes large.

Now, let us see what happens when the water surface approaches the channel bottom, i.e., $y \to 0$. If we use Chezy's formula for the friction losses, then

$$S_f = \frac{Q^2}{C^2 A^2 R} \tag{5-10}$$

in which $C = $ Chezy constant and $R = $ hydraulic radius.

For a very wide rectangular channel, $R \simeq y$. By substituting Eq. 5-10 into Eq. 5-5, replacing R by y, and simplifying the resulting equation, we obtain

$$\frac{dy}{dx} = \frac{gB\left(S_o C^2 B^2 y^3 - Q^2\right)}{C^2\left(gBy^3 - \alpha B Q^2\right)} \tag{5-11}$$

It follows from this equation that when $y \to 0$,

$$\lim_{y \to 0} \frac{dy}{dx} = \frac{g}{\alpha C^2} \tag{5-12}$$

Therefore, as $y \to 0$, the slope of the water-surface profile is finite, has a positive value, and is a function of the Chezy constant, C, and the velocity-head coefficient, α.

If we use Manning formula instead of Chezy formula, we will find that $dy/dx \to \infty$ as $y \to 0$. This is left as an excercise for the reader to prove (see Problem 5-5).

To illustrate the use of the above general remarks, let us consider water-surface profiles in a channel having a mild slope. As we discussed before, $y_n > y_c$ if the slope is mild. Therefore, the flow depth, y, in the three zones is as follows:

- Zone 1: $y > y_n > y_c$
- Zone 2: $y_n > y > y_c$
- Zone 3: $y_n > y_c > y$

The qualitative characteristics of the water-surface profiles may be studied as follows.

Zone 1 (M1 Profile).   Since $y > y_n$ in zone 1, $S_f < S_o$. Therefore, the numerator of Eq. 5-7 is positive. Similarly, we have $\mathbf{F}_r < 1$, since $y > y_c$. Therefore, the denominator of Eq. 5-7 is positive as well. Hence, it follows from Eq. 5-7 that

$$\frac{dy}{dx} = \frac{S_o - S_f}{1 - \mathbf{F}_r^2} = \frac{+}{+} = +$$

This means that y increases with distance x. As discussed previously, $y \rightarrow y_n$ asymptotically, and the water surface becomes almost horizontal as y becomes large.

Zone 2 (M2 Profile).   In this case, $S_f > S_o$, since $y < y_n$. Therefore, the numerator of Eq. 5-7 is negative. However, the denominator is positive, since $\mathbf{F}_r < 1$ because $y > y_c$. Hence, it follows from Eq. 5-7 that

$$\frac{dy}{dx} = \frac{S_o - S_f}{1 - \mathbf{F}_r^2} = \frac{-}{+} = -$$

Thus, y decreases as x increases. As discussed previously, $y \rightarrow y_n$ asymptotically; and $y \rightarrow y_c$ almost vertically.

Zone 3 (M3 Profile).   In Zone 3, $S_f > S_o$, since $y < y_n$. Therefore, the numerator of Eq. 5-7 is negative. The denominator is negative as well, since $\mathbf{F}_r > 1$ because $y < y_c$. Hence, it follows from Eq. 5-7 that

$$\frac{dy}{dx} = \frac{S_o - S_f}{1 - \mathbf{F}_r^2} = \frac{-}{-} = +$$

Thus, y increases as x increases.

As discussed previously, $y \rightarrow y_c$ almost vertically while the water surface profile approaches the channel bottom at a finite positive slope.

By using the preceding qualitative conclusions, the water-surface profiles in each region may be sketched as shown in Fig. 5-3. Note that the profiles are shown by dashed lines as they approach the critical-depth line and as they approach the channel bottom to indicate uncertainty in their shapes.

Proceeding similarly, the qualitative characteristics of surface profiles for other types of channel bottom slopes may be studied. In general, the procedure is as follows: We first determine the signs of the numerator and of the denominator of Eq. 5-7 and hence determine the sign of dy/dx. Then, by utilizing the qualitative remarks made in the previous paragraphs, we may sketch the water-surface profiles as they approach the normal- and critical-depth lines and the channel bottom.

The characteristics and shapes of various profiles and the situation in which they occur in real life are presented in Fig. 5-4.

Note that H1 and A1 profiles do not exist, since there is no zone 1 in both cases. In addition, profile C2 actually represents uniform flow rather than gradually varied flow.

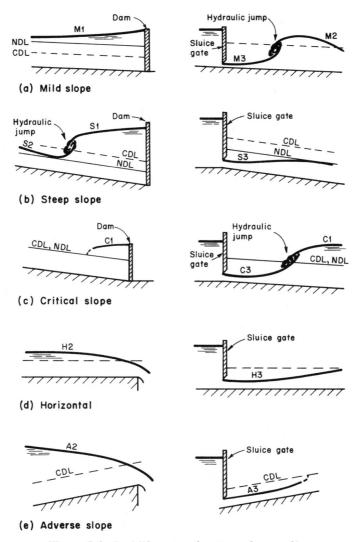

Figure 5-4 Real-life cases of water-surface profiles.

5-5 Sketching Water-Surface Profiles

A channel section at which there is a unique relationship between the depth and discharge is referred to as a *control*. The properties of surface profiles we discussed in the previous two sections are for a prismatic channel having a control section either at the upstream or at the downstream end. In real life, however, many situations arise in which there may be several control sections. In addition, a channel system having variable cross section or bottom slope may be divided into several prismatic channels. To qualitatively sketch the profiles in these cases, a number of guidelines are outlined in the following paragraphs, and two examples are included for illustrative purposes.

For the specified discharge, roughness coefficient, and channel cross section, compute the normal and critical depths in each channel. Now, using an exaggerated vertical scale, plot the channel bottom and the normal- and critical-depth lines. Then, on this diagram, mark the locations of controls—i.e., the locations where the water surface profile passes through critical depth ($y = y_c$) and identify the channel reaches where the flow is expected to be uniform ($y = y_n$).

A downstream control governs if the flow is subcritical and an upstream control governs if the flow is supercritical. It is possible to have situations where part of the channel is governed by the upstream control and part of the channel is governed by the downstream control. In addition, a control at an intermediate location (e.g., a weir, sluice gate, or spillway) may act as a control for both the upstream and downstream directions from the control location.

At a channel entrance, the surface profile passes through the critical depth if the lake or reservoir level is higher than the critical-depth line, and the channel bottom slope at the channel entrance is steep. To allow for the velocity head and losses at the channel entrance, the water surface at the upstream end of the channel may be slightly lower than the water level in the reservoir.

At a free overfall, the water surface passes through the critical-depth line approximately three to four times the critical depth upstream of the fall if the flow depth upstream of the fall is greater than the critical depth.

A hydraulic jump is formed whenever the flow changes from supercritical to subcritical flow. The exact location of the jump is determined by detailed calculations, as discussed in Chapter 8. However, an approximate location of the jump may be estimated while sketching the water-surface profiles.

The following examples illustrate these procedures.

Example 5-1

Sketch the water-surface profile in the channels connecting the two reservoirs, as shown in Fig. 5-5(a). The bottom slope of channel 1 is steep and that of channel 2 is mild.

Solution   Compute the critical and normal depths for each channel. Then plot the critical-depth line (marked as CDL in Fig. 5-5(b)) and the normal-depth line (marked as NDL in Fig. 5-5(b)).

The water depth at the channel entrance is equal to the critical depth, since the water level in the upstream reservoir is above the CDL of channel 1. Let us mark this water level at the channel entrance by a dot. The water level at the downstream end is lower than the CDL at the downstream end of channel 2. Therefore, the water surface passes through the CDL approximately three to four times the critical depth upstream of the entrance to the downstream reservoir. Let us again mark this water level at the downstream end by a dot, as shown in Fig. 5-5(b).

In channel 1, the water surface at the entrance passes through the critical depth and then it tends to the normal depth. Thus, we have an S2 profile in channel 1. The flow decelerates downstream of the junction of channels 1 and 2 because of mild slope. Hence, the flow depth keeps on increasing until it intersects the CDL. Approximately at this location, a hydraulic jump is formed. The water surface follows the M2 profile downstream of the jump. Detailed calculations are required to determine the exact location of the jump.

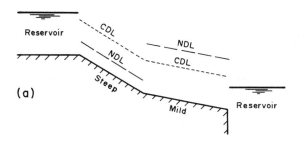

(a)

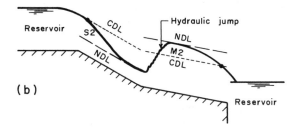

(b)

Figure 5-5 Water-surface profiles for Example 5-1.

Example 5-2

Sketch all possible water-surface profiles in the channel of Fig. 5-6. The channel is long and has a steep slope. Consider two different cases of gate opening.

Solution   Several different water-surface profiles are possible depending upon the location of the control section. We may divide these profiles into the following two categories:

    1. Control at the channel entrance
    2. Control at the gate

Let us consider each of these cases.

    If the control is at the channel entrance, then the channel discharge is not controlled by the sluice gate. This is because we have supercritical flow in part of the channel length upstream of the sluice gate. Depending upon the gate opening, several water-surface profiles are possible upstream of the gate, as shown in Fig. 5-6(a) and 5-6(b). The water-surface profile asymptotically approaches the NDL, since the channel is long. Depending upon the gate opening, S1, S2, or S3 profiles are possible, as shown in Fig. 5-6.

    The sluice gate controls the channel discharge only if the backwater from the gate extends to the channel entrance. In such a case, we have an S1 profile upstream of the gate in all cases (Fig. 5-6). Downstream of the gate, however, the type of the surface profile depends upon the gate opening, as shown in Fig. 5-7.

    In this figure, a small drop in the water level is shown at the channel entrance to account for the entrance losses and the velocity head.

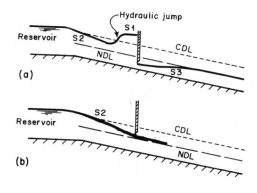

Figure 5-6 Water-surface profiles for control at channel entrance.

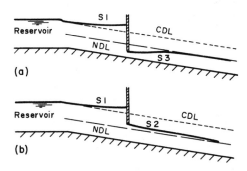

Figure 5-7 Water-surface profiles for control at gate.

5-6 Discharge From a Reservoir

In the discussion for sketching the water-surface profiles in the previous sections, we assumed that the discharge is known. However, this may not always be the case, as the following example illustrates.

Let us consider a channel-reservoir system as shown in Fig. 5-8. The reservoir is large so that the flow velocity in the reservoir approaches zero. In addition, the reservoir water level is known and remains constant independent of the discharge in the channel. The channel cross section, entrance-loss coefficient, k, Manning n, and channel-bottom slope, S_o, are specified. We want to determine the flow depth, y, and the discharge, Q, in the channel.

Referring to Fig. 5-8, H_o, S_o, n, and the properties of the channel section are known and we want to determine y and Q.

For the specified flow variables and channel parameters, the bottom slope may be

1. Steep;
2. Critical; or
3. Mild.

The flow depth at the channel entrance is critical if the bottom slope is critical or steep and the reservoir water level is higher than the critical-depth line. However,

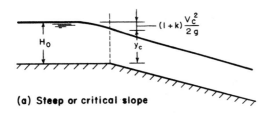

(a) Steep or critical slope

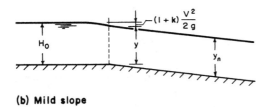

(b) Mild slope

Figure 5-8 Discharge from a reservoir.

normal depth occurs just downstream of the channel entrance if the bottom slope is mild.

To determine the type of bottom slope, let us first determine the critical slope, S_c. The following two equations describe the relationships between different flow variables if the flow depth is critical at the channel entrance and $\alpha = 1$.

$$\frac{Q^2}{2gA^2} = \frac{D}{2} \tag{5-13}$$

and

$$H_o = y_c + (1+k)\frac{Q^2}{2gA^2} \tag{5-14}$$

in which k = entrance loss coefficient and D = hydraulic depth. Note that both D and A are functions of y_c. We can solve these two equations for Q and y_c. If the slope of the channel bottom is equal to the critical slope, S_c, then the flow at this depth and discharge will be uniform. By utilizing this fact, we can determine the value of S_c from the Manning equation

$$Q = \frac{1}{n}AR^{2/3}S_c^{1/2} \tag{5-15}$$

The channel bottom slope is critical if $S_o = S_c$; it is steep if $S_o > S_c$; and, it is mild if $S_o < S_c$.

The discharge and the flow depth we determined above are correct if the slope is critical; while only the discharge is correct if the bottom slope is steep. The flow depth may now be computed, starting with the critical depth at the entrance. However, if the slope is mild, then we solve the following two equations simultaneously to determine y and Q:

$$Q = \frac{1}{n} A R^{2/3} S_o^{1/2} \tag{5-16}$$

and

$$
\begin{aligned}
H_o &= y + \frac{V^2}{2g} + k \frac{V^2}{2g} \\
&= y + \frac{1+k}{2g} \left(\frac{Q}{A} \right)^2
\end{aligned}
\tag{5-17}
$$

Eliminating Q from Eqs. 5-16 and 5-17, we obtain

$$H_o = y + \frac{1+k}{2gn^2} R^{4/3} S_o \tag{5-18}$$

Solution of this equation gives the flow depth in the channel. The discharge corresponding to this depth can now be determined from Eq. 5-16.

The following example illustrates this procedure.

Example 5-3

A 10-m-wide, rectangular, concrete-lined channel ($n = 0.013$) has a bottom slope of 0.01 and a constant-level reservoir at the upstream end. The reservoir water level is 6.0 m above the channel bottom at entrance. Assuming the entrance losses and the approach velocity in the reservoir to be negligible, determine the channel discharge and qualitatively sketch the water surface profile.

Given:

$n = 0.013$
$S_o = 0.01$
$B = 10.0$ m
$H_o = 6.0$ m

Entrance losses are negligible.

Determine:

$Q = ?$
Water-surface profile = ?

Solution: Let us assume the control is at the channel entrance, i.e., the bottom slope is steep or critical. Then,

$$
\begin{aligned}
y_c &= \frac{2}{3} H_o \\
&= \frac{2}{3} \times 6 \\
&= 4 \text{ m}
\end{aligned}
$$

For critical flow, unit discharge,

$$q = \sqrt{g y_c^3}$$
$$= \sqrt{9.81(4)^3}$$
$$= 25.06 \, \text{m}^3/\text{s/m}.$$

Hence,

$$Q = Bq$$
$$= 10 \times 25.06$$
$$= 250.6 \, \text{m}^3/\text{s}$$

Let us now determine the critical slope, S_c. This is the bottom slope for which we will have critical flow in the channel for $Q = 250.6 \, \text{m}^3/\text{s}$. Now, the Manning equation may be written as

$$Q = \frac{1}{n} A R^{2/3} S_c^{1/2}$$

or

$$S_c = \frac{n^2 Q^2}{A^2 R^{4/3}}$$
$$= \frac{(0.013)^2 (250.6)^2}{(10 \times 4)^2 [40/(10 + 8)]^{4/3}}$$
$$= 0.00229$$

Since, $S_c < S_o$, the slope of the channel bottom is steep and the channel discharge is 250.6 m³/s.

To sketch the water-surface profile, we first determine the normal depth. This may be done using the procedure we presented in Chapter 4. The flow area, A, and the hydraulic radius, R, corresponding to the normal depth satisfy the following equation

$$A R^{2/3} = \frac{nQ}{\sqrt{S_o}}$$

The substitution of the values of n, Q, and S_o and the expressions for A and R in terms of y_n into this equation gives

$$10 y_n \left[\frac{10 y_n}{10 + 2 y_n} \right]^{2/3} = \frac{0.013 \times 250.6}{\sqrt{.01}} = 32.57$$

The solution of this equation by trial and error yields

$$y_n = 2.37 \, \text{m}$$

The water surface at the entrance will be at the critical depth and then it will asymptotically approach the normal depth, as shown in Fig. 5-9.

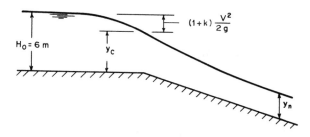

Figure 5-9 Water-surface profile for Example 5-3.

5-7 Profiles in Compound Channels

The discussion of water-surface profiles in the previous sections is applicable only to channels having regular cross sections. However, water-surface profiles in compound channels (channels having a compound cross section) should be given a special treatment because there may be more than one critical depth, as we discussed in Chapter 3. Quintela (1982) briefly discussed the shapes of flow profiles in a compound channel for two different cases of steep and mild slopes. He illustrated the occurrence of rapidly varied flow when the slope of the channel bottom changes from mild to steep. He tacitly assumed that the normal depth, y_n, is greater than the highest critical depth, y_{c_3}, if the slope is mild and that y_n is less than the lowest critical depth, y_{c_1}, if the slope is steep. However, another situation is possible when $y_{c_1} < y_n < y_{c_3}$, as pointed out by Chaudhry and Bhallamudi (1988). In this case, critical depth, y_{c_2} becomes important (Fig. 5-10).

In this section, two examples are presented to show how the possibility of multiple critical depths affects the water-surface profile in a compound channel. First, a long channel with a free overfall at the downstream end is considered and then a long channel with a reservoir at the upstream end.

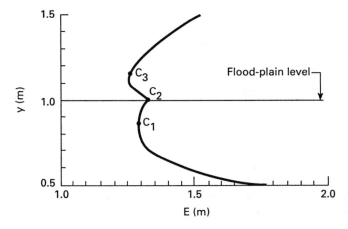

Figure 5-10 Specific energy versus depth curves for compound channel.

Example 5-4

A long channel with a compound cross section (Fig. 5-11) has a free overfall at the downstream end. The channel discharge is 2.5 m³/s, and the Manning n for the main channel and for the flood plain are 0.013 and 0.0144, respectively. Discuss and sketch the water surface profiles for the following four channel bottom slopes, $S_o = 0.0094, 0.0049, 0.0029,$ and 0.001.

The variation of specific energy and Froude number with depth for the section and for the specified discharges are plotted in Fig. 5-12. There are three critical depths: $y_{c_1} = 0.86$ m, $y_{c_2} = 1.002$ m and $y_{c_3} = 1.12$ m.

Let us consider each of the channel bottom slopes one by one.

$S_o = 0.0094$.   For this bottom slope, $y_n = 0.75$ m (determined by solving the Manning equation by trial and error). As shown in Fig. 5-12, $y_n < y_{c_1}$ and the Froude number F_{rc} is greater than 1. Therefore, the flow is supercritical, control is at the upstream end, and the free overfall at the downstream end does not affect the flow. Once the flow depth approaches the normal depth, it changes only slightly at the downstream end. The water-surface profile is shown in Fig. 5-13(a).

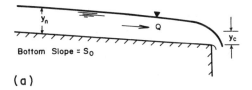

(a)

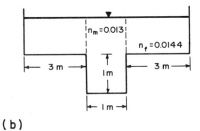

(b)

Figure 5-11 Compound channel with a free fall.

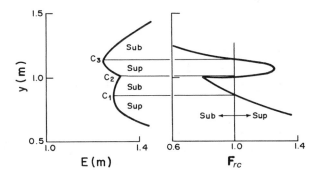

Figure 5-12 Variation of E and $\mathbf{F}_{rc}$ with depth for channel of Fig. 5-11.

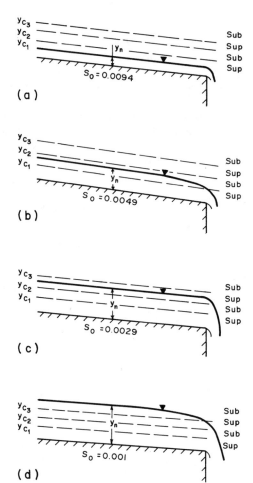

Figure 5-13 Water-surface profiles for channel of Fig. 5-11.

$S_o = 0.0049$.   For this slope, $y_n = 0.97$ m; i.e., $y_{c_1} < y_n < y_{c_2}$ and $\mathbf{F}_{rc} < 1$. Therefore, the flow is subcritical (Fig. 5-12) and the control is at the downstream end. The water-surface profile may be computed starting near the downstream end with a depth equal to y_{c_1}. The water-surface profile is shown schematically in Fig. 5-13(b).

$S_o = 0.0029$.   For this slope, $y_n = 1.07$ m; i.e., $y_{c_2} < y_n < y_{c_3}$. Figure 5-12 shows that the Froude number for this normal depth is greater than 1. Therefore, the flow is supercritical control is at the upstream end and the free overfall at the downstream end does not significantly affect the flow in the channel. The water-surface profile is shown schematically in Fig. 5-13(c).

$S_o = 0.001$.   For this slope, $y_n = 1.2$ m; i.e., $y_n > y_{c_3}$ and the flow is subcritical, as indicated by Fig. 5-12. The water-surface profile may be computed by starting at the downstream end with depth equal to y_{c_3}. The depth varies from normal depth, y_n, at some upstream point to the critical depth, y_{c_3}, near the downstream end. The flow varies rapidly from one critical depth, y_{c_3}, to the other, y_{c_1}, at the downstream end. The flow profile is schematically shown in Fig. 5-13(d).

Example 5-5

Water flows from a reservoir into a long channel with a compound cross section, as shown in Fig. 5-14. The reservoir water surface is 1.2 m above the channel bottom at the channel entrance. Sketch various possible water-surface profiles.

The method presented in the previous section to determine the channel discharge is followed. First, the channel-bottom slope corresponding to the critical flow (critical slope) is calculated. If the actual channel bottom slope is steeper than the critical slope, then the flow is supercritical. The flow is subcritical if the channel-bottom slope is less than the critical slope. Since it is possible to have more than one critical slope for a compound channel, it is necessary to consider them in the analysis. The following discussion illustrates this point.

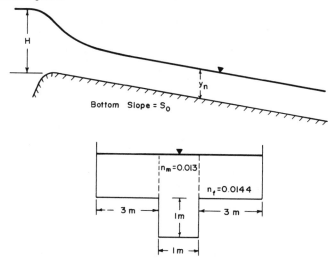

Figure 5-14 Channel of Example 5-5.

Figure 5-15 shows the variation of discharge and Froude number with depth for the given reservoir level, $H = 1.2$ m. The discharge, Q, is calculated by solving the following energy equation:

$$Q = \left[2(H - y) \frac{g A^2}{\alpha} \right]^{1/2} \tag{5-19}$$

in which y is the flow depth in the channel. The entrance losses and the velocity of approach are neglected in this equation.

Critical conditions occur when $\mathbf{F}_{rc} = 1$. It is clear from the discharge-versus-depth diagram (Fig. 5-15) that the discharge is not necessarily a local maximum at critical conditions. For example, it is actually minimum for y_{c_2}. As can be seen, there are three critical depths: $y_{c_1} = 0.8$ m, $y_{c_2} = 1.03$ m, and $y_{c_3} = 1.10$ m. Corresponding to these critical depths, there are three critical discharges, $Q_{c_1} = 2.241$ m^3/s, $Q_{c_2} = 1.960$ m^3/s, and $Q_{c_3} = 2.0$ m^3/s. Critical bottom slopes, S_c, corresponding to these critical discharges

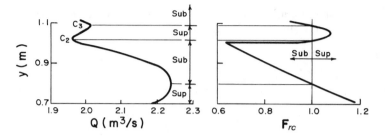

Figure 5-15 Variation of Q and $\mathbf{F}_{rc}$ with depth.

may be solved from the Manning formula as $S_{c_1} = 0.0064$, $S_{c_2} = 0.0024$ and $S_{c_3} = 0.0015$, respectively.

Depending upon the channel bottom slope, S_o, the following four cases are possible.

$S_o > S_{c_1}$.   This bottom slope is steeper than all three critical slopes and the flow in the channel is supercritical. Flow depth varies rapidly from the reservoir level to the critical depth, y_{c_3}, passes through the other two critical depths, y_{c_2} and y_{c_1}, and approaches the normal depth, as shown in Fig. 5-16(a). The discharge in the canal is equal to the discharge corresponding to y_{c_1}, i.e., $Q = Q_{c_1} = 2.24$ m^3/s. We will have the same discharge for all values of slope, $S_o > S_{c_1}$.

$S_{c_1} > S_o > S_{c_2}$.   For this bottom slope, y_n is greater than y_{c_1} but less than y_{c_2}. Therefore, as indicated by Fig. 5-15, the Froude number is less than 1 and the flow is subcritical. The discharge in the channel depends on the channel bottom slope and may be determined by solving the energy equation simultaneously with the Manning equation for uniform flow. For $S_o = 0.0035$, the discharge is equal to 2.08 m^3/s and the corresponding depth is equal to 0.96 m. The flow profile is shown schematically in Fig. 5-16(b).

$S_{c_2} > S_o > S_{c_3}$.   For this bottom slope, the normal depth, y_n, is greater than y_{c_2} but less than y_{c_3}. Therefore, as indicated by Fig. 5-15, the Froude number is greater than 1 and the flow is supercritical. Flow depth passes through the critical depth y_{c_3} and then approaches the normal depth. The discharge in the canal is equal to the discharge corresponding to y_{c_3}, i.e., 2.0 m^3/s. The flow profile is shown schematically in Fig. 5-16(c).

$S_o < S_{c_3}$.   In this case, the channel bottom slope is less than all critical slopes and the flow is subcritical. The channel discharge may be calculated as discussed previously for case 2. The flow profile is shown schematically in Fig. 5-16(d).

5-8 Summary

In this chapter, an equation describing the spatial variation of flow depth in gradually varied flow was derived. The classification of water-surface profiles was discussed, and several general remarks were made on the properties of water surface profiles. Procedures for sketching the water surface profiles in a channel were then discussed. A procedure for determining the discharge from a reservoir was presented. The properties of water surface profiles in a compound channel were outlined.

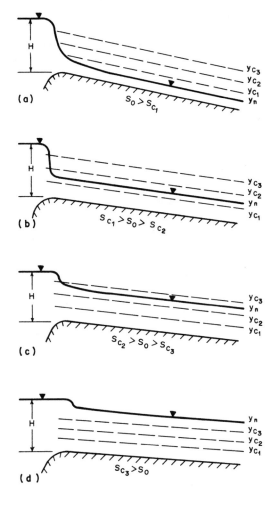

Figure 5-16 Water-surface profiles.

Problems

5-1. Prove that the gradually varied flow equation for a wide rectangular channel may be written as

$$\frac{dy}{dx} = S_o \frac{1 - (y_n/y)^{10/3}}{1 - (y_c/y)^3}$$

if the Manning formula is used and as

$$\frac{dy}{dx} = S_o \frac{1 - (y_n/y)^3}{1 - (y_c/y)^3}$$

if the Chezy formula is used for the friction losses.

5-2. Derive the gradually varied flow equation for a prismatic channel having lateral flow of q per unit length. Assume that the lateral flow enters the channel perpendicular to the flow direction. Will this equation be valid if we have lateral outflow instead of lateral inflow?

5-3. Sketch the water-surface profiles in the channel system of Fig. 5-17. In this figure, NDL and CDL denote normal and critical depth lines, respectively.

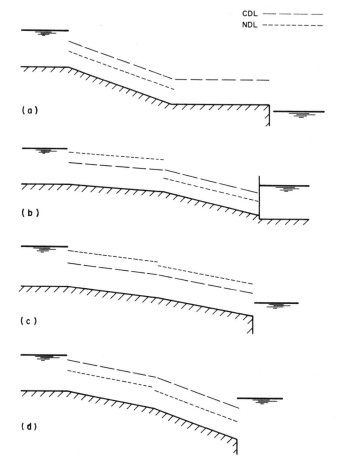

Figure 5-17 Channel systems for Prob. 5-3.

5-4. If a sluice gate is used to control flow from a lake, should the gate be located near to the lake outlet or at a long distance? Why? If the gate is located at a sufficient distance from the lake outlet, is there a situation in which the outflow from the lake does not depend upon the gate opening? Sketch all possible flow situations assuming the channel-bottom slope to be

    i. Mild;
    ii. Steep.

5-5. If the Manning formula is used to compute S_f, prove that the slope of the water surface in gradually varied flow, dy/dx, approaches ∞ as $y \to 0$.

5-6. A 5-m-wide rectangular concrete-lined canal takes off from a lake having a constant water level of 2 m above the channel bottom at the entrance. The channel is long and has a bottom slope of 0.004, and $n = 0.013$.

  i. If the head losses at the entrance are negligible, determine the rate of discharge in the canal.

  ii. Compute the rate of discharge if the bottom slope is changed to 0.001 and the entrance losses are $0.1 V^2/(2g)$.

5-7. Lakes A and B are connected by a 10-m-wide rectangular channel, as shown in Fig. 5-18. If n for the flow surfaces is 0.013, sketch the water-surface profile in the channel if the Lake B is at

  i. El. 155.0;

  ii. El. 161.0.

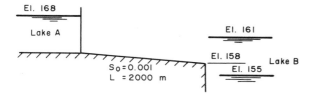

Figure 5-18 Lake system for Prob. 5-7.

5-8. A 15-m-wide, 15-km-long, concrete-lined channel ($n = 0.013$) is planned for conveying water from reservoir X to reservoir Y. The water level in reservoir X is at El. 129.65 m and the level of the channel bottom at the entrance is at El. 121.4 m. Determine the channel discharge and sketch and label the type of water-surface profile in each case.

  i. The slope of the channel bottom is 0.001 and the water level in reservoir Y is at El. 109 m.

  ii. The slope of the channel bottom is 0.008 and the water level in reservoir Y is at El. 7 m.

Assume the entrance losses are negligible in both cases.

5-9. Prove that the following equation describes the gradually varied flow in a channel having variable cross section along its length:

$$\frac{dy}{dx} = \frac{S_o - S_f + (V^2/gA)\partial A/\partial x}{1 - (BV^2)/(gA)}$$

5-10. For a wide rectangular channel, derive expressions for the channel bottom slope to be mild, steep, and critical.

5-11. A chute spillway is blasted through rock and is not lined. The bottom drops 1.5 ft in 20 ft. Determine the flow depth and the rate of discharge in the chute if the reservoir water level is 10 ft above the channel bottom at the entrance.

5-12. Name the water-surface profiles shown in Fig. 5-19.

5-13. Sketch the water-surface profiles in the channels shown in Fig. 5-20.

5-14. The bottom slope of a long trapezoidal channel (bottom width = 15 ft, side slopes = 1:1) is suddenly changed from 0.0005 to 0.05. The flow in the channel is 800 ft³/sec. and the Manning n is 0.028. Compute the critical and normal flow depths in each channel reach and sketch the water-surface profile.

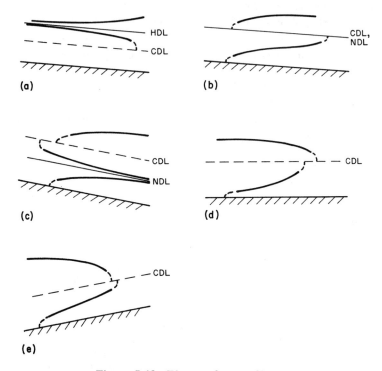

Figure 5-19 Water-surface profiles.

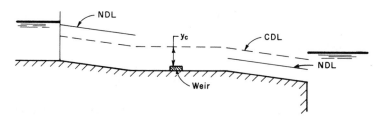

Figure 5-20 Channels for Prob. 5-13.

5-15. Sketch the water-surface profile for the channel of Prob. 5-14 if the channel downstream of the slope change is long and has a bottom slope of 0.0003. The bottom slope of the upstream channel is 0.05.

5-16. Sketch and label the types of water surface profiles in the channel of Fig. 5-21.

5-17. Sketch and label the type of water surface profiles in the channel shown in Fig. 5-22.

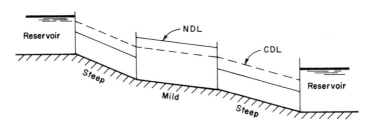

Figure 5-21   Channel for Prob. 5-16.

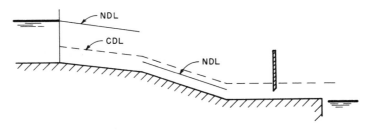

Figure 5-22   Channel of Prob. 5-17.

REFERENCES

Chaudhry, M. H., and Bhallamudi, S. M.   1988. Critical depth in compound channels, *Jour. Hydraulic Research*, Inter. Assoc. for Hydraulic Research, 26, no. 4:377–95.

Chow, V. T.   1959. *Open-channel hydraulics*, McGraw-Hill, New York, NY.

Gill, H. A.   1976. Exact solution of gradually varied flow, *Jour. Hydraulics Division*, Amer. Soc. Civil Engrs., 102, no. 9:1353-64

Henderson, F. M.   1966. *Open channel flow*, Macmillan, New York, NY.

Quintela, A. C.   1982. Discussion of Blalock, M.E. and Sturm, T.W. Minimum specific energy in compound channel, *Jour. Hydraulics Division*, Amer. Soc. Civil Engrs., 108:729–31. (See also closure, vol. 109, (1983):483–86).

Sturm, T. W.; Skolds, D. M.; and Blalock, M. E.   1985. Water surface profiles in compound channels, *Proc.,* Hydraulics and Hydrology in Computer Age, Amer. Soc. Civil Engrs., 1:569–74.

6 Computation of Gradually Varied Flow

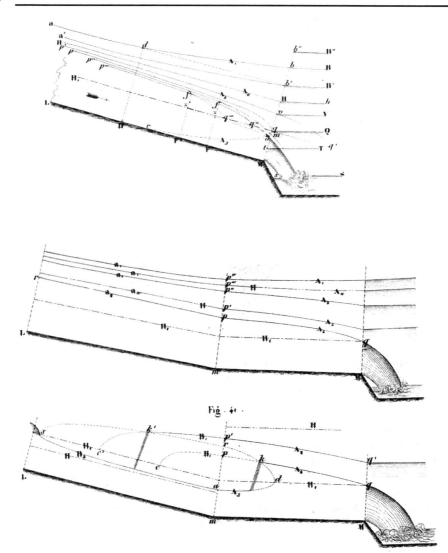

Gradually varied flow profiles in a single, and in two series channels (After Boudine, 1861).

6-1 Introduction

In the last chapter, we discussed how to qualitatively sketch water-surface profiles in channels having gradually varied flows. For many engineering applications, however, it is necessary to compute the flow conditions in these flows. These computations, generally referred to as water-surface profile calculations, determine the water-surface elevations along the channel length for a specified discharge. The water-surface elevations are required for the planning, design, and operation of open channels so that the effects of the addition of engineering works and the channel modifications on water levels may be assessed. The addition of a dam, for example, raises water levels upstream of the dam. It is necessary to know the flow depths to determine the extent of flooding.

In addition, steady-state flow conditions are needed to specify proper initial conditions for the computation of unsteady flows.* Improper initial conditions introduce false transients into the simulation, which may lead to incorrect results. It is possible to use unsteady-flow algorithms directly to determine the initial conditions by computing for long simulation time. However, such a procedure is computationally inefficient and may not converge to the proper steady-state solution if the finite-difference scheme is not consistent.

In this chapter, methods to compute gradually varied flows are presented. Preference is given to the methods suitable for a computer solution. Two traditional methods—commonly referred to as the direct- and standard-step methods—are first presented. The computations progress step by step in these methods. Then higher-order accurate methods to numerically integrate the governing differential equation are introduced. A procedure is then presented to compute the flow conditions at all specified locations of a channel system simultaneously instead of computing them from one section to the next.

6-2 General Remarks

The continuity, momentum, and energy equations describe the relationships among various flow variables, such as the flow depth, discharge, and flow velocity. Therefore, we solve these equations to determine the flow conditions throughout a specified channel length. These analyses yield the change in flow depth in a given distance or compute the distance in which a specified change in flow depth will occur. The channel cross

---

*These are discussed in Chapters 11 through 16.

section, Manning n, bottom slope, and the rate of discharge are usually known for these steady-state-flow computations.

The rate of change of flow depth in gradually varied flows is usually small. Therefore, the assumption of hydrostatic pressure distribution is valid. In addition, by introducing the velocity-head coefficient, α, we may use the mean flow velocity to compute the velocity head at a channel section. For a channel having no lateral inflows or outflows, the continuity equation between sections 1 and 2 (Fig. 6-1) may be written as

$$Q = V_1 A_1 = V_2 A_2 \tag{6-1}$$

in which V = mean flow velocity; A = flow area; Q = rate of discharge; and the subscripts 1 and 2 refer to the variables for sections 1 and 2, respectively. Similarly, the energy equation between sections 1 and 2 of a channel with small bottom slope may be written as

$$z_1 + y_1 + \alpha_1 \frac{V_1^2}{2g} = z_2 + y_2 + \alpha_2 \frac{V_2^2}{2g} + h_l \tag{6-2}$$

in which z = elevation of the channel bottom above a specified datum; y = flow depth; and h_f = head losses between sections 1 and 2. The head losses comprise the friction and form losses between these two sections.

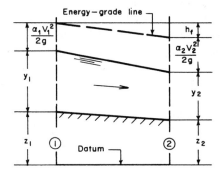

Figure 6-1 Definition sketch.

We derived the following equation for gradually varied flows in Chapter 5 by differentiating the energy equation

$$\frac{dy}{dx} = \frac{S_o - S_f}{1 - (\alpha Q^2 B)/(g A^3)} \tag{6-3}$$

in which x = distance along the channel (measured positive in the downstream direction); S_o = longitudinal slope of the channel bottom; S_f = slope of the energy line; B = top water-surface width; and g = acceleration due to gravity. If the momentum coefficient β is equal to unity, then this equation may also be obtained by applying Newton's second law of motion to a volume of water in a short channel length (see Problem 6-1).

Equation 6-3 is a first-order ordinary differential equation in which x is the independent variable and y is the dependent variable. This equation describes the rate of variation of flow depth, y, with respect to distance, x. A close look at the terms of the right-hand side of this equation shows that this rate is a function of the properties of the channel section, flow depth, and the rate of discharge. For a given channel section, the channel properties (e.g., top water-surface width, B, and the flow area, A) are functions of y only. The bottom slope, S_o, Manning n, and the rate of discharge, Q, are known for a given problem. Therefore, for a specified discharge in a given channel, we may say that the right-hand side of Eq. 6-3 is a function of the flow depth, y, which in turn is a function of x. Let us designate this function as $f(x, y)$. Then we may write Eq. 6-3 as

$$\frac{dy}{dx} = f(x, y) \tag{6-4}$$

in which

$$f(x, y) = \frac{S_o - S_f}{1 - (\alpha B Q^2)/(g A^3)} \tag{6-5}$$

We integrate this equation to determine the flow depth along a channel length. A closed-form solution of this equation is not available except for very simplified cases because $f(x, y)$ is a nonlinear function. Therefore, numerical methods are used for its integration. These methods yield the flow depth at discrete locations. Let us first discuss the basis of these methods.

Let the flow depth, y, be known at a given distance, x. Let us denote these known values by y_1 and x_1, respectively. To determine the water-surface profile, we may follow either of the following two procedures: Determine y_2 at a specified location, x_2, or determine the location x_2 where specified flow depth y_2 will occur. Let us discuss each of them in more detail, starting with determining the value of y_2 at distance x_2.

By multiplying both sides of Eq. 6-4 by dx, and integrating, we obtain

$$\int_{y_1}^{y_2} dy = \int_{x_1}^{x_2} f(x, y) \, dx \tag{6-6}$$

This equation, upon applying the limits of integration, may be written as

$$y_2 = y_1 + \int_{x_1}^{x_2} f(x, y) \, dx \tag{6-7}$$

Computations progress in the downstream direction if dx is positive and they progress in the upstream direction if dx is negative. We can determine y_2 by numerically evaluating the integral term of this equation. Then, by successive application of this equation, we may compute the water-surface profile in the desired channel length.

To determine x_2 where the flow depth will be y_2, we may proceed as follows. Equation 6-3 may be written as

$$\frac{dx}{dy} = F(x, y) \tag{6-8}$$

in which

$$F(x, y) = \frac{1 - (\alpha B Q^2)/(g A^3)}{S_o - S_f} \tag{6-9}$$

By multiplying both sides of this equation by dy and integrating

$$x_2 = x_1 + \int_{y_1}^{y_2} F(x, y)\, dy \qquad (6\text{-}10)$$

The value of x_2 may be determined by numerically evaluating the integral term.

Instead of the differential equation, Eq. 6-3, we may use the energy equation, Eq. 6-2, between two sections for computing the water-surface profile. The main difficulty in the use of this equation is in determining the head losses, h_f. After selecting an expression to approximate h_f, we solve a nonlinear algebraic equation to determine the flow depth at a specified location or to determine the location where a specified flow depth will occur. These are discussed in the next two sections. Of course, similar difficulty arises if we use the differential equation. The average value of S_f between two sections may be used if the distance between the sections is short. This is usually a satisfactory approximation since short step lengths are used to numerically integrate Eq. 6-3.

Several procedures to compute the water-surface profiles have been developed (Bakhmeteff 1932; Chow 1959; Henderson 1966; Eichert 1970; Prasad 1970; McBean and Perkins 1975; Chaudhry and Schulte 1985; and Schulte and Chaudhry 1987). Some of the earlier procedures used various varied-flow functions developed by integrating the differential equation (Eq. 6-3) describing the gradually varied flows. Several graphical and mathematical methods were developed for the integration of this equation or for solving the energy equation between two sections (Bakhmeteff 1932; Chow 1959; Henderson 1966). Some of these methods have been used in various general-purpose computer programs for computing water-surface profiles (Soil Conservation Service 1976; U.S. Geological Survey 1976; U.S. Army Corps of Engineers 1982).

We discussed in Chapters 3 and 5 that subcritical flow has a downstream control and the supercritical flow has an upstream control. To compute the water-surface profile, we start the computations at a location where the flow depth for the specified discharge is known. Consequently, we start the computations at a downstream control section if the flow is subcritical and proceed in the upstream direction. In supercritical flows, however, we start at an upstream control section and compute the profile in the downstream direction. Unfortunately, this fact has been incorrectly attributed in many well-known publications to mean that computations become unstable or yield incorrect results if this convention is not followed. Other than the fact that the flow depth is known at a control section, there appears to be no reason why we should proceed in either the upstream or downstream direction. This is because all we are doing is either numerically solving a differential equation for the specified initial condition or solving a nonlinear algebraic equation. Whether we proceed in the positive or negative x direction should make little difference.

6-3 Direct Step Method

In the previous section, we discussed how we may compute the location where a specified depth will occur. Let us discuss this in more detail to develop a systematic procedure for the computations. Following Chow (1959), we call this procedure the *direct step method.*

Referring to Fig. 6-2, let us say that we know the flow depth at section 1 and we want to determine the location of section 2, where a specified flow depth, y_2, will occur in a given channel for a specified discharge, Q. In other words, the statement of our problem is as follows: The flow depth, y_1, at distance x_1 (i.e., section 1 in Fig. 6-2) is known; determine distance x_2 where a specified flow depth y_2 will occur. The properties of the channel section, S_o, Q, and n, are known.

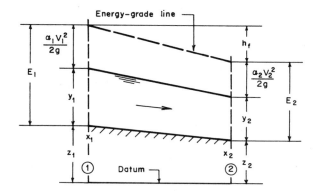

Figure 6-2 Computation of distance for specified depth.

If S_o = slope of the channel bottom, then, referring to Fig. 6-2,

$$z_2 = z_1 - S_o(x_2 - x_1) \tag{6-11}$$

In addition, the specific energy

$$E_1 = y_1 + \frac{\alpha_1 V_1^2}{2g}$$
$$E_2 = y_2 + \frac{\alpha_2 V_2^2}{2g} \tag{6-12}$$

The slope of the energy-grade line (we will refer to it as the friction slope in the following discussion for simplicity) in gradually varied flow may be computed with negligible error by using the corresponding formulas for friction slopes in uniform flow (Chow 1959; Henderson 1966). However, since the flow depth, y, varies with distance, x, the friction slope S_f is a function of x as well. The following approximations have been used (U.S. Army Corps of Engineers 1982) to select a representative value of S_f for the channel length between sections 1 and 2.

Average friction slope

$$\bar{S}_f = \frac{1}{2}(S_{f_1} + S_{f_2}) \tag{6-13a}$$

Geometric mean friction slope

$$\bar{S}_f = \sqrt{S_{f_1} S_{f_2}} \tag{6-13b}$$

Harmonic mean friction slope

$$\bar{S}_f = \frac{2S_{f_1}S_{f_2}}{S_{f_1} + S_{f_2}} \tag{6-13c}$$

By expanding the right-hand side of the preceding approximations in a Taylor series, we can prove (Prob. 6-15) that these three formulations for the approximation of the friction slope give identical results if the terms of the order $(\Delta S_f / S_{f1})^2$ and higher are neglected. In this expression, $\Delta S_f = S_{f_2} - S_{f_1}$.

Laurenson (1986) showed that the average slope (Eq. 6-13a) gives the lowest maximum error, although it is not always the smallest error. If the distance between sections 1 and 2 is short or the flow depths y_1 and y_2 are not significantly different, then Eq. 6-13(a) yields satisfactory results, in addition to being the simplest of the three approximations. Therefore, its use is recommended, and we will use it herein. Hence an expression for h_f may be written as

$$h_f = \frac{1}{2}(S_{f_1} + S_{f_2})(x_2 - x_1) \tag{6-14}$$

Substitution of Eqs. 6-12 and 6-14 into Eq. 6-2 yields

$$z_1 + E_1 = z_2 + E_2 + \frac{1}{2}(S_{f_1} + S_{f_2})(x_2 - x_1) \tag{6-15}$$

By substituting the expression for z_2 from Eq. 6-11 into Eq. 6-15 and canceling out z_1, we obtain

$$E_2 - E_1 = S_o(x_2 - x_1) - \frac{1}{2}(S_{f_1} + S_{f_2})(x_2 - x_1) \tag{6-16}$$

This equation may be written as

$$x_2 = x_1 + \frac{E_2 - E_1}{S_o - \frac{1}{2}(S_{f_1} + S_{f_2})} \tag{6-17}$$

Now, the location of section 2 is known. This is the starting value for the next step. Then, by successively increasing or decreasing the flow depth and determining where these depths will occur, the water-surface profile in the desired channel length may be computed. In Eq. 6-17, the direction of computations is automatically taken care of if proper sign is used for the numerator and the denominator.

The main *disadvantages* of this method are as follows: (1) The flow depth is not computed at predetermined locations. Therefore, interpolations may become necessary if the flow depths are required at specified locations. Similarly, the cross-sectional information has to be estimated if such information is available only at the given locations. This may not yield accurate results in addition to requiring additional effort. (2) It is cumbersome to apply to nonprismatic channels.

The following example should help in understanding this computational procedure.

Example 6-1

A trapezoidal channel having a bottom slope of 0.001 is carrying a flow of 30 m³/s. The bottom width is 10.0 m and the side slopes are 2H to 1V. A control structure is built at the downstream end which raises the water depth at the downstream end to 5.0 m. Compute the water surface profile. Manning n for the flow surfaces is 0.013 and α = 1.

Given:

>   Bottom slope $S_o = 0.001$
>   Discharge $Q = 30$ m³/s
>   Channel width $B_o = 10.0$ m
>   Manning $n = 0.013$
>   Depth at the downstream end (i.e., at $x = 0$) = 5.0 m
> $\alpha = 1$

Determine:

>   Water-surface profile in the channel

Solution: We computed the normal depth, y_n, for this channel in Example 4-1 as 1.16 m. The water surface approaches this depth asymptotically.

We start the computations with a known depth of 5.0 m at the control structure and proceed in the upstream direction. Let us call the location at the control structure as $x = 0$. Since we are considering the distance in the downstream flow direction as positive, the values of x we determine from Eq. 6-17 are negative.

Table 6-1 shows the calculations. The following explanatory remarks should be helpful to understand these calculations. In this discussion, we will call the depth for the step under consideration as the current depth and the depth for the previous step as the previous depth.

Column 1, y.   The flow depth approaches the normal depth asymptotically at an infinite distance. Therefore, the computation of the surface profile may be stopped when the flow depth is within about 1 percent of the normal depth. To conserve space, we will continue the calculations in this example until $y = 1.1y_n = 1.1 \times 1.1 = 1.21$ m. We first use large increments of y and decrease their size as the rate of variation of y with x becomes small.

Column 2, A.   This is the flow area for the depth of column 1.

Column 3, R.   The hydraulic radius, $R = A/P$, where P = wetted perimeter for the flow depth of column 1.

Column 4, V.   Flow velocity, V, is computed by dividing the specified rate of discharge, Q, by the flow area, A, of column 2.

Column 5, S_f.   By using the specified value of Manning n and the computed values of V of column 4 and R of column 3, this column is computed from the equation, $S_f = n^2 V^2/(C_o^2 R^{1.33})$.

Column 6, $\bar{S}_f$.   This is the average of S_f for the current depth and for the previous depth. This column is left blank for the first line, since there is no previous depth when we start the computations. To indicate that this is an average slope, we list it between the lines corresponding to the current and the previous depths.

Column 7, $S_o - \bar{S}_f$.   This is obtained by subtracting $\bar{S}_f$ of column 6 from the specified value of S_o.

Column 8, E.   The specific energy, E, is computed for the selected value of y of column 1 and the corresponding computed value of V of column 4, i.e., $E = y + \alpha V^2/(2g)$.

Table 6-1 Direct step method

$$Q = 30 \text{ m}^3/\text{s}; \quad B_o = 10 \text{ m}; \quad s = 2; \quad S_o = 0.001; \quad n = 0.013; \quad \alpha = 1.0; \quad C_o = 1.0$$

y (1)	A (2)	R (3)	V (4)	S_f (5)	$\bar{S}_f$ (6)	$S_o - \bar{S}_f$ (7)	E (8)	ΔE (9)	Δx (10)	x_2 (11)
5.00	100.0	3.09	0.30	0.000003			5.00459			0.0
					0.000004	0.000996		−0.49831	−500.5	
4.50	85.5	2.84	0.35	0.000005			4.50627			−500.4
					0.000007	0.000993		−0.49743	−500.8	
4.00	72.0	2.58	0.42	0.000008			4.00885			−1001.1
					0.000010	0.000990		−0.33743	−340.8	
3.66	63.4	2.40	0.47	0.000012			3.67142			−1342
					0.000014	0.000986		−0.32651	−331.3	
3.33	55.5	2.23	0.54	0.000017			3.34490			−1673
					0.000021	0.000979		−0.32499	−332.0	
3.00	48.0	2.05	0.63	0.000025			3.01991			−2005
					0.000030	0.000970		−0.24466	−252.3	
2.75	42.6	1.91	0.70	0.000035			2.77525			−2257.4
					0.000043	0.000957		−0.24263	−253.5	
2.50	37.5	1.77	0.80	0.000050			2.53262			−2511
					0.000063	0.000937		−0.23952	−255.5	
2.25	32.6	1.63	0.92	0.000075			2.29310			−2766
					0.000095	0.000905		−0.23459	−259.5	
2.00	28.0	1.48	1.07	0.000115			2.05851			−3025
					0.000142	0.000858		−0.18196	−212.1	
1.80	24.5	1.36	1.23	0.000169			1.87655			−3238
					0.000214	0.000786		−0.17371	−220.9	
1.60	21.1	1.23	1.42	0.000258			1.70284			−3459
					0.000337	0.000663		−0.15999	−241.4	
1.40	17.9	1.10	1.67	0.000416			1.54285			−3700
					0.000479	0.000521		−0.07188	−137.8	
1.30	16.4	1.04	1.83	0.000541			1.47097			−3838.4
					0.000629	0.000371		−0.06379	−171.9	
1.20	14.9	0.97	2.02	0.000717			1.40718			−4010.4

Column 9, $\Delta E = E_2 - E_1$. This column is obtained by subtracting E for the current depth from E for the previous depth. Since this column is the difference of E values corresponding to the current and the previous depths, we list its value at the middle of the lines for these depths.

Column 10, $\Delta x = x_2 - x_1$. The distance increment is computed from the equation $\Delta x = (E_2 - E_1)/(S_o - \bar{S}_f)$, i.e., dividing column 9 by column 7.

Column 11, x_2. This is the distance where depth y will occur. It is obtained by algebraically adding Δx of column 10 to the x_2 value for the previous depth.

6-4 Standard Step Method

The procedure described in the previous section is not suitable if we want to determine the flow depth at specified locations or if the channel cross sections are available only at some specified locations. In such cases, the procedure described in this section may be used. Following Chow (1959), we will call this method the *standard step method*, since this name has been widely used. A very popular computer program called HEC-2, developed by Hydrologic Engineering Center, U. S. Army Corps of Engineers (1982) is based on this method.

Referring to Fig. 6-3, the flow depth, y_1, for the specified discharge, Q, in the given channel at section 1 (distance x_1) is known; and we want to determine the flow depth at distance x_2 (section 2). Let us assume that the values of the velocity-head coefficient, α, at sections 1 and 2 are either known or we have determined their values, as discussed in Secs. 1-5 and 4-7. Since y_1 is known, we can determine the flow velocity, V_1, at section 1 for the specified discharge, Q, from the continuity equation. Hence, the total head, H, at section 1,

$$H_1 = z_1 + y_1 + \frac{\alpha_1 V_1^2}{2g} \tag{6-18}$$

is known. According to the energy equation, the total head at section 2 is

$$H_2 = H_1 - h_f \tag{6-19}$$

in which h_f = head losses (sum of the friction and form losses) between sections 1 and 2. By substituting the expression for h_f from Eq. 6-14 into Eq. 6-19, we obtain

$$H_2 = H_1 - \frac{1}{2}(S_{f_1} + S_{f_2})(x_2 - x_1) \tag{6-20}$$

Substituting an expression for H_2 (similar to that for H_1 in Eq. 6-18) into Eq. 6-20 and transposing all terms to the left-hand side, we obtain

$$y_2 + \frac{\alpha_2 Q^2}{2g A_2^2} + \frac{1}{2}S_{f_2}(x_2 - x_1) + z_2 - H_1 + \frac{1}{2}S_{f_1}(x_2 - x_1) = 0 \tag{6-21}$$

In this equation, A_2 and S_{f_2} are functions of y_2 and all other quantities are either known or have already been calculated at section 1. Hence, y_2 may be determined by solving the following nonlinear algebraic equation:

$$F(y_2) = y_2 + \frac{\alpha_2 Q^2}{2g A_2^2} + \frac{1}{2}S_{f_2}(x_2 - x_1) + z_2 - H_1 + \frac{1}{2}S_{f_1}(x_2 - x_1) = 0 \tag{6-22}$$

Equation 6-22 may be solved for y_2 by a trial-and-error procedure or by using the Newton or bisection methods. We discuss only the use of the Newton method.

For this method, we need an expression for dF/dy_2. This expression may be obtained by differentiating Eq. 6-22 with respect to y_2, i.e.,

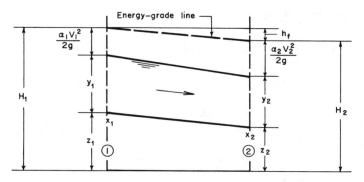

Figure 6-3 Computation of depth at specified location.

$$\frac{dF}{dy_2} = 1 - \frac{\alpha_2 Q^2}{g A_2^3} \frac{dA_2}{dy_2} + \frac{1}{2}(x_2 - x_1)\frac{d}{dy_2}\left(\frac{Q^2 n^2}{C_o^2 A_2^2 R_2^{4/3}}\right) \qquad (6\text{-}23)$$

The last term of this equation may be evaluated as follows:

$$
\begin{aligned}
\frac{d}{dy_2}\left(\frac{Q^2 n^2}{C_o^2 A_2^2 R_2^{4/3}}\right) &= \frac{-2Q^2 n^2}{C_o^2 A_2^3 R_2^{4/3}}\frac{dA_2}{dy_2} - \frac{4}{3}\frac{Q^2 n^2}{C_o^2 A_2^2 R_2^{7/3}}\frac{dR_2}{dy_2} \\
&= \frac{-2Q^2 n^2}{C_o^2 A_2^2 R_2^{4/3}}\frac{B_2}{A_2} - \frac{4}{3}\frac{Q^2 n^2}{C_o^2 A_2^2 R_2^{4/3}}\frac{1}{R_2}\frac{dR_2}{dy_2} \qquad (6\text{-}24)\\
&= -2\left(S_{f_2}\frac{B_2}{A_2} + \frac{2}{3}\frac{S_{f_2}}{R_2}\frac{dR_2}{dy_2}\right)
\end{aligned}
$$

Note that we have replaced $dA_2/dy_2 = B_2$ in this equation. By substituting Eq. 6-24 into Eq. 6-23, we obtain

$$\frac{dF}{dy_2} = 1 - \frac{\alpha_2 Q^2 B_2}{g A_2^3} - (x_2 - x_1)\left(S_{f_2}\frac{B_2}{A_2} + \frac{2}{3}\frac{S_{f_2}}{R_2}\frac{dR_2}{dy_2}\right) \qquad (6\text{-}25)$$

The derivative dR_2/dy_2 of the last term in this equation may be evaluated as follows.

$$
\begin{aligned}
\frac{dR_2}{dy_2} &= \frac{d}{dy_2}\left(\frac{A_2}{P_2}\right) \\
&= \frac{1}{P_2}\frac{dA_2}{dy_2} + A_2\frac{d}{dy_2}\left(\frac{1}{P_2}\right) \qquad (6\text{-}26)\\
&= \frac{B_2}{P_2} - \frac{A_2}{P_2^2}\frac{dP_2}{dy_2}
\end{aligned}
$$

For a rectangular channel, $dP_2/dy_2 = 2$, and for a trapezoidal channel, $dP_2/dy_2 = 2\sqrt{1 + s^2}$, in which s = side slope of the channel (s horizontal to 1 vertical).

A step-by-step procedure for computing y_2 by using the Newton method is as follows:

1. Calculate H_1 at section 1 from Eq. 6-18 for the known values of y_1 and z_1.
2. Estimate the flow depth at section 2. Let us designate this estimated flow depth and other quantities corresponding to this estimated depth by superscript *. At the beginning of the calculations, the rate of variation of y at x_1 may be determined from Eq. 6-3 by using $y = y_1$, i.e., $dy/dx = f(x_1, y_1)$. Then, the flow depth, y_2^*, can be computed from the equation $y_2^* = y_1 + f(x_1, y_1)(x_2 - x_1)$. During subsequent steps, however, y_2^* may be determined by extrapolating the change in flow depth computed during the preceding step.
3. By using the estimated value of flow depth, y_2^*, at section 2, compute B_2^*, A_2^*, R_2^*, and $S_{f_2}^*$. The value of z_2 is either given in the available data or it may be computed from the channel bottom slope and the known value of z_1.
4. Compute the value of $F(y_2^*)$ from Eq. 6-22 by using y_2^*, B_2^*, A_2^*, R_2^*, and $S_{f_2}^*$.
5. Compute dF/dy_2 from Eq. 6-25 using y_2^* and the corresponding values of A_2^*, R_2^*, and $S_{f_2}^*$, etc.
6. Then, a better estimate for y_2 can be computed from the equation

$$y_2 = y_2^* - \frac{F(y_2^*)}{[dF/dy_2]^*}$$

7. If $|y_2 - y_2^*|$ is less than a specified tolerance, ϵ, then y_2^* is the flow depth, y_2, at section 2; otherwise, set $y_2^* = y_2$ and repeat steps 3 to 7 until a solution is obtained.

The following example should help in understanding this procedure.

Example 6-2

A trapezoidal channel having a bottom slope of 0.001 is carrying a flow of 30 m^3/s. The bottom width is 10.0 m and the side slopes are 2H to 1V. At the downstream end, a control structure raises the water depth to 5.0 m. Determine the water-surface levels at 1, 2, and 4 km upstream of the control structure. The Manning n for the flow surfaces is 0.013, $\alpha = 1.0$, and the elevation of the channel bottom at the downstream end is 0.0.

Given:

       Bottom slope $S_o = 0.001$
       Discharge $Q = 30$ m³/s
       Channel width $B_o = 10.0$ m
       Manning $n = 0.013$
       Depth at downstream end (i.e., at $x = 0$) = 5.0 m
       $\alpha = 1.0$

Determine:

       Water-surface levels at 1, 2, and 4 km upstream of the control structure

Solution: Let us call the location at the control structure $x = 0$. Since the distance in the downstream flow direction is considered positive, the upstream distances where we want to determine the flow depths are negative.

      Table 6-2 lists the calculations using a trial-and-error procedure. A computer program using the Newton method is presented in Appendix D-2.

Table 6-2 Standard step method

x (1)	y (2)	A (3)	P (4)	R (5)	$R^{1.33}$ (6)	V (7)	$V^2/(2g)$ (8)	z (9)	H (10)	S_f (11)	$\bar{S}_f$ (12)	Δx (13)	h_f (14)	H (15)	(16)	(17)
0	5.000	100.0	32.36	3.090	4.484	0.300	0.0046	0.0	5.0046	0.00000379						
−1000	4.001	72.00	27.89	2.582	3.542	0.417	0.0088	1.0	5.0095	0.00000828	0.00000583	−1000	0.00583	5.0104	0.0009	
−1000	4.002	72.04	27.90	2.583	3.543	0.416	0.0088	1.0	5.0104	0.00000827	0.00000583	−1000	0.00583	5.0104	0.0000	ok
−2000	3.003	48.10	23.43	2.052	2.607	0.624	0.0198	2.0	5.0233	0.00002525	0.00001676	−1000	0.01676	5.0272	0.0038	
−2000	3.007	48.16	23.45	2.054	2.611	0.623	0.0198	2.0	5.0271	0.00002512	0.00001670	−1000	0.01670	5.0271	0.0000	ok
−3000	2.014	28.30	19.01	1.487	1.697	1.062	0.0574	3.0	5.0719	0.00011224	0.00006868	−1000	0.06868	5.0958	0.0239	
−3000	2.038	28.68	19.11	1.501	1.718	1.046	0.0558	3.0	5.0935	0.00010763	0.00006638	−1000	0.06638	5.0935	0.0000	ok
−4000	1.078	13.10	14.82	0.884	0.848	2.290	0.2674	4.0	5.3450	0.00104508	0.00057635	−1000	0.57635	5.6698	0.3249	
−4000	1.232	15.40	15.51	0.990	0.986	1.954	0.1947	4.0	5.4263	0.00065429	0.00038096	−1000	0.38096	5.4744	0.0481	
−4000	1.263	15.83	15.65	1.011	1.015	1.896	0.1832	4.0	5.4465	0.00059837	0.00035300	−1000	0.35300	5.4465	0.0000	ok

136

The following explanatory remarks should be helpful in understanding the computations of Table 6-2. For each distance where we want to determine the water level, a flow depth is estimated. We then check whether this estimated flow depth satisfies Eq. 6-22 or not. If it does not satisfy this equation, then we discard the calculations corresponding to this depth and estimate another value. This process is continued until a solution is obtained.

Column **1,** *x* **.**   This is the specified location at which flow depth, y, is to be computed.

Column **2,** *y*.   This is the estimated flow depth.

Column **3, A.**   This is the flow area, A, for the flow depth of column 2.

Column **4, P.**   This is the wetted perimeter, P, for the flow depth of column 2.

Column **5, R.**   This is the hydraulic radius, R, corresponding to the flow depth of column 2 obtained by dividing column 3 by column 4.

Column **6, R$^{4/3}$.**   This column lists the value of R raised to the power $\frac{4}{3}$.

Column **7, V.**   Flow velocity, $V = 30/A$, where A is listed in column 3. Velocity head corresponding to this depth is shown in column 8.

Column **9, z.**   This is the elevation of the channel bottom. It is computed from the known bottom elevation ($z_d = 0$) at the downstream end ($x_d = 0$) and the known channel bottom slope, S_o, of 0.001; i.e., $z = z_d - S_o(x - x_d)$.

Column **10, H.**   Total head of column 10 corresponds to the flow depth of column 2, velocity head of column 8, and the channel bottom level of column 9, i.e., $H = z + y + \alpha V^2/(2g)$.

Column **11, S$_f$.**   This is the slope of the energy-grade line. It is computed using the velocity of column 7, $R^{4/3}$ of column 6, and the known value of Manning n from the equation $S_f = n^2 V^2/(C_o^2 R^{4/3})$.

Column **12, $\bar{S}_f$.**   This is the average of the S_f value for the flow depth at the current distance and that for the flow depth at the previous distance.

Column **13, Δx.**   This is the distance between the current location where we want to determine the flow depth and the location where we determined the flow depth during the previous step.

Column **14, h$_f$.**   The head losses in distance Δx are computed from the equation $h_f = \bar{S}_f \Delta x$, where $\bar{S}_f$ is given in column 12 and Δx is given in column 13.

Column **15, H.**   This is the elevation of the energy-grade line, computed by adding the head losses (h_f of column 14) to the elevation of the energy grade line (i.e., H of column 10 for the previous step) at the location where the flow depth was computed during the previous step.

Column **16.**   If the values of H in columns 10 and 15 are within an acceptable tolerance, then the estimated depth of column 2 is the flow depth at the location under consideration. We then compute the flow depth at the next desired location. However, if these values are not within the specified tolerance, we discard the values corresponding to this flow depth and start with another value for the estimated depth.

6-5 Integration Of Differential Equation

In Section 6-2, we discussed the computation of the water-surface profile by integrating the differential equation, Eq. 6-3. We also mentioned that the integration has to be done numerically, since $f(x, y)$ is a nonlinear function. In the following sections, we present several numerical methods that may be used for this purpose. Some of these methods have been used in the past, but others are being introduced for computing the water-surface profiles. We may divide these methods into the following categories (McCracken and Dorn 1964; Chapra and Canale 1988):

1. Single-step methods
2. Predictor-corrector methods

Single-step methods are just like the step methods discussed in the previous two sections. The unknown depth at a section is expressed in terms of a function $f(x, y)$, at a neighboring point where the flow depth is either initially known or has been computed during the previous step. In the predictor-corrector method, a value of the unknown depth is first predicted by using the available information from the previous step. This predicted value is then refined by an iterative procedure during the corrector part until a solution is obtained with a specified accuracy. The details of both methods are presented in the following sections.

6-6 Single-step Methods

There are several single-step methods (McCracken and Dorn 1964). However, only four of these are presented here:

1. Euler method
2. Modified Euler method
3. Improved Euler method
4. Fourth-order Runge-Kutta method

Let us now discuss how we can use these methods to compute the water-surface profiles.

Referring to Fig. 6-4, let us say that we know the flow depth, y_i, at distance x_i and that we want to determine the flow depth at distance x_{i+1}. Let $y = y(x)$ be the exact solution of the differential equation (Eq. 6-4). Then the curve $y = y(x)$ represents the variation of y with respect to x.

Euler Method

We can evaluate the rate of variation of y with respect to x at distance x_i from Eq. 6-4, i.e.,

$$y_i' = \frac{dy}{dx}\bigg|_i = f(x_i, y_i) \tag{6-27}$$

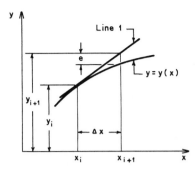

Figure 6-4 Geometrical representation of the Euler method.

in which the subscript i refers to quantities at distance x_i, a prime on y indicates a derivative of y with respect to x, and

$$f(x_i, y_i) = \frac{S_o - S_{fi}}{1 - Q^2 B_i/(g A_i^3)} \tag{6-28}$$

Since all the variables on the right-hand side of this equation are known, we can compute $f(x_i, y_i)$, which is the rate of variation of y at the point (x_i, y_i). If we assume that this rate of variation, y_i', is constant in the interval x_i to x_{i+1}, then we can determine the flow depth at x_{i+1} from the equation

$$y_{i+1} = y_i + y_i' \Delta x \tag{6-29}$$

in which $\Delta x = x_{i+1} - x_i$. The substitution of Eq. 6-27 into Eq. 6-29 yields

$$y_{i+1} = y_i + f(x_i, y_i)\Delta x \tag{6-30}$$

This is referred to as the Euler's method. Now, y_{i+1} is known and we can determine y_{i+2} at distance x_{i+2} by repeating the same procedure.

Let us briefly discuss the accuracy of the Euler method. We may expand y_{i+1} in a Taylor series as

$$y_{i+1} = y_i + y_i'\Delta x + O\left[(\Delta x)^2\right] \tag{6-31}$$

in which $O(\Delta x)^2$ means that the remaining terms are of the order of $(\Delta x)^2$ or smaller. A comparison of Eqs. 6-30 and 6-31 shows that we are including in our solution terms up to the first power of Δx. Therefore, this method is referred to as *first-order* accurate.

Equation 6-29 is the equation of a straight line, line 1, shown in Fig. 6-4. Since y' is not constant in the interval x_i to x_{i+1}, we introduce an error, e, at each step. Note that if the flow surface is a straight line, then $e = 0$. Because of this error at each step, the numerically computed value may diverge from the correct solution. This method is usually unstable; i.e., a small error—round-off or truncation—is magnified as the value of x increases.

In the Euler method, we used the slope at only one point (x_i, y_i) to compute the value of y_{i+1}. By using the slope at more than one point, we may improve the accuracy of this method. Two such methods, improved Euler and modified Euler, are presented in the following paragraphs.

Improved Euler Method

Let us call the flow depth at x_{i+1} obtained by using Euler method as y_{i+1}^*, i.e.,

$$y_{i+1}^* = y_i + y_i'\Delta x \qquad (6\text{-}32)$$

By using this value, we can compute the slope of the curve $y = y(x)$ at $x = x_{i+1}$, i.e., $y_{i+1}' = f(x_{i+1}, y_{i+1}^*)$. Let us use the average value of the slopes of the curve at x_i and x_{i+1}. Then we can determine the value of y_{i+1} from the equation

$$y_{i+1} = y_i + \frac{1}{2}(y_i' + y_{i+1}')\Delta x \qquad (6\text{-}33)$$

This equation may be written as

$$y_{i+1} = y_i + \frac{1}{2}[f(x_i, y_i) + f(x_{i+1}, y_{i+1}^*)]\Delta x \qquad (6\text{-}34)$$

This method, called the *improved* Euler method, is *second-order* accurate. A geometrical representation of this method is shown in Fig. 6-5. In this figure, line 1 is tangent at (x_i, y_i) and has a slope of y_i' whereas line 2 is tangent at (x_{i+1}, y_{i+1}) and has a slope of y_{i+1}'. Line 3 is drawn through point (x_i, y_i) with a slope of $\frac{1}{2}(y_i' + y_{i+1}')$.

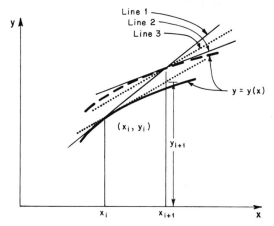

Figure 6-5 Geometrical representation of the improved Euler method.

Modified Euler Method

We may also improve the accuracy of the Euler method by using the slope of the curve $y = y(x)$ at $x = x_{i+1/2}$ and $y = y_{i+1/2}$, in which $x_{i+1/2} = \frac{1}{2}(x_i + x_{i+1})$ and $y_{i+1/2} = y_i + \frac{1}{2}y_i'\Delta x$. Let us call this slope $y_{i+1/2}'$. Then

$$y_{i+1} = y_i + y_{i+1/2}'\Delta x$$

or

$$y_{i+1} = y_i + f(x_{i+1/2}, y_{i+1/2})\Delta x \qquad (6\text{-}35)$$

This method, called the *modified* Euler method, is *second-order* accurate. Figure 6-6

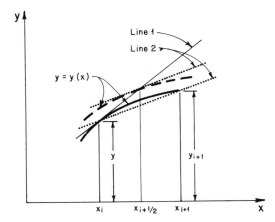

Figure 6-6 Geometrical representation of the modified Euler method.

shows a geometrical representation of this method. In this figure, line 1 is tangent at (x_i, y_i) and has a slope of y_i', whereas line 2 is tangent at $(x_{i+1/2}, y_{i+1/2})$ and has a slope of $y_{i+1/2}'$. Line 3 is drawn through point (x_i, y_i) with a slope of $y_{i+1/2}'$.

By expanding the numerical solution in Taylor series, we can show that both the modified and the improved Euler methods are second-order accurate.

Fourth-Order Runge-Kutta Method

In the fourth-order Runge-Kutta method, the slope of the curve $y = y(x)$ is determined from the following equations:

$$
\begin{aligned}
k_1 &= f(x_i, y_i) \\
k_2 &= f(x_i + \frac{1}{2}\Delta x, y_i + \frac{1}{2}k_1\Delta x) \\
k_3 &= f(x_i + \frac{1}{2}\Delta x, y_i + \frac{1}{2}k_2\Delta x) \\
k_4 &= f(x_i + \Delta x, y_i + k_3\Delta x)
\end{aligned}
\tag{6-36}
$$

Then

$$
y_{i+1} = y_i + \frac{1}{6}(k_1 + 2k_2 + 2k_3 + k_4)\Delta x
\tag{6-37}
$$

As the name implies, this method is *fourth-order* accurate. Humpidge and Moss (1971) developed a general-purpose computer program based on this method to compute the water-surface profiles.

6-7 Predictor-Corrector Methods

In the numerical methods discussed in the previous section, we used the known information at point x_i and, to improve the accuracy, we used the value of the function $f(x, y)$ at more than one point, e.g., at x_i, x_{i+1}, and $x_{i+1/2}$. In the predictor-corrector method, we do not compute the function at several points but rather predict the unknown flow depth first, correct this predicted value, and then recorrect this corrected value. This iterative procedure is continued until a solution of a desired accuracy is obtained.

There are several predictor-corrector methods. However, to conserve space, we present only one of these methods.

In the predictor part, let us use the Euler method to predict the value of y_{i+1}, i.e.,

$$y_{i+1}^{(0)} = y_i + f(x_i, y_i)\Delta x \tag{6-38}$$

in which the superscript enclosed in parentheses indicates the number of the iteration (zero iteration is the initially estimated or predicted value). Then, we may correct it using the following equation:

$$y_{i+1}^{(1)} = y_i + \frac{1}{2}[f(x_i, y_i) + f(x_{i+1}, y_{i+1}^{(0)})]\Delta x \tag{6-39}$$

Now, we may recorrect $y_{i+1}^{(1)}$ again to obtain a better value:

$$y_{i+1}^{(2)} = y_i + \frac{1}{2}[f(x_i, y_i) + f(x_{i+1}, y_{i+1}^{(1)})]\Delta x \tag{6-40}$$

Thus, the jth iteration is

$$y_{i+1}^{(j)} = y_i + \frac{1}{2}[f(x_i, y_i) + f(x_{i+1}, y_{i+1}^{(j-1)})]\Delta x \tag{6-41}$$

We continue this iterative procedure until $|y_{i+1}^{(j)} - y_{i+1}^{(j-1)}| \leq \epsilon$, where ϵ = specified tolerance. A similar method is used by Prasad (1970) to compute water-surface profiles, except that he compared the derivative, y'_{i+1}, between two successive iterations instead of the depths.

6-8 Simultaneous Solution Procedure

The procedures presented in the previous sections are suitable for single or series channels. However, to compute the flow conditions in parallel channels or in channel networks (Fig. 6-7), manual trial-and-error methods have to be employed in conjunction with the use of computer programs for the computation of water-surface profiles. As an example, let us consider the analysis of the parallel-channel system shown in Fig. 6-7(a). We first assume a discharge distribution, Q_1 and Q_2, in both channels so that the continuity equation is satisfied, i.e., $Q = Q_1 + Q_2$. Then, the water-surface profiles are computed in channel 1 for Q_1 and in channel 2 for Q_2 from the point of separation (point E) to the point of union (point F). The elevation of the energy grade line in the three channels at junction E must be the same for the computed water levels and corresponding flow velocities. This corresponds to identical water levels in all channels at the junction if the junction losses and the difference in the velocity heads in different channels at the junction are neglected. If the elevation of the energy-grade line is not the same, then other values of Q_1 and Q_2 are selected, and the entire procedure is

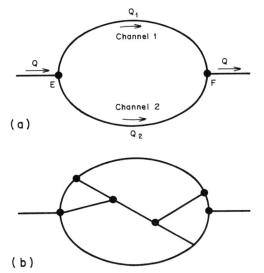

(a)

(b)   **Figure 6-7** Channel network.

repeated. It is clear that this is a time-consuming process; for a complex network it is very difficult, if not impossible to apply.

We may compute gradually varied flows in single or series channels or in channel networks directly by using the simultaneous solution approach presented in this section. This approach utilizes the Newton-Raphson iterative procedure for the solution of a system of nonlinear equations (Epp and Fowler 1970) and computes the flow conditions in the entire network simultaneously. Based on this method, Wylie (1972) presented an algorithm for channel systems. Chaudhry and Schulte (1986) used this method to analyze systems having two parallel channels; then Schulte and Chaudhry (1987) extended this concept for application to channel networks. This solution procedure is presented in this section. Unlike the algorithm presented by Wylie, the governing equations are in terms of the commonly used variables, namely, flow depths and discharges. Therefore, this formulation is easier to understand and apply. In addition, a procedure is presented to number the nodes of a parallel system such that a banded matrix is obtained. This reduces the computational time and storage requirements and improves the accuracy of the computed results.

Governing Equations

Let us first present the notation we will use in the following discussion. We will use two subscripts to designate variables at different channel sections: The first subscript refers to the number of the channel, whereas the second subscript refers to the section number on that channel. For example, $y_{i,j}$ refers to the flow depth at section j of channel i. The only exception to this rule is the head loss term, $h_{f_{j,j+1}}$, which implies the losses between sections j and $j + 1$.

Referring to the longitudinal profile of a channel shown in Fig. 6-8, the energy equation for the channel length (commonly termed a reach) between sections j and

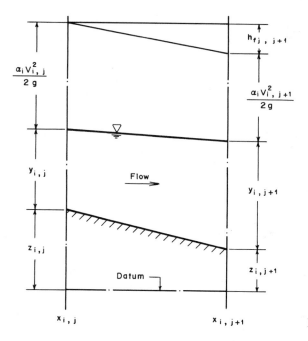

Figure 6-8 Definition sketch.

$j + 1$ of channel i may be written as

$$z_{i,j} + y_{i,j} + \alpha_{i,j} \frac{Q_{i,j}^2}{2g A_{i,j}^2} = z_{i,j+1} + y_{i,j+1} + \alpha_{i,j+1} \frac{Q_{i,j+1}^2}{2g A_{i,j+1}^2} + h_{f_{j,j+1}} \qquad (6\text{-}42)$$

As an approximation, the head loss between sections j and $j + 1$ of channel i may be computed by using the average of the friction slopes at sections j and $j + 1$. Expressing the flow velocity as a function of the discharge and the flow area, Eq. 6-42 becomes

$$z_{i,j} + y_{i,j} + \alpha_{i,j} \frac{Q_{i,j}^2}{2g A_{i,j}^2} = z_{i,j+1} + y_{i,j+1} + \frac{Q_{i,j+1}^2}{2g A_{i,j+1}^2}$$
$$+ \frac{1}{2}(x_{i,j+1} - x_{i,j}) \left(\frac{Q_{i,j+1}^2 n_{i,j+1}^2}{C_o^2 A_{i,j+1}^2 R_{i,j+1}^{1.33}} + \frac{Q_{i,j}^2 n_{i,j}^2}{C_o^2 A_{i,j}^2 R_{i,j}^{1.33}} \right)$$
$$(6\text{-}43)$$

The second governing equation is the continuity equation

$$Q_{i,j} = Q_{i,j+1} \qquad (6\text{-}44)$$

Equation 6-44 is valid if there is no lateral inflow or outflow between sections j and $j + 1$ (Fig. 6-8). Although this equation may appear to be trivial at this stage, its inclusion in the system of governing equations becomes important while computing the water-surface profile in a channel network. This will become apparent when we later discuss the analysis of branching systems and channel networks.

Single and Series Channels

In a series channel system, a number of channels are connected such that the outflow of one channel is the inflow of the next. Each channel may have different cross section, Manning n, bottom slope, etc. To facilitate an understanding of the computational procedure, we first consider a single channel and then a series channel system before studying the case of channel networks.

Figure 6-9 shows the longitudinal profile of channel i. The channel is subdivided into N_i reaches, where i refers to the channel number. If the first section is numbered 1, then the last section will be $N_i + 1$. For a single channel, the continuity equations need not be included in the system of equations, since the discharge at all sections is the same, i.e.,

$$Q_{i,1} = Q_{i,2} = \cdots = Q_{i,N_i+1} = Q_i \tag{6-45}$$

The values of α and n are generally the same at different sections of a particular channel, although they may be different for different channels. Therefore, we use only one subscript representing the channel number with these variables.

By writing the energy equation (Eq. 6-43) for each of the N_i reaches, we obtain the following system of equations:

$$F_{i,1} = y_{i,2} - y_{i,1} + z_{i,2} - z_{i,1} + \frac{1}{2g}\left(\frac{\alpha_i Q_i^2}{A_{i,2}^2} - \frac{\alpha_i Qi^2}{A_{i,1}^2}\right)$$

$$+ \frac{1}{2}(x_{i,2} - x_{i,1})\left(\frac{Q_i^2 n_i^2}{C_o^2 A_{i,2}^2 R_{i,2}^{1.33}} + \frac{Q_i^2 n_i^2}{C_o^2 A_{i,1}^2 R_{i,1}^{1.33}}\right) = 0$$

$$F_{i,2} = y_{i,3} - y_{i,2} + z_{i,3} - z_{i,2} + \frac{1}{2g}\left(\frac{\alpha_i Q_i^2}{A_{i,3}^2} - \frac{\alpha_i Qi^2}{A_{i,2}^2}\right)$$

$$+ \frac{1}{2}(x_{i,3} - x_{i,2})\left(\frac{Q_i^2 n_i^2}{C_o^2 A_{i,3}^2 R_{i,3}^{1.33}} + \frac{Q_i^2 n_i^2}{C_o^2 A_{i,2}^2 R_{i,2}^{1.33}}\right) = 0 \tag{6-46}$$

$$\cdot$$
$$\cdot$$
$$\cdot$$

$$F_{i,N_i} = y_{i,N_i+1} - y_{i,N_i} + z_{i,N_i+1} - z_{i,N_i} + \frac{1}{2g}\left(\frac{\alpha_i Q_i^2}{A_{i,N_i+1}^2} - \frac{\alpha_i Qi^2}{A_{i,N_i}^2}\right)$$

$$+ \frac{1}{2}(x_{i,N_i+1} - x_{i,N_i})\left(\frac{Q_i^2 n_i^2}{C_o^2 A_{i,N_i+1}^2 R_{i,N_i+1}^{1.33}} + \frac{Q_i^2 n_i^2}{C_o^2 A_{i,N_i}^2 R_{i,N_i}^{1.33}}\right) = 0$$

Since A and R are functions of the properties of the channel cross section and the flow depth, the preceding equations are functions only of the flow depth. However, we have N_i equations in $N_i + 1$ unknowns. It is, therefore, necessary to have one more

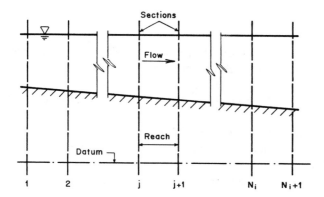

Figure 6-9 Channel reaches.

equation to obtain a unique solution of the system of equations. This additional equation is provided by the *end condition*. For subcritical flows, the end condition is the specified flow depth, y_d, at the downstream end of the channel, i.e.,

$$F_{i,N_i+1} = y_{i,N_i+1} - y_d = 0 \qquad (6\text{-}47)$$

Similarly, the end condition for supercritical flow is a specified flow depth, y_u, at the upstream end of the channel, i.e.,

$$F_{i,1} = y_{i,1} - y_u = 0 \qquad (6\text{-}48)$$

For brevity, only the solution of a system having subcritical flow is discussed in the following paragraphs; i.e., Eqs. 6-46 and 6-47 describe the flow conditions in the channel system. These nonlinear equations may now be solved simultaneously using the Newton-Raphson method as follows.

We are interested in determining corrections $\Delta y_{i,j}$ such that $y_{i,j}^{(1)} = y_{i,j}^{(0)} + \Delta y_{i,j}$ is a better estimate for the flow depth at section (i, j), where $y_{i,j}^{(0)}$ ($j = 1, 2, ..., N_i + 1$) are the initial estimates for the flow depths (The superscript in parentheses indicates the number of the iteration).

By expanding Eqs. 6-46 and 6-47 in Taylor series and writing the system of equations in a matrix form, we obtain

$$
\begin{bmatrix}
\dfrac{\partial F_{i,1}}{\partial y_{i,1}} & \dfrac{\partial F_{i,1}}{\partial y_{i,2}} & \cdots & \dfrac{\partial F_{i,1}}{\partial y_{i,N_i+1}} \\[2ex]
\dfrac{\partial F_{i,2}}{\partial y_{i,1}} & \dfrac{\partial F_{i,2}}{\partial y_{i,2}} & \cdots & \dfrac{\partial F_{i,2}}{\partial y_{i,N_i+1}} \\[2ex]
\vdots & \vdots & \ddots & \vdots \\[2ex]
\dfrac{\partial F_{i,N_i+1}}{\partial y_{i,1}} & \dfrac{\partial F_{i,N_i+1}}{\partial y_{i,2}} & \cdots & \dfrac{\partial F_{i,N_i+1}}{\partial y_{i,N_i+1}}
\end{bmatrix}^{(0)}
\begin{pmatrix}
\Delta y_{i,1} \\[2ex]
\Delta y_{i,2} \\[2ex]
\vdots \\[2ex]
\Delta y_{i,N_i+1}
\end{pmatrix}
= -
\begin{pmatrix}
F_{i,1} \\[2ex]
F_{i,2} \\[2ex]
\vdots \\[2ex]
F_{i,N_i+1}
\end{pmatrix}^{(0)}
$$

$$(6\text{-}49)$$

In this equation, the superscript 0 within the parentheses indicates that the functions $F_{i,j}$ and their partial derivatives are evaluated for the estimated flow depth, $y_{i,j}^{(0)}$.

The Jacobian matrix (matrix of partial derivatives) of the preceding system has an important characteristic. For each energy equation, all the partial derivatives are zero, except the partial derivative with respect to the flow depth at the section under consideration and with respect to the depth at the next section. For example, only the partial derivatives with respect to $y_{i,j}$ and $y_{i,j+1}$ of the energy equation for reach i between section j and $j+1$ are not zero. These nonzero partial derivatives are

$$\frac{\partial F_{i,j}}{\partial y_{i,j}} = -1 + Q_i^2 \left(\frac{\alpha_i B_{i,j}}{g A_{i,j}^3} - \frac{2n_i^2 (x_{i,j+1} - x_{i,j})}{3 C_o^2 A_{i,j}^2 R_{i,j}^{2.33}} \frac{dR_{i,j}}{dy_{i,j}} \right.$$

$$\left. - \frac{n_i^2 B_{i,j}(x_{i,j+1} - x_{i,j})}{C_0^2 A_{i,j}^3 R_{i,j}^{1.33}} \right) \qquad (6\text{-}50)$$

$$\frac{\partial F_{i,j}}{\partial y_{i,j+1}} = 1 - Q_i^2 \left(\frac{\alpha_i B_{i,j+1}}{g A_{i,j+1}^3} + \frac{2n_i^2 (x_{i,j+1} - x_{i,j})}{3 C_0^2 A_{i,j}^2 R_{i,j+1}^{2.33}} \frac{dR_{i,j+1}}{dy_{i,j+1}} \right.$$

$$\left. + \frac{n_i^2 B_{i,j+1}(x_{i,j+1} - x_{i,j})}{C_0^2 A_{i,j+1}^3 R_{i,j+1}^{1.33}} \right) \qquad (6\text{-}51)$$

where the value of dR/dy depends on the shape of the channel cross section, as discussed in Section 6-4.

It is clear from the preceding discussion that all nonzero partial derivatives lie on or near the principal diagonal of the Jacobian matrix. Therefore, the resulting matrix is banded, as shown in the following equation:

$$\begin{bmatrix} \frac{\partial F_{i,1}}{\partial y_{i,1}} & \frac{\partial F_{i,1}}{\partial y_{i,2}} & 0 & 0 & \cdots & 0 & 0 \\ 0 & \frac{\partial F_{i,2}}{\partial y_{i,2}} & \frac{\partial F_{i,2}}{\partial y_{i,3}} & 0 & \cdots & 0 & 0 \\ 0 & 0 & \frac{\partial F_{i,3}}{\partial y_{i,3}} & \frac{\partial F_{i,3}}{\partial y_{i,4}} & \cdots & 0 & 0 \\ \vdots & \vdots & \vdots & \vdots & \ddots & \vdots & \vdots \\ 0 & 0 & 0 & 0 & \cdots & \frac{\partial F_{i,N_i+1}}{\partial y_{i,N_i}} & \frac{\partial F_{i,N_i+1}}{\partial y_{i,N_i+1}} \end{bmatrix} \qquad (6\text{-}52)$$

This particular Jacobian has two diagonal nonzero elements, and it is referred to as a Jacobian of bandwidth two.

The advantage of having such a banded matrix is that the computer memory required to store its elements and computational time required to invert it are significantly reduced. In addition, most computer systems have standard subroutines for the inversion of banded matrices.

The solution algorithm is as follows: The functions $F_{i,j}$ (Eq. 6-46) and the partial derivatives of the banded Jacobian (Eqs. 6-50 and 6-51) are computed for the estimated flow depths. Instead of the Jacobian of Eq. 6-49, the banded Jacobian of Eq. 6-52 is used, and the system is solved for the corrections $\Delta y_{i,j}$, ($j = 1, 2, ..., N_i + 1$). Then, better estimates of the flow depths are

$$y_{i,j}^{(1)} = y_{i,j}^{(0)} + \Delta y_{i,j} \qquad (6\text{-}53)$$

If the absolute value of each of these corrections, $\Delta y_{i,j}$, is less than a specified tolerance, then the flow depths, $y_{i,j}^{(1)}$ computed from Eq. 6-53 are the desired solution. Otherwise, $y_{i,j}^{(0)}$ are set equal to $y_{i,j}^{(1)}$, and the previously described procedure is repeated until an acceptable solution is obtained. Good estimates of the initial flow depths are desirable for a rapid convergence of the iterations. The depth specified as the end condition may be specified as the initial estimate for the flow depths.

Let us now discuss how to analyze a series system having M channels in series (Fig. 6-10a). First, we write the governing equations for all M channels and then solve them simultaneously by following the preceding procedure.

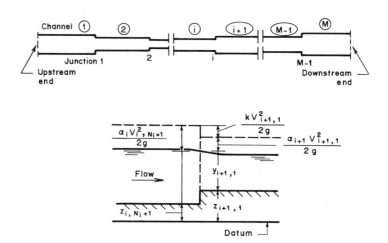

Figure 6-10   Series channels.

We have $\sum_{i=1}^{M}(N_i + 1)$ sections on M channels. Therefore, we need as many equations to determine the depths at these sections. Since the discharge has the same value at all sections, we do not include it as an unknown. Therefore, we do not have to include the continuity equation in the governing equations. By writing the energy equation for all reaches of the system, we will have $\sum_{i=1}^{M} N_i$ equations. In addition, there are $M - 1$ channel junctions. For each of these junctions, we may write the energy equation as well. For example, this equation for the junction of channel i and $i + 1$ is

$$z_{i,N_i+1} + y_{i,N_i+1} + \frac{V_{i,N_i+1}^2}{2g} = z_{i+1,1} + y_{i+1,1} + (1 + k)\frac{V_{i+1,1}^2}{2g} \qquad (6\text{-}54)$$

in which k = coefficient of head losses at the junction. If the junction losses and the difference in the velocity heads at the junction are small, they may be neglected in this equation.

Thus, the energy equations for all reaches of M channels, the energy equations for the $M - 1$ channel junctions, and the end condition provide the necessary number of equations. These may be solved simultaneously to determine the depths at all sections of the system.

Channel Networks

Let us now discuss how to analyze channel networks (Fig. 6-7). We will consider only subcritical flow in the following discussion. For supercritical flow, additional constraints arise from the channel geometries at branching nodes. The analysis of such a situation is beyond the scope of this section. Let us consider the channel networks shown in Fig. 6-7. The flow in all channels is subcritical. In addition to the flow depths, the discharges in the individual channels are not known as well. Therefore, the continuity equation (Eq. 6-44) for each channel reach is also included to obtain the necessary number of equations.

Let us write the energy equation (Eqs. 6-46) and the continuity equation (Eq. 6-44) for N_i reaches of channel i. The resulting system of equations is

$$F_{i,1} = y_{i,2} - y_{i,1} + z_{i,2} - z_{i,1} + \frac{\alpha_i}{2g}\left(\frac{Q_{i,2}^2}{A_{i,2}^2} - \frac{Q_{i,1}^2}{A_{i,1}^2}\right)$$

$$+ \frac{1}{2}(x_{i,2} - x_{i,1})\left(\frac{Q_{i,2}^2 n_{i,2}^2}{C_o^2 A_{i,2}^2 R_{i,2}^{1.33}} + \frac{Q_{i,1}^2 n_{i,1}^2}{C_o^2 A_{i,1}^2 R_{i,1}^{1.33}}\right) = 0$$

$$F_{i,2} = Q_{i,2} - Q_{i,1} = 0$$

$$F_{i,3} = y_{i,3} - y_{i,2} + z_{i,3} - z_{i,2} + \frac{\alpha_i}{2g}\left(\frac{Q_{i,3}^2}{A_{i,3}^2} - \frac{Q_{i,2}^2}{A_{i,2}^2}\right)$$

$$+ \frac{1}{2}(x_{i,3} - x_{i,2})\left(\frac{Q_{i,3}^2 n_{i,3}^2}{C_o^2 A_{i,3}^2 R_{i,3}^{1.33}} + \frac{Q_{i,2}^2 n_{i,2}^2}{C_o^2 A_{i,2}^2 R_{i,2}^{1.33}}\right) = 0$$

$$\tag{6-55}$$

$$F_{i,4} = Q_{i,3} - Q_{i,2} = 0$$

$$\vdots$$

$$F_{i,2N_i-1} = y_{i,N_i+1} - y_{i,N_i} + z_{i,N_i+1} - z_{i,N_i} + \frac{\alpha_i}{2g}\left(\frac{Q_{i,N_i+1}^2}{A_{i,N_i+1}^2} - \frac{Q_{i,N_i}^2}{A_{i,N_i}^2}\right)$$

$$+ \frac{1}{2}(x_{i,N_i+1} - x_{i,N_i})\left(\frac{Q_{i,N_i+1}^2 n_{i,N_i+1}^2}{C_o^2 A_{i,N_i+1}^2 R_{i,N_i+1}^{1.33}} + \frac{Q_{i,N_i}^2 n_{i,N_i}^2}{C_o^2 A_{i,N_i}^2 R_{i,N_i}^{1.33}}\right) = 0$$

$$F_{i,2N_i} = Q_{i,N_i+1} - Q_{i,N_i} = 0$$

For illustration purposes, let us consider the simplest type of network having two parallel channels, as shown in Fig. 6-11. Writing the energy and the continuity equations for the remaining three channels, $i+1$, $i+2$, and $i+3$, of this system in the same manner as Eqs. 6-55 gives a total of $2(N_i + N_{i+1} + N_{i+2} + N_{i+3})$ equations (here the subscripts refer to the channel number). Since the flow depth and the rate of discharge are the two unknowns for each section, we have $2(N_i + N_{i+1} + N_{i+2} + N_{i+3} + 4)$ unknowns. Therefore, we need eight additional equations for a unique solution. Two of these

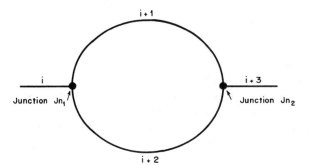

Figure 6-11 Parallel-channel system.

equations are given by the *end conditions*. These end conditions for subcritical flows are the specified flow depth, y_d, and the specified discharge, Q_d, at the downstream end of channel $i + 3$, i.e.,

$$F_{i+3,2N_{i+3}+1} = y_{i+3,N_{i+3}+1} - y_d = 0 \tag{6-56}$$

$$F_{i+3,2N_{i+3}+2} = Q_{i+3,N_{i+3}+1} - Q_d = 0 \tag{6-57}$$

In addition, three equations are provided by the energy equations at the junction of channels i, $i + 1$, and $i + 2$. Similarly, three equations are available at the junction of channels $i + 1$, $i + 2$, and $i + 3$ (see Fig. 6-12).

Assuming no lateral inflow, the continuity equation and the two energy equations at junction, jn_1, of channels i, $i + 1$, and $i + 2$ (Fig. 6-12a) may be written as

$$F_{jn_1,1} = Q_{i,N_i+1} - Q_{i+1,1} - Q_{i+2,1} = 0 \tag{6-58}$$

$$F_{jn_1,2} = z_{i,N_i+1} + y_{i,N_i+1} + \frac{Q_{i,N_i+1}^2}{2g A_{i,N_i+1}^2}$$

$$- z_{i+1,1} - y_{i+1,1} - (1+k)\frac{Q_{i+1,1}^2}{2g A_{i+1,1}^2} = 0 \tag{6-59}$$

$$F_{jn_1,3} = z_{i,N_i+1} + y_{i,N_i+1} + \frac{Q_{i,N_i+1}^2}{2g A_{N_i+1}^2}$$

$$- z_{i+2,1} - y_{i+2,1} - (1+k)\frac{Q_{i+2,1}^2}{2g A_{i+2,1}^2} = 0 \tag{6-60}$$

where the subscript jn_1 identifies the upstream junction. The junction losses and the differences in the velocity heads at the junction may be neglected if they are small.

Similarly, the following three equations are available at the downstream junction (Fig. 6-12b), in which we use subscript jn_2 to denote values for the junction.

$$F_{jn_2,1} = Q_{i+3,1} - Q_{i+1,N_{i+1}+1} - Q_{i+2,N_{i+2}+1} = 0 \tag{6-61}$$

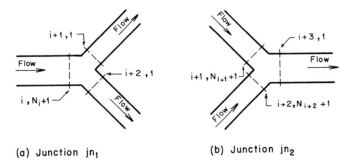

Figure 6-12   Channel junctions.

$$F_{jn_2,2} = z_{i+1,N_{i+1}+1} + y_{i+1,N_{i+1}+1} + \frac{Q_{i+1,N_{i+1}+1}^2}{2gA_{i+1,N_{i+1}+1}^2}$$
$$- z_{i+3,1} - y_{i+3,1} - (1+k)\frac{Q_{i+3,1}^2}{2gA_{i+3,1}^2} = 0 \tag{6-62}$$

$$F_{jn_2,3} = z_{i+2,N_{i+2}+1} + y_{i+2,N_{i+2}+1} + \frac{Q_{i+2,N_{i+2}+1}^2}{2gA_{i+2,N_{i+2}+1}^2}$$
$$- z_{i+3,1} - y_{i+3,1} - (1+k)\frac{Q_{i+3,1}^2}{2gA_{i+3,1}^2} = 0 \tag{6-63}$$

For complex networks (Fig. 6-7b), Eqs. 6-58 through 6-60 or Eqs. 6-61 through 6-63 are included for the branching junction of three channels; Eq. 6-54 is included for a series junction of two channels.

Now, the system of equations is solved simultaneously as follows. We first estimate $y_{l,j}^{(0)}$ as well as $Q_{l,j}^{(0)}$, ($l = i, i+1, \ldots, i+3$ and $j = 1, 2, \ldots, N_i + 1$). Reasonable initial estimated values for the flow depths may be obtained by setting all of them equal to the downstream depth specified as the end condition. It is desirable to satisfy continuity at each node by choosing appropriate discharge estimates as well as the respective flow directions. However, a correct solution is obtained even if the initial estimates do not satisfy the continuity condition, although convergence in this case is slower. To account for the reverse flow—i.e., flow directions opposite to the assumed ones—the energy equation may be written as

$$F_{i,k} = y_{i,j+1} - y_{i,j} + z_{i,j+1} - z_{i,j}$$
$$+ \frac{\alpha_i}{2g}\left(\frac{Q_{i,j+1}|Q_{i,j+1}|}{A_{i,j+1}^2} - \frac{Q_{i,j}|Q_{i,j}|}{A_{i,j}^2}\right)$$
$$+ \frac{1}{2}(x_{i,j+1} - x_{i,j})\left(\frac{n_{i,j+1}^2 Q_{i,j+1}|Q_{i,j+1}|}{C_o^2 A_{i,j+1}^2 R_{i,j+1}^{1.33}} + \frac{n_{i,j}^2 Q_{i,j}|Q_{i,j}|}{C_o^2 A_{i,j}^2 R_{i,j}^{1.33}}\right) = 0 \tag{6-64}$$

Replacing $Q_{i,j}^2$ by $Q_{i,j}|Q_{i,j}|$ yields the correct sign for the head-loss term.

By expanding in Taylor series, a matrix system similar to Eq. 6-49 is obtained. For each energy equation, there are now four nonzero partial derivatives, namely, the partial derivatives with respect to the flow depth and with respect to the discharge at the section under consideration as well as the partial derivatives with respect to the corresponding variables for the adjacent section. Thus, for an energy equation $F_{i,k}$ between sections j and $j+1$ of channel i, the following nonzero partial derivatives are obtained:

$$\frac{\partial F_{i,k}}{\partial y_{i,j}} = -1 + Q_{i,j}^2 \left(\frac{\alpha_i B_{i,j}}{g A_{i,j}^3} - \frac{2n_i^2 (x_{i,j+1} - x_{i,j})}{3 C_o^2 A_{i,j}^2 R_{i,j}^{2.33}} \frac{d R_{i,j}}{d y_{i,j}} - \frac{n_i^2 B_{i,j}(x_{i,j+1} - x_{i,j})}{C_o^2 A_{i,j}^3 R_{i,j}^{1.33}} \right)$$

$$\frac{\partial F_{i,k}}{\partial Q_{i,j}} = 2 Q_{i,j} \left(-\frac{\alpha_i}{2g A_{i,j}^2} + \frac{n_i^2 (x_{i,j+1} - x_{i,j})}{2 C_o^2 A_{i,j}^2 R_{i,j}^{1.33}} \right)$$

$$\frac{\partial F_{i,k}}{\partial y_{i,j+1}} = 1 - Q_{i,j+1}^2 \left(\frac{\alpha_i B_{i,j+1}}{g A_{i,j+1}^3} + \frac{2n_i^2 (x_{i,j+1} - x_{i,j})}{3 C_o^2 A_{i,j+1}^2 R_{i,j+1}^{2.33}} \frac{d R_{i,j+1}}{d y_{i,j+1}} \right.$$
$$\left. + \frac{n_i^2 B_{i,j+1}(x_{i,j+1} - x_{i,j})}{C_o^2 A_{i,j+1}^3 R_{i,j+1}^{1.33}} \right) \tag{6-65}$$

$$\frac{\partial F_{i,k}}{\partial Q_{i,j+1}} = 2 Q_{i,j+1} \left(\frac{\alpha_i}{2g A_{i,j+1}^2} + \frac{n_i^2 (x_{i,j+1} - x_{i,j})}{2 C_o^2 A_{i,j+1}^2 R_{i,j+1}^{1.33}} \right)$$

Here the subscript k refers to the equation number, and its value is not identical to that of j. Similarly, the nonzero partial derivatives for any continuity equation $F_{i,k+1}$ are those with respect to the discharges of the adjacent sections, i.e.,

$$\frac{\partial F_{i,k+1}}{\partial Q_{i,j}} = -1$$
$$\frac{\partial F_{i,k+1}}{\partial Q_{i,j+1}} = 1 \tag{6-66}$$

Similar equations may be written for the remaining three channels. The partial derivatives at the junctions are (values in the parentheses apply if the junction losses and the difference in the velocity heads at the junction are neglected):

$$\frac{\partial F_{jn}}{\partial y_{i,j}} = -1 + \frac{Q_{i,j}^2 B_{i,j}}{g A_{i,j}^3} \qquad (\text{or} - 1)$$

$$\frac{\partial F_{jn}}{\partial y_{i+1,1}} = 1 - \frac{(1+k) Q_{i+1,1}^2 B_{i+1,1}}{g A_{i+1,1}^3} \qquad (\text{or } 1)$$

$$\frac{\partial F_{jn}}{\partial Q_{i,j}} = -\frac{Q_{i,j}}{g A_{i,j}^2} \qquad (\text{or } 0) \tag{6-67}$$

$$\frac{\partial F_{jn}}{\partial Q_{i+1,1}} = \frac{(1+k) Q_{i+1,1}}{g A_{i+1,1}^2} \qquad (\text{or } 0)$$

Note that the preceding equations are written assuming the flow direction from section (i, j) to section $(i + 1, 1)$.

If the equations for a channel network are arbitrarily arranged, then all the nonzero elements of the Jacobian matrix may not necessarily lie on or near the principal diagonal, as do those for a series system. This results in increased storage requirements, increased computer time, and, most probably, reduced accuracy. For the parallel-channel network shown in Fig. 6-11, these limitations may be avoided by arranging the governing equations as discussed in the following paragraphs.

For each reach of channel i, the energy and continuity equations are written consecutively from section 1 to section $N_i + 1$. Then, Eqs. 6-58 through 6-60, i.e., $F_{jn_1,k}$ ($k = 1, 2, 3$), are written for the upstream junction. The energy equation between sections 1 and 2 of channel $i + 1$ is written, followed by the energy equation between sections 1 and 2 of channel $i + 2$. Then, the continuity equation between sections 1 and 2 of channel $i + 1$ is written, followed by the continuity equation between sections 1 and 2 of channel $i + 2$. These four equations are repeated in the same alternating sequence for the remaining reaches of the parallel channels, $i + 1$ and $i + 2$. In order to have such a numbering system, it is necessary that the number of sections on the parallel channels be the same. Next, Eqs. 6-61 through 6-63 are written for the downstream junction, jn_2. Finally, the energy and continuity equations for channel $i + 3$ are written, similar to those for channel i. If the governing equations are arranged in this manner, then the resulting Jacobian is banded with a bandwidth of seven. The remainder of the solution algorithm is identical to that described previously. For a parallel-channel system with M parallel channels, this arrangement of equations results in a Jacobian of bandwidth $3M + 1$.

For complex channel networks, however, no generalized procedure is available for arranging the governing equations to produce a Jacobian matrix of minimum bandwidth. This is because the system is asymmetric. However, by manually numbering each channel and each node from the upstream end to the downstream end in a semisystematic manner, a banded matrix of small bandwidth may be obtained. It may not, however, have the minimum band width. A number of procedures are available to reduce the bandwidth of nonsymmetric matrices (Berztiss 1971; Deo 1974). However, they are generally very complex and the effort required to apply them exceeds the additional time and storage required for the solution of a nonminimized Jacobian. Therefore, the use of these methods appears to be unjustified, and they are not presented here.

Example 6-3

Figure 6-13 *shows a channel network. The channel data are listed in Table* 6-3. *All channels have subcritical flow and rectangular cross sections. The flow depth at the downstream end (node F) is 5 m for a flow of* $250 \, m^3/s$. *The form loss coefficient, k, for all junctions is 0.0 and the velocity-head coefficient, α, for all channels is 1. Compute the flows and depths in different channels of the system.*

Table 6-3. Channel data

Channel	Length (m)	Width (m)	Reach (m)	n	S_o
1	200.0	30.0	50.0	0.013	0.0005
2	200.0	40.0	50.0	0.013	0.0005
3	200.0	20.0	50.0	0.012	0.0005
4	100.0	20.0	25.0	0.014	0.0005
5	100.0	20.0	25.0	0.013	0.0005
6	100.0	25.0	25.0	0.013	0.0005
7	100.0	30.0	25.0	0.014	0.0005
8	300.0	50.0	75.0	0.014	0.0005

Solution: The channels and nodes are numbered as shown in Fig. 6-13. The iterative procedure is started by assuming all flow depths equal 5.0 m. The discharges are given random values without satisfying continuity at the branching junctions. The assumed flow directions in different channels are shown by the arrows of Fig. 6-13. A tolerance of 0.0005 for both $\Delta y_{i,j}$ and $\Delta Q_{i,j}$ is specified for the convergence of the solution procedure. The iterations converge after only six iterations. The computed discharges and flow depths are listed in Table 6-4.

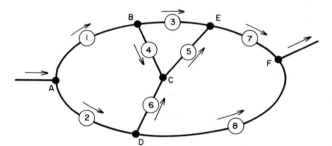

Figure 6-13 Network of Example 6-3.

These values were verified by computing the flow profiles using the fourth-order Runge-Kutta method. Instead of using graphical or other manual methods to determine the discharge in various channels, the values calculated by the procedure of this section were used directly. The water levels computed by the Runge-Kutta method were identical (within the specified tolerance) to those listed in Table 6-4. The computational time for each individual channel was approximately the same for both methods. It is not possible to estimate the number of trial-and-error iterations required to determine the discharge in the Runge-Kutta method. Assuming m such iterations, the computational time for the Runge-Kutta method is at least m times that required by the procedure presented in this section.

Practical Applications

Parallel-channel systems occur frequently in nature, mainly in the form of flow around islands. Channel networks are not as common as parallel systems, although they occur

Table 6-4. Network of example 6-3

Section	Channel 1 $Q = 95.748$ m³/s		Channel 2 $Q = 154.252$ m³/s		Channel 3 $Q = 55.093$ m³/s		Channel 4 $Q = 40.655$ m³/s	
	Depth (m)	Distance (m)	Depth (m)	Distance (m)	Depth (m)	Distance (m)	Depth (m)	Distance (m)
1	4.754	0.0	4.754	0.0	4.853	0.0	4.853	0.0
2	4.779	50.0	4.779	50.0	4.877	50.0	4.865	25.0
3	4.803	100.0	4.803	100.0	4.902	100.0	4.977	50.0
4	4.828	150.0	4.828	150.0	4.927	150.0	4.890	75.0
5	4.853	200.0	4.852	200.0	4.951	200.0	4.902	100.0

Section	Channel 5 $Q = 52.669$ m²/s		Channel 6 $Q = 12.014$ m²/s		Channel 7 $Q = 107.762$ m²/s		Channel 8 $Q = 142.238$ m²/s	
	Depth (m)	Distance (m)	Depth (m)	Distance (m)	Depth (m)	Distance (m)	Depth (m)	Distance (m)
1	4.902	0.0	4.852	0.0	4.951	0.0	4.852	0.0
2	4.914	25.0	4.864	25.0	4.963	25.0	4.889	75.0
3	4.927	50.0	4.877	50.0	4.976	50.0	4.926	150.0
4	4.939	75.0	4.889	75.0	4.988	75.0	4.963	225.0
5	4.952	100.0	4.902	100.0	5.000	100.0	5.000	300.0

in a braided river system, e.g., in a delta. A common design problem is to provide cutoff channels in a meandering stream for flood control. The allowable flow rate and water levels in the old stream dictate the design of the new channel. By using the algorithm presented in this section, different designs may be modeled efficiently.

The algorithm of this section may also be used to determine other channel parameters, such as the roughness coefficients. For example, if the flow depths in a channel system are known for a specified discharge, then the governing equations may be solved simultaneously to directly determine Manning n instead of using a trial-and-error procedure.

6-9 Computer Programs

To demonstrate computer applications of this chapter, four sample computer programs are presented in Appendix D. The direct-step method is used in the program of Appendix D-1, the standard step method is used in Appendix D-2, the modified Euler method is used in Appendix D-3, and the simultaneous solution procedure is employed in Appendix D-4.

6-10 Summary

A number of procedures to compute steady-state, gradually varied flows were presented in this chapter. The first two methods—direct and standard step—solve the energy equation between two consecutive channel sections. Then, a number of numerical methods were presented to integrate the governing differential equation. These methods do not allow a direct solution of parallel-channel systems or complex channel networks. Therefore, trial-and-error procedures have to be used. For the analysis of these channel systems, a simultaneous solution algorithm based on the Newton-Raphson method was presented.

Problems

6-1. Derive Eq. 6-3 by using the principle of conservation of momentum. (Hint: Apply Newton's second law of motion to a short channel length, Δx.)

6-2. A 10-m-wide, rectangular, concrete-lined canal has a bottom slope of 0.01 and a constant-level lake at the upstream end. The water level in the lake is 6.0 m above the bottom of the canal at the entrance. If the entrance losses are negligible, determine

　i. The flow depth 800 m downstream of the canal entrance; and
　ii. The distance from the lake where the flow depth is 2.5 m.

6-3. A trapezoidal channel having a bottom slope of 0.001 is carrying a flow of 75 m³/s. The channel bottom is 50 m wide and the channel side slopes are 1V : 1.5H. If a control structure is built at the downstream end that raises the water depth at the downstream end to 12 m, determine the amount by which the channel banks must be raised along its length. Assume the channel had uniform flow prior to the construction of the control structure and $n = 0.025$.

6-4. A 5-km-long canal is lined and has a free overfall at the lower end and a constant-level reservoir at the upper end. If the critical depth at the fall is 4 m, determine the minimum water level in the lake assuming $n = 0.013$ and the head losses at the entrance are $0.2V^2/(2g)$. The canal has a bottom width of 8.0 m and side slopes of 1V : 1.5H and a bottom slope of 0.0001.

6-5. Figure 6-14 shows the cross section of a natural stream. The channel bottom slope is 0.0002, $n = 0.035$, and the discharge is 80 m³/s. If the flow depth at a bridge crossing is 8.0 m, determine the flow depth 3.0 km upstream of the bridge.

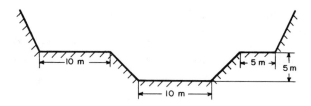

Figure 6-14　Channel cross section for Prob. 6-5.

6-6. Develop computer programs to compute the water-surface profile in a trapezoidal channel having a free overfall at the downstream end. To compute the profile, use the following methods:

  i. Direct step method
  ii. Standard step method
  iii. Euler method
  iv. Modified Euler method
  v. Fourth-order Runge-Kutta method

6-7. By using the computer programs of Prob. 6-6, compute, plot, and compare the surface profile in a trapezoidal channel having the following data:

Bottom width = 20.0 m
Side slopes = 2H : 1V
Manning $n = 0.013$
Discharge = 30 m^3/s
Channel bottom slope = 0.00015
A free overfall at the downstream end.

Select appropriate values for the flow depths in method (i) and appropriate distance locations in methods (ii) – (v) of Prob. 6-6.

6-8. Investigate the sensitivity of the computed water level at a distance of 5 km upstream from the outfall by using different increments for the flow depth and different distance locations in Prob. 6-4.

6-9. Write a computer program to compute the water-surface profile in a trapezoidal channel having a steep bottom slope. The water level in a constant-level lake located at the upper end is 1.0 m above the normal depth in the channel. Neglect the entrance losses.

6-10. Use the computer program of Prob. 6-9 to determine the flow depth 0.5 km downstream from the lake outlet.

6-11. Verify the validity of the computer programs developed in Prob. 6-6 by comparing the computed results with those obtained from the following equation for a very wide channel derived by Bresse (Bresse 1860; Rouse 1950):

$$x = \frac{y}{S_o} - y_n \left[\frac{1}{S_o} - \frac{C^2}{g} \right] \phi$$

In this equation, the Bresse function is

$$\phi = \frac{1}{6} \ln \left[\frac{w^2 + w + 1}{(w - 1)^2} \right] - \frac{1}{\sqrt{3}} \tan^{-1} \left[\frac{\sqrt{3}}{2w + 1} \right] + C_1$$

$w = y/y_n$; C_1 = a constant of integration; y = flow depth at distance x; y_n = normal depth; S_o = bottom slope; and C = Chezy constant.
(Hint: Solve the following problem and compare the results: $B = 500$ m; $Q = 500$ m^3/s; $S_o = 0.0008$; $C = 100$. Assume critical depth occurs 10 m upstream of the free overfall.)

6-12. By using the Euler, modified Euler, and fourth-order Runge-Kutta methods, compute the flow depth 1 km upstream of the fall of Prob. 6-11 by using $\Delta x = 25$, 50, 100, and 200 m. By assuming the results of the Bresse equation to be exact, compute and plot the error versus Δx.

6-13. A meandering river (Fig. 6-15) has a bottom slope of 1.0 m/km. The stations as marked are distances, in kilometers, along the river from point A. To reduce flooding, it is planned to provide cutoffs as shown. The river bottom at point F is at El. 500 and the water level at F for a flow of 500 m³/s is at El. 505. Compute the water level at point A if

   i. There are no cutoffs;

  ii. Only cutoff BD is provided; and

 iii. Both cutoffs BD and CE are provided.

Assume Manning n for the river channel and for the cutoff are 0.050 and 0.020, respectively. The river channel is 500 m wide and the cutoffs are 150 m wide.

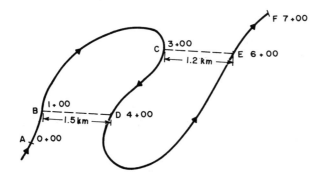

Figure 6-15 Meandering river and cutoffs.

6-14. Figure 6-16 shows the tailrace system of a hydroelectric power plant. If the water level at the downstream weir is at El. 504.00 for a flow of 1688 m³/s, determine the water levels in each manifold.

6-15. Prove that all three formulations for the approximation of the friction slope (Eq. 6-13) give identical results if the terms of the order $(\Delta S_f / S_{f_1})^2$ and higher are neglected. In this expression, $\Delta S_f = S_{f_2} - S_{f_1}$.

6-16. In order to reduce the height of the Pont du Gard for the Roman Aqueduct of Nimes (Fig. 4-9), the bottom slope of the upstream part was increased (Hauck and Novak, 1987). This resulted in reducing the available bottom slope downstream of the crossing. There has been heavy incrustation of the aqueduct due to low velocities and due to some other factors. If you were the designer rehabilitating the aqueduct, list the modifications you would propose to maximize the flow capacity with minimum structural modification to the channel cross sections (you may raise or lower the channel). Analyze and compare the effect of these modifications on the flow capacity, costs, and available freeboard along the length. Which modification do you prefer and why?

    By assuming that the incrustation progressed as shown in Fig. 6-17, compute and plot the water levels in the aqueduct for flows of 210 and 450 l/s.

6-17. Plot the water surface profile in the outlet of Prob. 3-18.

6-18. Debris accumulates at a bridge, which raises the water level to 12 ft. The flood channel is trapezoidal in shape, 20 ft wide at the bottom with side slopes of 2H : 1V, $n = 0.025$ and with the bottom slope of 0.0003. How far will the effect of clogging extend for a flow of 800 ft³/sec.

6-19. A 10-ft-square box culvert is 150 ft long and is laid at a slope of 0.01. The flow depth upstream of the entrance is 15 ft. The accumulation of debris at a channel crossing 0.5 mi

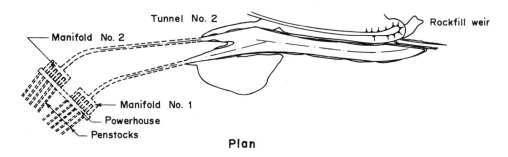

Plan

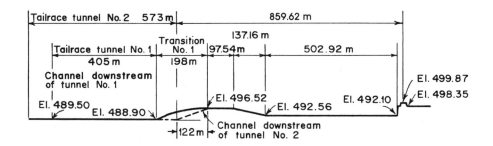

Center line profile

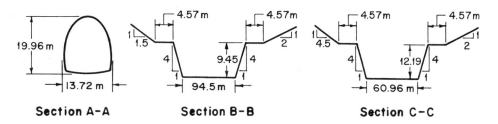

Section A-A   Section B-B   Section C-C

Figure 6-16 Tailrace system of a hydropower plant.

downstream of the culvert raises the water level by 5 ft at the crossing. The channel is trapezoidal in shape, is 10 ft wide at the bottom, and has side slopes of 1V : 1.5H. If the channel bottom level drops 1.2 ft from the culvert exit to the crossing, compute and plot the water-surface profiles in the channel and inside the culvert. Assume the channel had uniform flow prior to the accumulation of debris.

6-20. A 10-m-wide, rectangular, concrete-lined channel ($n = 0.013$) has a bottom slope of 0.01. There is a constant-level lake at the upstream end with the lake water surface 6 m above the channel bottom. If the flow depth at the channel entrance is critical, determine the locations where the flow depth is 3.9, 3.7, 3.5, 3.3, and 3.0 m.

6-21. A rectangular canal is 10 m wide and carries a flow of 50 m³/s. The bottom and sides of the canal are concrete lined, the longitudinal bottom slope is 0.0006, and the canal ends in

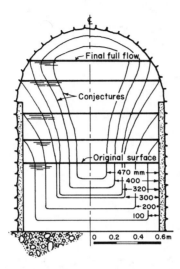

Figure 6-17 Possible scenarios for progressive incrustation (After Hauck and Novak, 1987).

a free outfall. What is the depth of flow 2 km upstream of the fall? Assume the concrete lining has deteriorated somewhat due to weathering.

     If the flow depth is critical at a distance of $4y_c$ upstream of the fall, compute the water-surface profile in the canal.

6-22. A trapezoidal channel with bottom width of 10 m and side slopes of 1V : 1.5H is carrying a flow of 80 m³/s. The channel bottom slope is 0.002 and $n = 0.015$. A dam is planned that will raise the flow depth to 10 m. Compute the flow depth in the channel 250, 500, and 750 m upstream of the dam.

6-23. The normal depth in a 10-m-wide rectangular channel having a bottom slope of 0.001 is 2 m. The Manning n for the flow surfaces may be assumed to be 0.020. The construction of a bridge raises the upstream water level by 1 m. Determine the distance from the bridge where the flow depth in 2.5 m.

REFERENCES

Bakhmeteff, B. A. 1932. *Hydraulics of open channels*, McGraw-Hill, New York, NY.

Berztiss, A. T. 1971. *Data structures—theory and practice,* Academic Press, New York, NY.

Boudine, E. J. 1861. De l'axe hydraulique des cours d'eau contenus dans un lit prismatique et des dispositifs réalisant, en pratique, ses formes diverses (The flow profiles of water in a prismatic channel and actual dispositions in various forms), *Annales des travaux publiques de Belgique*, Brussels, 20, 397-555.

Bresse, J. A. C. 1860. *Cours de mecanique appliquee, hydraulique,* 2nd ed. Mallet-Bachelier, Paris, France.

Chapra, S. C., and Canale, R. P. 1988. *Numerical methods for engineers,* 2nd ed., McGraw-Hill, New York, NY.

Chaudhry, M. H., and Schulte, A. 1986. Computation of steady-state, gradually varied flows in parallel channels, *Canadian Jour. of Civil Engineering*, 13, no. 1:39–45.

Chow, V. T. 1959. *Open-channel hydraulics*, McGraw-Hill, New York, NY.

Davis, D. W., and Burnham, M. W. 1987. Accuracy of Computed Water Surface Profiles, *Proc.*, National Hydraulic Engineering Conference, Amer. Soc. Civil Engrs., pp. 818–23.

Deo, N. 1974. *Graph theory with applications to engineering and computer science,* Prentice-Hall, Englewood Cliffs, NJ.

Eichert, Bill S. 1970. Survey of Programs for Water-Surface Profiles, *Jour. Hydraulics Division*, Amer. Soc. Civil Engrs., 96, no. 2:547–63.

Epp, R., and Fowler, A. G. 1970. Efficient code for steady-state flows in networks, *Jour. Hydraulics Division*, Amer. Soc. Civil Engrs., 96, no. HY1 January:43–56.

Hauck, G. F., and Novak, R. A. 1987. Interaction of flow and incrustation in the Roman aqueduct of Nimes, *Jour. Hydraulic Engineering,* Amer. Soc. Civil Engrs., 113, no 2:141–57.

Henderson, F. M. 1966. *Open channel flow*, Macmillan, New York, NY.

Humpidge, H. B., and Moss, W. D. 1971. The development of a comprehensive computer program for the calculation of flow profiles in open channels, *Proc.*, Institution of Civil Engrs. (London) 50 (September):49–65.

Kumar, A. 1978. Integral solutions of the gradually varied equations for rectangular and triangular channels, *Proc.*, Institution Civil Engrs. (London) 65, pt. 2 (September):509–15.

Kumar, A. 1979. Gradually varied surface profiles in horizontal and adversely sloping channels, *Proc.*, Inst. Civil Engrs. (London) 67, pt. 2 (June):435–52.

Laurenson, E. M. 1986. Friction slope averaging in backwater calculations, *Jour. Hydraulic Engineering,* Amer. Soc. Civil Engrs., 112, no. 12:1151–63.

McBean, E., and Perkins, F. 1975. Numerical errors in water profile computation, *Jour. Hydraulics Div.*, Amer. Soc. Civil Engrs. 101, no. 11:1389–1403.

McCracken, D. D., and Dorn, W. S. 1964. *Numerical methods and FORTRAN programming*, John Wiley, New York, NY.

Molinas, A., and Yang, C. T. 1985. Generalized water surface profile computations, *Jour. Hydraulic Engineering,* Amer. Soc. Civil Engrs., 111, no. 3:381–97.

Prasad, R. 1970. Numerical method of computing flow profiles, *Jour. Hydrulics Division*, Amer. Soc. Civil Engrs., 96, no. 1:75-86.

Rouse, H. (ed.). 1950. *Engineering Hydraulics*, John Wiley, New York, NY.

Schulte, A. M., and Chaudhry, M. H. 1987. Gradually varied flows in open channel networks, *Jour. of Hydraulic Research*, Inter. Assoc. for Hydraulic Research, 25, no. 3:357–71.

Soil Conservation Service, 1976. *WSP-2 Computer Program*, Technical Release No. 61.

U. S. Army Corps of Engineers, 1982. *HEC-2, Water-surface Profiles, User's Manual*, Hydrologic Engineering Center, Davis, California.

U. S. Geological Survey, 1976. Computer applications for step-backwater and floodway analysis, *Open File Report* 76-499.

Wylie, E. B. 1972. Water surface profiles in divided channels, *Jour. Hydraulic Research* Inter. Assoc. for Hydraulic Research. 10, No. 3:325–41.

7 Rapidly Varied Flow

Stilling basins of the Mangla dam spillway, Pakistan. The dissipator is designed for 31150 m³/s and 100-m fall and comprises two basins in series. The upper basin has 12.2-m high end sill and the lower basin has a row of 15, 9.1-m high baffle blocks. (Courtesy of Harza Engineering Company, Chicago, Ill.).

7-1 Introduction

The streamlines in the uniform and gradually varied flows we considered in the previous chapters are either parallel or may be assumed as parallel. Therefore, the acceleration in these flows is negligible, and the pressure distribution may be assumed as hydrostatic. The analyses in which the pressure distribution is hydrostatic is referred to as *shallow-water theory*. However, as we discussed in Chapters 1 and 5, many times the streamlines have sharp curvatures, thereby making the assumption of hydrostatic pressure distribution invalid. In addition, the flow surface may become discontinuous if the flow depth changes rapidly such that the surface profile breaks, e.g., in a hydraulic jump.

Due to nonhydrostatic pressure distribution, rapidly varied flow can not be analyzed by using the same approach as that for parallel flow. In the past, these flows have been mainly investigated experimentally; and empirical relationships and other information in the form of charts and diagrams have been developed. Each particular phenomenon has been studied more or less in isolation, and a considerable amount of information is available for typical design applications.

The rapidly varied flows have been analyzed based on the Boussinesq and Fawer assumptions. In the Boussinesq assumption, the vertical-flow velocity is assumed to vary linearly from zero at the channel bottom to the maximum at the free surface. In the Fawer assumption, this variation is assumed to be exponential.

The rapidly varied flow usually occurs in a short distance. Therefore, the losses due to shear at the channel boundaries are small and may be neglected in a typical analysis. However, because of sudden changes in the channel geometry, the flow may separate, and eddies and swirls may form. These phenomena complicate the flow pattern, and it becomes difficult to generalize the velocity distribution at a cross section. Even in those cases where it may be possible to approximate the velocity distribution, the energy coefficient, α, and the momentum coefficient, β, are difficult to quantify and usually have a value significantly higher than unity. Because of separation zones and formation of rollers and eddies, it becomes difficult to define the flow boundaries and to determine the average flow variables for a cross section.

In this and in the following chapter, we discuss rapidly varied flows. The material presented in this chapter follows the traditional approach. First, we discuss how the application of common conservation laws for the analysis of rapidly varied flows requires special consideration. Then, empirical information on the transitions, hydraulic jump, spillways, and energy dissipators is included for design purposes. Chapter 8 deals mainly with the numerical computation of these flows.

7-2 Application of Conservation Laws

As we discussed in the last section, the assumption of hydrostatic pressure distribution may not be valid in rapidly varied flows because the streamlines are not parallel, due to variation in the cross-sectional shape and size, due to change in the flow direction, or for other reasons. Therefore, special care has to be taken while applying the conservation laws of mass, momentum, and energy for the analysis of rapidly varied flows. To illustrate this, we discuss a number of typical situations in the following paragraphs (Ippen 1950).

Let us consider the flow conditions at three sections in a constant-width channel with a sudden step in its bottom (Fig. 7-1). Sections 1 and 3 are located at some distance upstream and downstream of the step, but section 2 is located just downstream of the step. Let us assume the flow is uniform at section 1 and becomes uniform again at section 3. The flow separates at the step due to sudden change in the bottom profile, and the velocity distribution at section 2 is as shown in this figure (Fig. 7-1). At sections 1 and 3, the pressure distribution may be assumed hydrostatic and the velocity in most of the cross sections is approximately equal to the mean velocity.

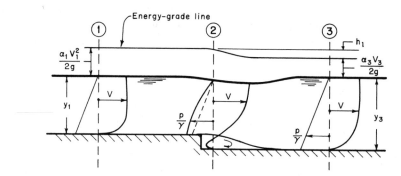

Figure 7-1 Definition sketch for abrupt drop.

Let us now consider the application of the three conservation laws one by one to the flow at sections 1, 2, and 3.

The volumetric rate of flow through a cross section may be written as

$$Q = \int_A \mathbf{v} \cdot d\mathbf{A} \qquad (7-1)$$

in which Q = volumetric rate of flow and $\mathbf{v}$ = flow velocity at any point in the cross section. Note that the dot product automatically takes care of whether vectors $\mathbf{v}$ and $\mathbf{A}$ are orthogonal or not. If the flow velocity v is normal to the flow area dA, then we may write this equation as

$$Q = \int_A v \, dA \qquad (7-2)$$

Assuming the mass density to be constant and the flow to be steady, the volumetric flow rate at section 1 is equal to the flow rate at section 3, i. e., $Q_1 = Q_3$, or $V_1 A_1 = V_3 A_3$, in which V = mean flow velocity at the section and the subscripts 1 and 3 refer to the variables for cross sections 1 and 3, respectively. This is commonly known as the *continuity equation*. However, note that the flow velocity at section 2 is negative in part of the channel depth. In this case, we cannot express the volumetric flow rate in terms of the mean flow velocity. Thus, it is necessary to know the velocity distribution to compute the rate of discharge at section 2.

For computing the rate of momentum flux at any section, we may write

$$\text{Rate of momentum flux in the } x \text{ direction} = \rho \int_A v_x (v \, dA) \tag{7-3}$$

in which v_x = component of the flow velocity in the x direction. To evaluate this integral, the distribution of flow velocity and its component in the x direction should be known. Similar equations may be written for the momentum flux in the y and z directions.

To account for the nonuniform velocity distribution, we may write

$$\text{Rate of momentum flux} = \rho Q (\beta V_x) \tag{7-4}$$

in which β = momentum coefficient as defined in Chapter 1 and V_x = mean flow velocity in the x direction. Equation 7-4 may be used for sections 1 and 3, but it is unsuitable for section 2, for which Eq. 7-3 must be used.

Note that the pressure distribution may not be hydrostatic in the region of curvilinear flow because of local accelerations. Therefore, the pressure distribution should be known to determine the force acting on a section. For example, the pressure force acting at sections 1 and 3 may be determined by assuming hydrostatic pressure distribution' at these sections. However, this is not the case at section 2, where the pressure distribution should be known. Such a distribution may not be known without measurements on the prototype or without tests on a hydraulic model. For this reason, the application of the momentum equation in the regions having rapidly varied flow is not simple and straightforward.

For the energy in curvilinear flow, we cannot write the expression

$$H = z + y + \alpha \frac{V^2}{2g} \tag{7-5}$$

for the total head in the usual sense, since the pressure distribution is not hydrostatic. In such a case, the energy flux at any section may be written in the power form as

$$P = \rho g \int_A (z + \frac{p}{\gamma} + \frac{v^2}{2g}) v \, dA \tag{7-6}$$

To evaluate this integral, the velocity and pressure distribution at the section should be known.

In summary, it is not usually possible to utilize the concept of mean quantities at a cross section in rapidly varied flows; instead, the distributions of velocity and pressure are needed to use the conservation laws properly. Such a distribution may not be available for universal applications. Two widely used velocity distributions at a cross section are the Boussinesq and Fawer assumptions. To simplify the analysis, however, we usually select sections away from the regions of rapidly varied flow and then apply the conservation laws to these sections.

7-3 Channel Transitions

A channel transition is a local change in the channel characteristics (usually area, shape, and/or direction) that results in changing the flow from one state to another. Typical examples of channel transitions are contractions, expansions, and bends. We will call a reduction in the cross-sectional area in the direction of flow a channel *contraction* and an increase in the area a channel *expansion*. A transition in which there is a unique relationship between the flow depth and the rate of discharge is referred to as a *control*. Critical flow occurs at a control, which may be natural or artificial. For example, structures such as spillways, weirs, etc., are artificial controls, whereas free falls are natural controls.

A transition is usually provided to change the channel alignment and/or cross section. The design requirements of a transition may be to minimize head losses, to dissipate energy, or to reduce flow velocities to prevent scour and erosion. Because of the unique relationship between the flow depth and the rate of discharge, a control may be used to measure the rates of discharge in a channel.

In this section, a number of general remarks on transitions are made. We first consider subcritical flow in contractions and expansions. The generation of shock waves in supercritical flows at changes in the channel geometries is then discussed.

General Remarks

The design and construction of transitions usually impose conflicting requirements. For example, for the ease and economy of construction, a transition should be simple and may thus have discontinuous boundaries. However, if the design objective is to minimize head losses in the transition, then its cross-sectional area, size, shape, and configuration should change gradually, and the boundaries should be streamlined. Such a design prevents eddy formation and flow separation and reduces the possibility of cavitation. Vittal and Chiranjeevi (1983) presented a procedure for the design of transitions having subcritical flow; and Sturm (1985) gave one for the design of a contraction having supercritical flow.

Because of large changes in the flow boundaries in a short distance, acceleration plays a dominant role in the flow through transitions as compared to the shear resistance at the channel boundaries. Therefore, the validity of the assumption of one-dimensional flow becomes questionable. In addition, the effects of curvilinear streamlines and the possibility of flow separation and cavitation has to be considered.

In flows having significant vertical acceleration, the flow velocity and pressure varies in the direction of flow as well as in the vertical direction. In such a situation, we may have to consider the flow as two- or three-dimensional. The energy losses in a transition are usually small and may be neglected. The flow may thus be assumed to be irrotational, and a flow net may be drawn to analyze flow condition in a transition. Such analyses help in identifying the regions of sudden pressure changes. A sudden pressure drop usually indicates the possibility of cavitation, whereas a sudden pressure rise shows the possibility of flow separation, instability, and vibrations.

In rapid transitions, the velocity distribution may not necessarily be uniform at a channel cross section. There may even be negative flow velocities in part of the channel depth. In such situations, it may be difficult to compute the total head at the section even though the flow depth at the section is known.

The boundary-layer theory may be utilized to predict flow separation. In such analyses, it is necessary to take into consideration the losses due to shear at the boundaries. If the pressure gradient at the boundary is negative—i.e., the flow is accelerating—then the momentum of the boundary layer is augmented, since the acceleration is greater than the boundary shear. For a positive pressure gradient, however, the momentum is further reduced. Therefore, a negative pressure gradient indicates a stable boundary layer and a positive pressure gradient usually leads to flow separation.

Now, let us discuss flow through transitions. In many cases, sharp water surface disturbances in the form of cross waves (like shock waves in gas dynamics) are present in supercritical flows. These require special consideration in the analysis. As a result, we will discuss subcritical and supercritical flows separately in the following paragraphs.

Subcritical Flow

We will first discuss channel expansions and then channel contractions.

Expansions.   A channel expansion may be due to an increase in the channel width, a drop in the channel bottom, or a combination of these two, as shown in Fig. 7-2. An expansion having abrupt geometrical changes is called a *sudden expansion* and it is called a *gradual expansion* if the changes occur over a finite distance.

Channel expansions are utilized in many hydraulic structures, such as flumes, outlets, siphons, and aqueducts. The design of a transition involves the selection of its shape so that flow does not separate and the form losses are minimized. To determine the optimum shape, experimental studies were conducted by several investigators (Smith and Yu 1966; Mazumdar 1967; Nashta and Garde 1988). The use of a number of devices to control separation has been investigated by Smith and Yu (1966), Rao and Seetharamiah (1969), Skogerboe et al. (1971), and Mazumdar (1967).

Several experimental investigations (Abbott and Kline 1962; Graber 1982) show that the flow downstream of an expansion becomes unsymmetrical for $B_2/B_1 \leq 1.5$, where B_1 = channel width upstream of the transition and B_2 = channel width downstream of the transition. It is found that the long and short eddy regions of the confined jet usually remain stable, although they may be interchanged by temporarily inserting a vane in the flow (Nashta and Garde 1988).

Experimental data show that the shape of the separating streamline downstream of an expansion does not depend on the Reynolds number; this shape may be determined

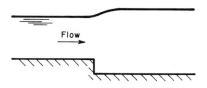

(a) Longitudinal section

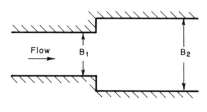

(b) Plan

Figure 7-2 Channel expansion.

from the following equation

$$\frac{B_x - B_1}{B_2 - B_1} = \frac{x}{L}\left[1 - \left(1 - \frac{x}{L}\right)^m\right] \tag{7-7}$$

in which B_x = twice the distance of the separating streamline from the transition centerline at distance x from the transition, L = distance of the point where the streamline meets the boundary, and m = an exponent that varies between 0.6 and 0.66. An expansion shaped according to this equation gives a smaller separation zone and thus reduces head losses.

Let us consider the flow through a sudden expansion (Fig. 7-3), in which the channel width increases from B_1 to B_2 and the channel sections 1, 2, and 3 are located as shown.

Let us make the following assumptions: (1) $E_2 = E_1$, (2) $F_{s1} = F_{s2}$, (3) $y_1 = y_2$, (4) the width of the jet at section 2 is equal to B_1, and (5) the Froude number, $\mathbf{F}_{r1}$ is small so that $\mathbf{F}_{r1}^4$ may be neglected. Based on these assumptions, Henderson (1966) showed that

$$E_1 - E_3 = \frac{V_1^2}{2g}\left[\left(1 - \frac{B_1}{B_2}\right)^2 + 2\mathbf{F}_{r1}^2(B_2 - B_1)\frac{B_1^3}{B_2^4}\right] \tag{7-8}$$

The second term inside the bracket is small if $\mathbf{F}_{r1} < 0.5$ or if $B_2/B_1 > 1.5$. In most practical situations, the former condition is satisfied. If $B_2/B_1 < 1.5$, then the overall energy loss in the transition is small, and as a result this term becomes insignificant. Hence, the head loss, H_l, in a sudden expansion may be written as

$$H_l = \frac{(V_1 - V_3)^2}{2g} \tag{7-9}$$

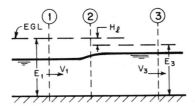

(a) Longitudinal section

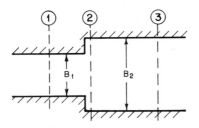

(b) Plan   **Figure 7-3** Definition sketch.

Experimental results of Formica (1955) show that Eq. 7-9 overestimates the head losses by about 10 percent.

The head losses may be reduced by providing a gradual expansion. Normally, a taper of 1 in 4 is recommended. The head loss in such a transition is

$$H_l = 0.3 \frac{(V_1 - V_3)^2}{2g} \tag{7-10}$$

More gradual tapering of walls than a taper of 1 in 4 does not significantly reduce the head losses in subcritical flow, although construction costs may be substantially increased.

Contractions.   The channel contraction may comprise a reduction in the channel width, raising the channel bottom, or a combination of the two (Fig. 7-4). An abrupt change in the cross section is called a *sudden* contraction, whereas if the change occurs over a distance, it is called a *gradual* contraction.

The head losses in channel contractions are less than those in channel expansions. According to tests conducted by Formica (1955), the head loss through a sudden or square-edged contraction is

$$H_l = 0.23 \frac{V_3^2}{2g} \tag{7-11a}$$

The head loss through a contraction when the edges are rounded is

$$H_l = 0.11 \frac{V_3^2}{2g} \tag{7-11b}$$

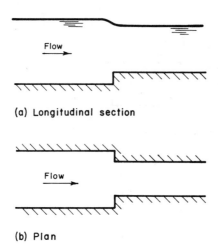

(a) Longitudinal section

(b) Plan   **Figure 7-4** Channel contractions.

in which V_3 is the velocity at a section in the narrower part, where the flow has almost become uniform. Yarnell's test results indicate (1934) higher loss coefficients: 0.35 for square and 0.18 for rounded-edge contractions, instead of those given in Eq. 7-11.

A contraction may choke the flow, as discussed in Chapter 2, if the channel width is reduced too much, since the energy may not be sufficient to pass the required amount of discharge per unit width.

7-4 Supercritical Flow

In supercritical flow through transitions, complications may arise due to the formation of shock waves on the flow surfaces. These waves are generated at a change in the flow boundary, as discussed in the following paragraphs. This treatment follows Henderson (1966) closely.

To illustrate the generation and formation of shock waves, let us consider the movement of an observer through a stationary fluid. Let us assume that the observer is traveling at velocity V and creates a disturbance (say by detonating an explosive charge) while arriving at different locations, marked as 1, 2, 3, ... in Fig. 7-5. Let us denote the celerity (i.e., velocity with respect to the medium in which the disturbance is traveling) of the disturbance by c. Then, depending upon the relative magnitude of the velocity of the observer V to that of celerity, c, three different situations are possible, as shown in Fig. 7-5. These are discussed in the following paragraphs.

The disturbance created by the observer at each location travels outward, as shown in Fig. 7-5. If $V < c$, the observer lags behind the front of the disturbance and the front of the disturbances generated at different locations forms a circular pattern (Fig. 7-5a). For the case of $V = c$, the disturbance moves outward from the point of generation. Because the celerity of the disturbance and the velocity of the observer are equal, the

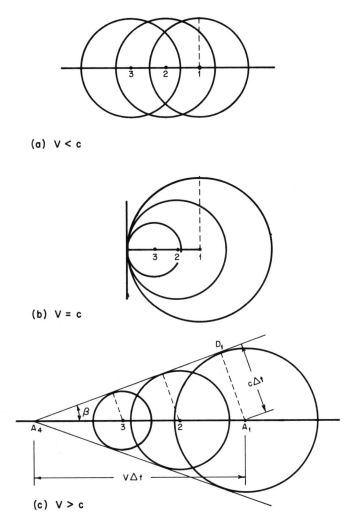

(a) V < c

(b) V = c

(c) V > c

Figure 7-5 Propagation of a disturbance.

waves generated at different locations combine and form a wave front (Fig. 7-5b). When $V > c$, the observer moves ahead of the front of the disturbance and when it reaches location 4, an envelope tangent to the fronts of the disturbance may be drawn. This forms an oblique wave front, as shown in Fig. 7-5(c).

Now, let us see whether there is a relationship between the wave celerity, c, the velocity of the observer, V, and the angle, β, between the wave fronts. Referring to Fig. 7-5(c), let the observer travel from location A_1 to A_4 in time Δt. During this time, the disturbance travels from A_1 to D_1. In terms of velocities and the distances traveled during time Δt, we may write

$$\sin \beta \ = \ \frac{A_1 D_1}{A_1 A_4}$$

$$= \ \frac{c\Delta t}{V\Delta t} \qquad\qquad (7\text{-}12)$$

$$= \ \frac{c}{V}$$

Note that in this case the location of the wave front varies from one time to the next. If the fluid is in motion and the observer is stationary at one location, then a standing oblique wave front is formed. The cause of the disturbance may be a change in the alignment of the side wall, an irregularity on the wall surface, etc.

If the disturbance may be assumed to be a long wave of small amplitude, then as we proved in Chapter 3, $c = \sqrt{gy}$, where y = flow depth. Based on this relationship, we may write Eq. 7-12 as

$$\sin \beta = \frac{c}{V} = \frac{1}{\mathbf{F}_r} \qquad\qquad (7\text{-}13)$$

in which $\mathbf{F}_r$ = Froude number. However, if the disturbance is either large in magnitude or cannot be assumed to be a long wave, then Eq. 7-13 cannot be used. In the following section, we consider a large disturbance produced by a wall deflecting inward to the flow.

Oblique Hydraulic Jump

Let the alignment of the vertical side wall of a channel change inward into flow by angle $\Delta\theta$, as shown in Fig. 7-6. This produces an oblique, positive wave front. Let the height of this standing wave be Δy and let it make an angle β with the wall.

Referring to Fig. 7-6, there is a component of velocity parallel to the wave front. Since there is no force acting parallel to and along the wave front, the tangential velocities on either side of the front should be the same, even though the flow depths differ by height Δy. Hence, we may write

$$V_1 \cos \beta = V_2 \cos(\beta - \Delta\theta) \qquad\qquad (7\text{-}14)$$

for the tangential components.

The components of flow velocity normal to the wave front are $V_1 \sin \beta$ and $V_2 \sin(\beta - \Delta\theta)$. Then it follows from the continuity equation that

$$y_1 V_1 \sin \beta = y_2 V_2 \sin(\beta - \Delta\theta) \qquad\qquad (7\text{-}15)$$

We may apply the momentum equation as we did for the hydraulic jump in Chapter 2, except that in this case we replace V_1 by the normal component, $V_1 \sin \beta$. Thus, Eq. 2-58 for the oblique jump may be written as

$$\frac{V_1^2 \sin^2 \beta}{g y_1} = \frac{1}{2} \frac{y_2}{y_1} \left(\frac{y_2}{y_1} + 1 \right) \qquad\qquad (7\text{-}16)$$

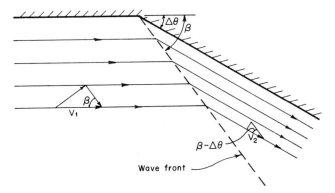

Wave front

Figure 7-6 Oblique hydraulic jump.

It follows from this equation that

$$\sin \beta = \frac{1}{\mathbf{F}_{r1}} \sqrt{\frac{1}{2} \frac{y_2}{y_1} \left(\frac{y_2}{y_1} + 1 \right)} \tag{7-17}$$

Note that for a small-amplitude wave, this equation becomes $\sin \beta = 1/\mathbf{F}_{r1}$, which is the same as Eq. 7-13.

Division of Eq. 7-15 by Eq. 7-14 and simplification of the resulting equation yields

$$\frac{y_2}{y_1} = \frac{\tan \beta}{\tan(\beta - \Delta \theta)} \tag{7-18}$$

By substituting $y_2 = y_1 + \Delta y$ and doing some algebraic manipulations, Eq. 7-18 may be reduced to

$$\frac{\Delta y}{y} = \frac{\sec^2 \beta \tan \Delta \theta}{\tan \beta - \tan \Delta \theta} \tag{7-19}$$

For small values of $\Delta \theta$, $\tan \Delta \theta \simeq \Delta \theta$ and $\tan \Delta \theta$ is very small as compared to $\tan \beta$. Then, in the limit as $\Delta \theta \to 0$, the preceding equation becomes

$$\frac{dy}{d\theta} = \frac{2y}{\sin 2\beta} \tag{7-20}$$

By combining Eqs. 7-13 and 7-20, we obtain

$$\frac{dy}{d\theta} = \frac{V^2}{g} \tan \beta \tag{7-21}$$

This equation defines the variation of flow depth along a curved wall. Thus, for a finite change in the wall angle, $\Delta \theta$, there is a change in the flow depth, Δy. This change in the flow depth may be assumed to be carried along an oblique front in the form of a shock wave even though the change in the flow depth is continuous.

The loss of energy in such a continuous change of flow depth is small and may be neglected. Thus, the specific energy, E, is constant. From the expression for E, it follows that $V = \sqrt{2g(E - y)}$. Substitution of this expression for V into Eq. 7-21 and elimination of β from the resulting equation and Eq. 7-13 yield

$$\frac{dy}{d\theta} = \frac{2(E - y)\sqrt{y}}{\sqrt{2E - 3y}} \tag{7-22}$$

Integration of this equation and substitution for E in terms of y and $\mathbf{F}_r$ give

$$\theta = \sqrt{3}\tan^{-1}\frac{\sqrt{3}}{\sqrt{\mathbf{F}_r^2 - 1}} - \tan^{-1}\frac{1}{\sqrt{\mathbf{F}_r^2 - 1}} + \theta_0 \tag{7-23}$$

where θ_0 is the integration constant obtained by substituting $\theta = 0$ for the initial depth $y = y_1$. This equation may be used to calculate the change in depth caused by the change in the direction of wall.

In the next four sections we present empirical and other information on weirs, hydraulic jumps, spillways, and energy dissipators.

7-5 Weirs

Weirs have been used in the laboratory and in the field for measuring discharge in open channels for more than 200 years. Wiers may be classified as sharp-crested and broad-crested. In the former, a thin vertical plate is fixed to the channel bottom and sides, whereas a broad-crested weir comprises a sudden rise in the channel bottom for some distance.

Sharp-Crested Weirs

A *sharp-crested weir* usually comprises a thin plate mounted perpendicular to the flow direction. The top of the plate has a beveled, sharp edge, which makes the nappe spring clear from the plate. These weirs may be rectangular or triangular in shape. The latter are used mainly for measuring small rates of discharge. There are no contractions in the lateral direction if the rectangular weir occupies the full width of the channel. This is called a *suppressed* weir. The basic theoretical development of weirs is based on the assumption that the pressure above and below the nappe is atmospheric. Therefore, it is necessary to vent the underside of the nappe so that the lower side of the nappe is at atmospheric pressure. If venting is not done, then the pressure under the jet may not be atmospheric, and assuming it as such in the discharge computations may produce erroneous results.

Let us consider the flow over a rectangular weir, as shown in Fig. 7-7, neglect the viscous losses, and assume the pressure at all points in the nappe is atmospheric. Then, the flow velocity at point a is $\sqrt{2gh}$, where $h =$ the vertical distance below the energy-grade line, measured positive in the downward direction. The discharge per unit width may then be determined from

$$q = \int_{h_o}^{H_o + h_o} \sqrt{2gh}\, dh = \frac{2}{3}\sqrt{2g}\left[(H_o + h_o)^{3/2} - h_o^{3/2}\right] \tag{7-24}$$

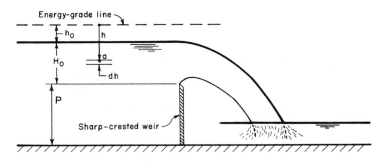

Figure 7-7   Sharp-crested weir.

in which $h_o = V_o^2/(2g)$. Theoretically speaking, it would be more appropriate to include the energy coefficient, α, in the velocity head to account for the nonuniform velocity distribution. However, experimental results show that α varies between 1.00 and 1.08 (Ranga Raju 1981) for flows approaching a weir, and we may thus safely assume it to be equal to 1.

To account for contraction and other effects, we may introduce a coefficient of discharge, C_d. This equation may then be written as

$$q = \frac{2}{3}C_d\sqrt{2g}\,H_o^{3/2} \qquad (7\text{-}25)$$

Based on Rehbock's experimental results (Henderson 1966), C_d may be approximated as

$$C_d = 0.611 + 0.08\frac{H_o}{P} \qquad (7\text{-}26)$$

in which P = height of the weir above the channel bottom (Fig. 7-7). Measurements by Rouse indicate that this formula is valid for $H_o/P < 5$, and it is approximate up to $H_o/P = 10$ when C_d is about 1.135 (Henderson 1966). For $H_o/P > 15$, the weir becomes a sill, and the discharge may be computed from the critical flow equation assuming $y_c = H_o$.

By proceeding similarly for a *triangular weir*, it may be shown that (Henderson 1966)

$$Q = \frac{8}{15}C_c\tan\frac{\alpha}{2}\sqrt{2g}\,H_o^{5/2} \qquad (7\text{-}27)$$

in which C_c = contraction coefficient and α = weir angle.

The contraction coefficient for the most commonly used 90° weir is approximately equal to 0.585. By substituting this value into Eq. 7-27 and simplifying, we obtain

$$Q = 2.5H_o^{5/2} \qquad (7\text{-}28)$$

An extensive amount of information on weirs is available in the literature. For details, see Bos (1976), Ackers et al. (1978), and Kabos (1984).

Broad-Crested Weirs

We showed in Chapter 3 that the flow may become critical if the channel bottom is raised by a specified amount. If the raised part of the channel is of sufficient length in the direction of flow, then the flow may be critical and the streamlines may be parallel to the weir. This may be utilized for flow measurement as follows.

Let us assume that the flow on the weir is critical and the losses between the weir and the location where the upstream flow depth is measured are negligible. Then we may write the energy equation between these two sections (Fig. 7-8) as

$$H + \frac{V^2}{2g} = \frac{3}{2}y_c \qquad (7\text{-}29)$$

in which H = upstream flow depth above the crest. By assuming the approach velocity V, to be negligible, this equation for the discharge per unit width, q, may be written as

$$q = \frac{2}{3}H\sqrt{\frac{2}{3}gH} \qquad (7\text{-}30)$$

However, a general equation for the discharge may be written as

$$Q = CB\sqrt{g}H^{3/2} \qquad (7\text{-}31)$$

in which B = channel width and C is a coefficient introduced to take into consideration the effects of various simplifying assumptions. If W is the height of the weir above the channel bottom, then $V = Q/[B(H + W)]$. Hence, it follows from Eqs. 7-29 and 7-31 that (Ranga Raju 1981)

$$\frac{H}{H + W} = \frac{\sqrt{3}(C^{2/3} - \frac{2}{3})^{1/2}}{C} \qquad (7\text{-}32)$$

It is unlikely in real life that the ideal situation—where the flow is critical as well as parallel to the crest—ever occurs (Henderson, 1966). Instead, there are two other possibilities: If the weir length in the direction of flow is short, then the flow depth on the crest varies with distance, the flow may be curvilinear, and it may even separate. Thus, errors may be introduced in the computation of discharge by assuming it to be critical parallel flow. Similarly, if the weir is too long, then the viscous effects are not negligible and a correction becomes necessary.

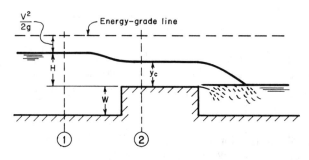

Figure 7-8 Broad-crested weir.

A weir may be assumed to be long if $L/H > 3$, where L = length of the weir in the direction of flow. In such a case, the value of H is reduced by δ^* to account for the viscous effects, where δ^* is the maximum displacement thickness of the boundary layer. For details, see Hall (1962).

The downstream submergence may affect the rate of discharge. A weir may be assumed to discharge freely if the tailwater level is lower than $0.8H$ above the weir crest (Henderson 1966).

7-6 Hydraulic Jump

As we discussed in Chapters 2 and 5, a hydraulic jump is formed whenever flow changes from supercritical to subcritical flow. In this transition from supercritical to subcritical flow, water surface rises abruptly, surface rollers are formed, intense mixing occurs, air is entrained, and a large amount of energy is usually dissipated. By utilizing these characteristics, a hydraulic jump can be used to dissipate energy, to mix chemicals, or to act as an aeration device.

In this section, we discuss different aspects of hydraulic jumps and present some empirical information. The computation of flows having a jump by using modern finite-difference techniques is outlined in the next chapter.

A jump in a horizontal, rectangular channel is referred to as a *classical jump*. Several experimental investigations have been conducted to determine the characteristics of such a jump, and an extensive amount of literature is available (see review articles by Rajaratnam 1967; McCorquodale 1986; and monograph by Hager 1991a).

Ratio of Sequent Depths

Equation 2-61 relates the flow depths on the upstream and downstream sides of a classical jump in terms of the Froude number at the entrance to the jump, $\mathbf{F}_{r1}$. We will call $\mathbf{F}_{r1}$ in the following discussion the approach Froude number. From this equation, it can be proved that for $\mathbf{F}_{r_1} > 2$,

$$y_r = \sqrt{2}\mathbf{F}_{r_1} - \frac{1}{2} \tag{7-33}$$

in which the ratio of the sequent depths is $y_r = y_2/y_1$. This is a linear relationship between the ratio y_r and the approach Froude number.

Note that we neglected the shear stress at the channel boundaries in the derivation of Eq. 2-61. Several experimental investigations show that this equation is valid in spite of this assumption, even for cases where the flow enters the jump at a steep angle. However, by utilizing extensive experimental data, Rajaratnam (1967) showed that y_r computed from an equation similar to Eq. 2-61 compares better with the experimental results if the boundary shear stress is included than that determined from an equation in which these losses are neglected. The viscous effects become important as the Reynolds number $\mathbf{R}_{e_1} = V_1 y_1 / \nu$, becomes small or as both $\mathbf{F}_{r_1}$ and y_1/B (B = channel width) become too large (Hager and Bremen 1989). Such a condition is possible on the scale models. As a rough guide, the equations derived by neglecting the viscous losses may be used with confidence for $\mathbf{F}_{r_1} < 12$ if $\mathbf{R}_{e_2} > 10^5$.

Length of a Jump

The length of a jump is needed to select the apron length and the height of the side walls of a stilling basin. To determine the length of a jump during laboratory investigations, it is difficult to mark the beginning and the end of a jump because of highly turbulent flow surface, formation of rollers and eddies, and air entrainment. In addition, the surface disturbances are of random nature, and the time-averaged quantities may not always give consistent results. The length of the roller may be taken to the point where the flow velocity at the top reverses and the jet continues.

For practical applications, experimental data have been summarized in a nondimensional form relating the approach Froude number, $\mathbf{F}_{r1}$, and L/y_1 or L/y_2, where L = length of the jump. Although satisfactory correlation has been observed for L/y_1, considerable disagreement exists between the data reported by different researchers for L/y_2. Figure 7-9 shows the curve recommended by the Bureau of Reclamation.

The following equation for the length of the roller, L_r, (Hager 1991a) gives good results if $y_1/B < 0.1$

$$\frac{L_r}{y_1} = -1.2 + 160 \tanh \frac{\mathbf{F}_{r_1}}{20} \tag{7-34}$$

In addition, by using the criterion that the turbulence has diminished at the end of the jump, Hager (1991a) developed the following equation for the length of the jump:

$$\frac{L}{y_1} = 220 \tanh \frac{\mathbf{F}_{r_1} - 1}{22} \tag{7-35}$$

or simply $L = 6y_2$ for $4 < \mathbf{F}_{r_1} < 12$.

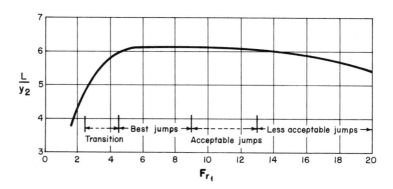

Figure 7-9 Length of hydraulic jump.

Jump Profile

The information on the jump profile is needed to determine the weight of water in a dissipator to counteract the uplift force if the basin floor is laid on a permeable foundation. Also, the height of the side walls may be varied for economic reasons if the water profile is known.

Figure 7-10 shows the jump profiles for different approach Froude numbers (Bakhmeteff and Matzke 1936). For design purposes, the vertical pressure on the basin floor may be assumed to be the same as that corresponding to the hydrostatic pressure for the profile depth. This has been confirmed by several experimental investigations.

Based on extensive laboratory experiments, Hager (1991a) developed the following empirical relationship for the flow depth, y, at distance, x, from the beginning of the jump

$$Y = \tanh(1.5X) \tag{7-36}$$

in which $X = x/L_r$, $Y = (y - y_1)/(y_2 - y_1)$, and L_r = length of the roller.

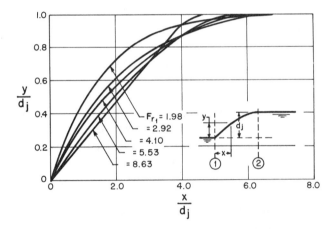

Figure 7-10 Jump profile (After Bakhmetef and Matzke, 1936).

Jump types

Hydraulic jump occurs in four distinct forms (Peterka 1958) depending upon the approach Froude number, F_{r_1}, as shown in Fig. 7-11. Each of these forms has a distinct flow pattern, formation of rollers and eddies, etc. The energy dissipation in the jump depends upon the flow pattern and the strength of the rollers. The range of the Froude number listed in the following paragraphs for various types of jumps is not precise, and there is some overlap from one type to the other.

Weak jump *(1 < F_{r_1} < 2.5)*.   For $1 < F_{r_1} < 1.7$, y_1 and y_2 are approximately equal to each other, and only a slight ruffle is formed on the surface. This undulation results in very little energy dissipation. However, as F_{r_1} approaches 1.7, a number of small rollers are formed on the water surface, although the downstream water surface remains smooth. The energy loss is again low in this jump.

(a) Pre-jump, very low energy loss (F_{r1} = 1.7 to 2.5)

(b) Transition, rough water surface (F_{r1} = 2.5 to 4.5)

(c) Good jump, least affected by tail water (F_{r1} = 4.5 to 9.0)

(d) Effective but rough (F_{r1} > 9.0)

Figure 7-11 Jump types (After Peterka, 1958).

Oscillating jump (2.5 < F_{r_1} < 4.5). The jet at the entrance to the jump oscillates from the bottom to the top at an irregular period. Turbulence may be near the channel bottom at one instant and at the water surface the next. These oscillations result in the formation of irregular waves, which may persist for a long distance downstream of the jump. They may cause considerable damage to the channel banks. Therefore, this range of F_{r_1} should be avoided while designing an energy dissipator.

Steady jump (4.5 < F_{r_1} < 9). For this range, the jump forms steadily at the same location, and the position of the jump is least sensitive to the downstream flow conditions. The jump is well balanced and the energy dissipation is considerable.

Strong jump (F_{r1} > 9). In this case the difference between the conjugate depths is large. At irregular intervals, slugs of water roll down the front of the jump face into high-velocity jet and generate additional waves. The jump action is very rough and the dissipation rate is high.

Energy loss

The difference between the total head upstream and downstream of the jump is the energy loss in the jump. For a horizontal channel bottom, this is the same as the difference in the specific energy upstream and downstream of the jump. We may derive an expression for this head loss, h_l, as follows:

$$h_l = E_1 - E_2 = (y_1 - y_2) + \frac{V_1^2}{2g} - \frac{V_2^2}{2g} \qquad (7\text{-}37)$$

in which the subscripts 1 and 2 refer to the quantities upstream and downstream of the jump. By substituting $q = V_1 y_1 = V_2 y_2$ and doing a number of algebraic manipulations, we may rewrite Eq. 7-37 as

$$h_l = \frac{(y_2 - y_1)^3}{4 y_1 y_2} \tag{7-38}$$

This is a theoretical expression for the head losses in a classical jump. Figure 7-12, based on experimental results, shows the energy dissipation in a jump for different values of the approach Froude number.

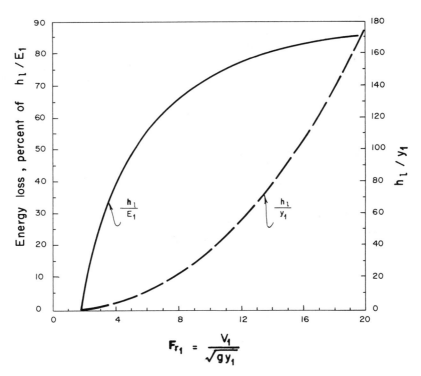

Figure 7-12   Energy dissipation in jump (After Peterka, 1958).

Jump Location

As we discussed in the previous paragraphs, a hydraulic jump is formed at a location where the flow depths upstream and downstream of the jump satisfy the equation for the sequent depth ratio (Eq. 2-61). We will illustrate the location of jump formation by considering the flow downstream of a sluice gate. Let the flow depth at the sluice outlet be y_1 and the sequent depth corresponding to this depth be y_2. There are several different possibilities for the formation of jump, depending upon the tailwater depth, y_d.

The jump is formed on the apron if the downstream depth, y_d, is equal to the depth y_2 required by Eq. 2-61 (Fig. 7-13a). If y_d is less than y_2, then the jump moves moves downstream to a point where the upstream depth y_1' is the sequent depth to y_d (Fig. 7-13(b)). In this figure, we have used a broken line to show the sequent depth y_2 required for the depth y_1 at the sluice outlet. However, if the tailwater depth is higher than the required amount, then the jump is pushed back, as shown in Fig. 7-13(c). This is called *submerged*, or *drowned, jump*.

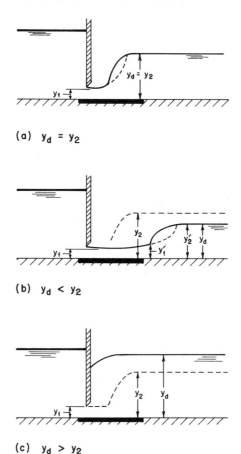

(a) $y_d = y_2$

(b) $y_d < y_2$

(c) $y_d > y_2$

Figure 7-13 Jump location.

Tailwater level plays a significant role in the formation of jump at a particular location. In most practical situations, the tailwater level depends upon the channel discharge, Q. A curve between Q and the tailwater level is referred to as the tailwater rating curve. Similarly, we may prepare a curve between y_2 and Q, which we will refer to as the jump curve. Depending upon these two curves, five different flow situations are possible (Leliavsky 1955). These are shown in Fig. 7-14, in which a solid line is used for the tailwater curve and a broken line is used for the jump curve.

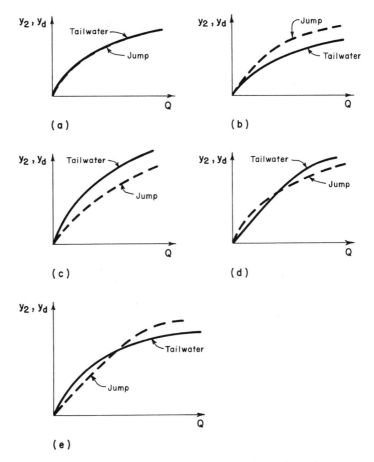

Figure 7-14 Effect of tailwater level on jump formation.

In (a) the tailwater rating curve and the jump curve coincide for all rates of discharge. The requirements for the sequent depth are always satisfied and the jump forms at the same location. This is an ideal situation that rarely occurs in nature.

The jump curve is always above the tailwater curve for (b). The downstream depth is less than the required sequent depth and the jump moves further downstream. To ensure jump formation on the apron, a sill may be used.

In (c) the tailwater curve is always above the jump curve. Thus the downstream depth is more than that required by the sequent depth. The jump moves upstream and may drown. The jump may be controlled at the desired location by providing a drop in the channel bottom or by letting the jump form on a sloping apron.

The tailwater curve is below the jump curve at low discharges and above it for higher discharges in (d). The stilling basin may be designed so that the jump is formed in the basin at low rates of discharges and the jump moves on to a sloping apron at higher discharges.

Case (e) is opposite to (d) in the sense that the tailwater curve is above the jump curve at low discharges and below the jump curve at high discharges. A stilling pool may be designed in this case to form the jump at high discharges.

Control of Jump

The location of a jump may be controlled by providing a number of appurtenances, such as baffle blocks, sills, drops, or rises in the channel bottom. The sill may be a sharp- or broad-crested weir. Typically, the flow in the vicinity of these appurtenances is rapidly varied, and the velocity distribution is not uniform. As a result, it becomes difficult to apply the momentum equation to analyze accurately the formation of jump. Therefore, laboratory experiments are done to develop empirical relationships for universal applications and model studies are conducted for specific projects.

In the following paragraphs, we discuss the control of jump by means of a sharp-crested weir and by an abrupt rise.

Sharp-crested weir.   Figure 7-15 shows the relationship between different variables for jump control by means of a sharp-crested weir. This diagram, developed by Forster and Skrinde (1950) from results of extensive laboratory tests, may be used to determine the effectiveness of the weir for jump formation provided the weir is not submerged. For the known value of the approach Froude number, F_{r_1}, the distance X between the toe of the jump and the weir may be determined from this figure. For different X/y_2 ratios, the values may be interpolated between the curves.

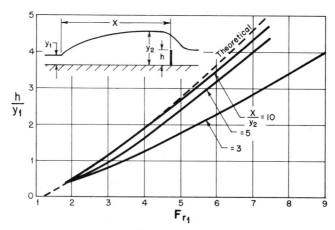

Figure 7-15 Jump control by a sharp-crested weir (After Forster and Skrinde, 1950).

Abrupt rise.   Based on laboratory experiments and theoretical analysis, Forster and Skrinde (1950) prepared Fig. 7-16 for the control of jump by means of an abrupt rise. The jump was formed at a distance $x = 5(h + y_3)$ upstream of the rise. This diagram may be used to predict the performance of an abrupt rise if the values of V_1, y_1, y_2, y_3, and h are known. The definition of these variables should be clear from Fig. 7-16.

An abrupt rise increases the drowning effect if a point corresponding to different flow variables lies above the line $y_2 = y_3$. The region between the lines $y_2 = y_3$ and $y_3 = y_c$ is further divided by the h/y_1 curves. A point lying on these curves between these two lines represents the condition when the jump is formed at $x = 5(h + y_3)$. A point above the h/y_1 curve shows a condition in which the jump is forced upstream and may be drowned. The condition where the rise is too low and the jump is forced downstream and may eventually be washed out is represented by points below the h/y_1 curves.

Points below the line $y_3 = y_c$ represent supercritical flow downstream of the rise. In this condition, the rise acts as a weir, and Fig. 7-15 may be used for the analysis.

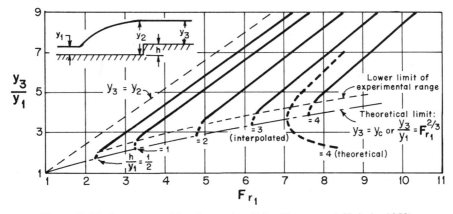

Figure 7-16 Jump control by abrupt rise (After Forster and Skrinde, 1950).

7-7 Spillways

A spillway is used to release surplus or floodwater or for other controlled releases, such as for irrigation, navigation, or environmental considerations. Spillways may be classified into different categories, using different criteria for such classification. For example, based on *function*, a spillway may be classified as service, auxiliary, or emergency; and based on the structural components, it may be called an overflow, chute, or tunnel spillway. Utilizing the type of inlet, a spillway may be classified as orifice, siphon, side channel, or morning glory.

The overflow spillway is one of the common types of spillway. Information on this spillway is presented here. For details on other types of spillways, see Chow (1959), Peterka (1958), and Davis and Sorenson (1969).

Overflow Spillway

An overflow spillway is used on concrete-gravity, arch, and buttress dams, where part of the dam length may be used as a spillway. Because of the shape, it is also called an ogee spillway.

An overflow spillway has three main parts: the crest, the sloping face, and the energy dissipator at the toe. The first two are discussed in this section, and the third is discussed in Sec. 7-8.

Crest shape.   The pressures on the crest are atmospheric if the crest shape is the same as the underside of the nappe of a jet over a sharp-crested weir. These pressures may be above atmospheric (positive) or subatmospheric (negative), depending upon the shape of the crest relative to the underside of the nappe over a sharp-crested weir (Fig. 7-17).

The shape of the crest is based on the design head, H_d, which is selected for a given site such that the minimum pressure at the crest is higher than -6 m in order to prevent cavitation. Figure 7-18 may be used to select H_d so that this condition for

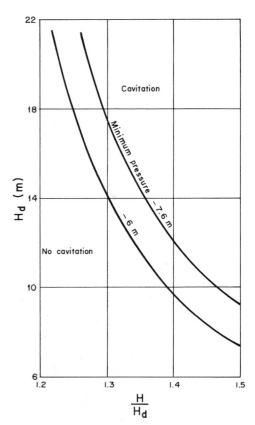

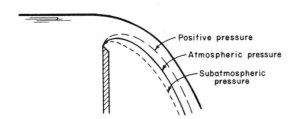

Figure 7-17 Pressure on a spillway crest.

Figure 7-18 Design head for overflow spillway.

the minimum pressure is satisfied. In this figure, H = maximum head on the crest = maximum upstream level − crest level. Usually, the design head is selected such that $1.3 < H/H_d < 1.5$. In this range, acceptable levels of subatmospheric pressures are produced on the spillway face. This results in an increase in the discharge capacity of the spillway and at the same time does not result in cavitation on the spillway. The crest shape may then be determined from the empirical relationships presented in Fig. 7-19. These relationships were developed by the U.S. Army Corps of Engineers based on extensive laboratory investigations.

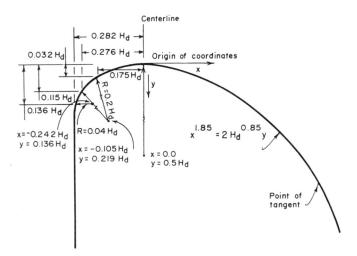

Figure 7-19 Crest shape (After U.S. Army Corps of Engrs., 1977).

Rating Curve.   A curve between the upstream reservoir level and the spillway discharge is called the *rating curve*. The relationship between discharge, Q, through an overflow spillway under a given total head on the spillway crest may be written as

$$Q = CL_e\sqrt{2g}H_e^{1.5} \tag{7-39}$$

in which L_e = effective spillway length; C = coefficient of discharge; and H_e = total head on the crest = $H + V_o^2/(2g)$. Normally the velocity of approach, V_o, for an overflow spillway is small. Therefore, the velocity head is negligible, and H_e may be taken as the difference between the upstream level and the crest level, i.e., $H_e = H$.

Through extensive laboratory tests, the U.S. Bureau of Reclamation has compiled information on the coefficient of discharge. The value of this coefficient, C_o, for the design head, H_d, is presented in Fig. 7-20(a), and its value C for other value of H is given in Fig. 7-20(b). To account for contractions at the piers and the abutments, the effective length, L_e, may be computed from the following equation:

$$L_e = L_n - 2(Nk_p + k_a)H_e \tag{7-40}$$

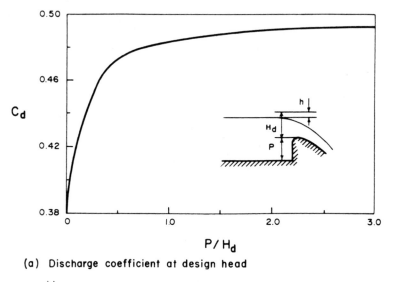

(a) Discharge coefficient at design head

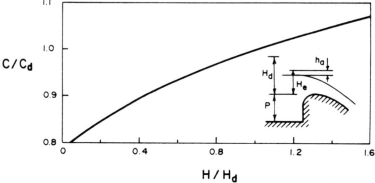

(b) Discharge coefficient at other heads

Figure 7-20 Discharge coefficient (After Roberson, et al. *Hydraulic Engineering*, Houghton Mifflin, Boston, MA. 1988).

in which L_n = net spillway length between the piers; k_p = pier coefficient; and k_a = abutment contraction coefficient. The pier coefficient may be determined from Fig. 7-21. For abutments having large radius of curvature, k_a is almost equal to zero and may be neglected. The values for k_a for unsymmetrical approach conditions may be determined from U.S. Army Corps of Engineers design criteria (1977).

Water-surface profile.   The water-surface profile is required to set the height of the side walls and to select the elevation of the trunion axis for the radial gates or for any other structure in the vicinity of the water surface. For preliminary design purposes, Fig. 7-22 presents data taken from the U.S. Army Corps of Engineers design criteria (1977). Hydraulic model studies are normally conducted for large spillways.

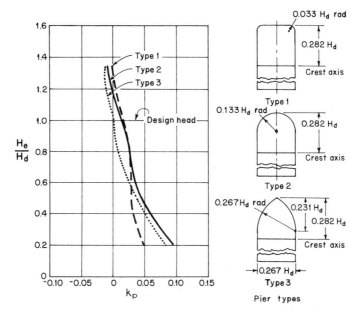

Figure 7-21 Water-surface profile (After U.S. Army Corps of Engrs., 1977).

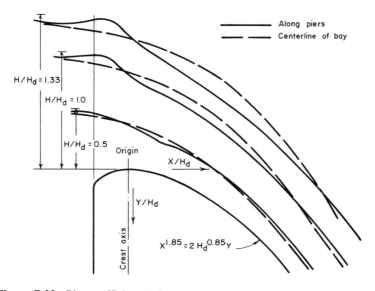

Figure 7-22 Pier coefficient (After U.S. Army Corps of Engrs., 1977).

Downstream face. The downstream face of a spillway has usually a very steep angle, such as $0.8\,H : 1\,\text{V}$ to $0.6\,H : 1\,\text{V}$. Because of this steepness, some difficulties arise in the analysis of flow conditions utilizing the procedures discussed in the earlier chapters. The development of a boundary layer starts at the crest, and the point where it intersects the flow surface is called the *transition point*. Downstream of the transition point, the turbulence is fully developed, a large amount of air is entrained, and bulking of flow occurs. This results in more flow depth than that if no air were entrained. The U.S. Army Corps of Engineers recommend increasing the flow depth by 20 percent to account for bulking (U.S. Army Corps of Engrs. 1965).

Several investigations have been conducted on the air entrainment or insufflation in high-velocity flows. A brief description is included on the topic in Chap. 10; for details, the reader should refer to Falvey (1980).

To determine the flow velocity at the toe of a spillway, no precise calculation procedures are presently available. Based on experience, computations, and experimental results, Paterka (1958) presented a graph for this purpose. Another viable procedure is to assume a value for the head losses on the spillway face (say 5 to 10 percent of the difference between the upstream reservoir level and the tailwater level) and then compute the theoretical velocity at the toe from the energy equation. This procedure is simpler and may be used for typical engineering applications.

7-8 Energy Dissipators

The flow velocity at the toe of a high-head spillway is usually high and may cause serious scour and erosion of the downstream channel if proper precautions are not taken. For this purpose, energy dissipators are provided to dissipate a sufficient amount of energy before water enters the downstream channel. In order to have an idea about the amount of energy dissipation, let us compute its value at the toe of the Grand Coulee dam, located on the Columbia River in the United States. The design discharge for the spillway is 28,320 m³/s, and the upstream and the tailwater levels for this flow are 393.8 m and 308.23 m, respectively. Assuming no losses on the spillway face, the amount of energy at the toe is $\rho g Q H$, where H is the difference between the headwater and tailwater levels. Substitution of the values of different variables into this expression yields an energy of 23 GW. This should give the reader an idea about the amount of energy involved and clearly shows that even excellent rock may be eroded if proper measures are not taken.

Three types of energy dissipators have been commonly used: stilling basins, flip buckets, and roller buckets (Mason 1982). Each dissipator has certain advantages and disadvantages and may be selected for a particular project, depending upon the site characteristics. A brief description of these dissipators is given in the following paragraphs.

Stilling Basins

The hydraulic jump is used for energy dissipation in a stilling basin. Typically, this basin may be used for heads less than 50 m. At higher heads, cavitation becomes a problem. A concrete apron is provided for the length of the jump and the invert level of the apron is set such that the downstream water level provides the necessary sequent depth for the flow depth and the Froude number at the entrance to the jump. Long apron lengths and low apron levels are needed for such a stilling basin. Low apron levels require large amounts of excavation and concrete. For economic reasons as well as to make the stilling basin operate efficiently over a wide range of flows, other devices and accessories may be provided to stabilize the jump, to reduce the length of the jump, and to permit the apron at a higher elevation. These devices include chute blocks, baffle blocks, and end sills (see Fig. 7-23).

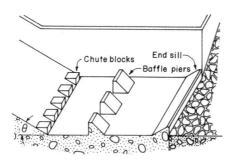

Figure 7-23 Stilling basin accessories (After Peterka, 1958).

The chute blocks serrate the flow entering the basin and lift up part of the jet. More eddies are produced, thus increasing energy dissipation, the jump length is decreased, and the tendency of the jump to sweep out of the basin is reduced. The baffle blocks stabilize the jump and dissipate energy due to impact. The sill mainly stabilizes the jump and inhibits the tendency of the jump to sweep out.

Based on laboratory tests and field experience, several standardized stilling basins have been developed. Notable among these are: St. Anthony Falls stilling basin; eight stilling basins developed by the U.S. Bureau of Reclamation (each suitable for a certain range of head), and a basin recommended by the U.S. Army Corps of Engineers (Murphy 1974). Only details of the last basin are given here; for information on the other basins, the reader should consult Chow (1959) and the monograph prepared by the U.S. Bureau of Reclamation (Peterka 1958).

Figure 7-24 shows a stilling basin (Murphy 1974) suitable for high-head installations. To design this basin, first the design and standard project flood are selected. The design flood may be less than the probable maximum flood, and the standard project flood may be less than the design flood. By assuming the head losses on the spillway face to be a certain percentage of the total head (say 5 to 10 percent), the values of V_1 and y_1 are computed for the design flood. Then, the values of Froude number, $\mathbf{F}_{r_1} = V_1/\sqrt{gy_1}$, and the sequent depth y_2 are determined. Similarly, the sequent depth y_2' is computed for the standard project flood. The dimensions of the basin and of the appurtenances are then determined from the following relationships:

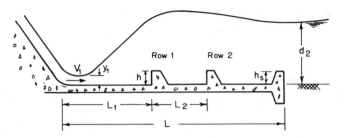

Figure 7-24 Corps of Engineers stilling basin.

$$L_1 = 1.5y_2 \qquad \text{for} \quad \mathbf{F}_{r_1} \le 4.6$$

$$L_1 = \left[1.5 + \frac{1}{11}(\mathbf{F}_{r_1} - 4.6)\right]y_2 \qquad \text{for} \quad \mathbf{F}_{r_1} > 4.6$$

$$L_2 = 2.5h$$

$$h = \frac{1}{6}y_2 \qquad \text{for} \quad \mathbf{F}_{r_1} \le 4.6$$

$$h = [1. + 0.13(\mathbf{F}_{r_1} - 4.6)]y_1 \qquad \text{for} \quad \mathbf{F}_{r_1} > 4.6$$

$$h_s = \frac{1}{2}h$$

$$d_2 \ge 0.85y_2$$

$$\simeq y_2'$$

$$L \ge L_1 + y_2$$

$$\ge 4y_2$$

Baffle block rows 1 and 2 are staggered and the baffle block width is less than or equal to h. The spacing between the blocks is to be at least equal to the baffle block width.

Flip Buckets

The flip bucket energy dissipator is suitable for sites where the tailwater depth is low (which would require a large amount of excavation if a hydraulic jump dissipator were used) and the rock in the downstream area is good and resistant to erosion. The flip bucket, also called a ski-jump dissipator, throws the jet at a sufficient distance away from the spillway, where a large scour hole may be produced. Initially, the jet impact causes the channel bottom to scour and erode. The scour hole is then enlarged by a ball-mill motion of the eroded rock pieces in the scour hole. A plunge pool may be excavated prior to the first spill for controlled erosion and to keep the plunge pool in a desired location.

A small amount of the energy of the jet is dissipated by the internal turbulence and the shearing action of the surrounding air as it travels in the air. However, most of the energy of the jet is dissipated in the plunge pool.

During the operation of a flip bucket, a large amount of spray is produced, which may be undesirable for roads, bridges, and electrical equipment, such as transmission lines or grid stations. In addition, large water-level fluctuations may be produced in the tailrace area by the plunging jet and by the associated return currents and eddies. These water-level oscillations near the draft-tube exits may impair the operation of Francis turbines, since these oscillations may result in load swings.

The horizontal throw distance, x_b, from the bucket lip to a point where the jet impinges the river bottom (Fig. 7-25) may be computed from the following equation:

$$\frac{x_b}{h_o} = \sin 2\alpha + 2\cos\alpha \sqrt{\sin^2\alpha + \frac{y_b}{h_o}} \tag{7-41}$$

in which y_b = the height of the bucket lip above the river bottom, α = bucket angle with the horizontal axis, and h_o = velocity head at the bucket lip.

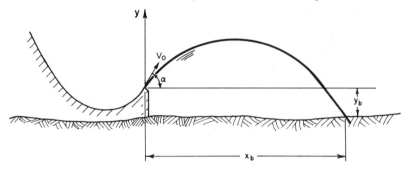

Figure 7-25 Definition sketch.

If the bucket lip is at the river bottom level, then $y_b = 0$, and the preceding equation simplifies as

$$\frac{x_b}{h_o} = 2\sin 2\alpha \tag{7-42}$$

The throw distance is normally less than the distance computed from this equation, since the air resistance and other losses are neglected in the derivation of the equation. For design purposes, this distance may be reduced by 20 percent.

It is clear from Eq. 7-42 that the throw distance is maximum for $\alpha = 45°$. However, experience has shown that a lip angle of 20° to 30° should be preferred, since a bucket having this angle produces scour holes of shallower depth due to the low angle of impact than that produced by a 45° bucket. The bucket radius is usually 10–20 m, and the height of the bucket lip is set approximately 10 percent of the bucket radius above the bucket invert.

Scour depth.   A plunge pool is said to be *stable* if the scour has reached equilibrium conditions. No procedures are presently available to accurately determine the maximum depth of the stable plunge pool downstream of a flip bucket. However, based on laboratory studies and field information obtained on several projects, a number of empirical formulas have been developed for predicting the maximum depth of scour of a stable plunge pool. Mason and Arumugam (1985) list and compare these formulas.

They also presented the following new formula which gives better comparison with both hydraulic-model and prototype results:

$$D_m = K \frac{q^a H^b h^{0.15}}{g^{0.3} d_m^{0.1}} \qquad (7\text{-}43)$$

in which D_m = maximum scour depth measured from the tailwater surface; q = flow intensity per unit width; h = tailwater depth above unscoured bottom level; H = head difference between the headwater and the tailwater levels; d_m = mean bed material size; and in SI units, $a = 0.60 - 300/H$; $b = 0.15 - H/200$; and $K = 6.42 - 3.1h^{0.1}$. Mason and Arumugam (1985) used a constant value for d_m of 0.25 m for prototypes. However, for rocky bottoms, d_m may be taken as the size of blocks into which the rock may be assumed to fracture.

Roller Buckets

A roller bucket may be used for energy dissipation if the downstream depth is significantly greater than that required for the formation of a hydraulic jump. In this dissipator, the dissipation is caused mainly by two rollers: a counterclockwise roller near the water surface above the bucket and a roller on the channel bottom downstream of the bucket. The movement of these rollers along with the intermixing of the incoming flows results in the dissipation of energy.

The Grand Coulee spillway on the Columbia River in the United States was the first spillway to have a roller bucket energy dissipator. Since then two types of roller buckets—solid and slotted—have been developed through hydraulic model studies and used successfully on several projects. In a solid bucket, the ground roller may bring the eroded material toward the bucket and deposit it in the bucket during periods of unsymmetrical operation. In a slotted bucket, part of the flow passes through the slots, spreads laterally, and is distributed over a greater area. Therefore, the flow concentration is less than that in a solid bucket. In addition, any material that might get into the bucket is washed out. The sweepout in a slotted bucket occurs at a slightly higher tailwater level than that in a solid bucket. Experience with both buckets indicates that a slotted bucket is preferable to a solid bucket.

The tailwater depth is the difference between the tailwater and the bucket invert level. The *minimum tailwater limit* is the tailwater depth just safely above the depth required for sweepout. At sweepout, a jet at the bottom scours the channel bottom and an unstable condition develops in the bucket, thereby causing excessive scour and erosion. Because of this, a bucket may not be designed for both submerged and flip action. A *safe lower limit*, T_{min}, has been established by adding a small factor of safety to the sweepout tailwater depth (Fig. 7-26a) determined from an extensive series of tests.

Similarly, an *upper tailwater limit* (Fig. 7-26b) was determined by raising the tailwater level until the flow dived from the apron lip.

Design procedure.   Figure 7-27 shows the typical parameters for a roller bucket. The steps for the design of a bucket for a specified discharge, Q, and the spillway width, B, are as follows:

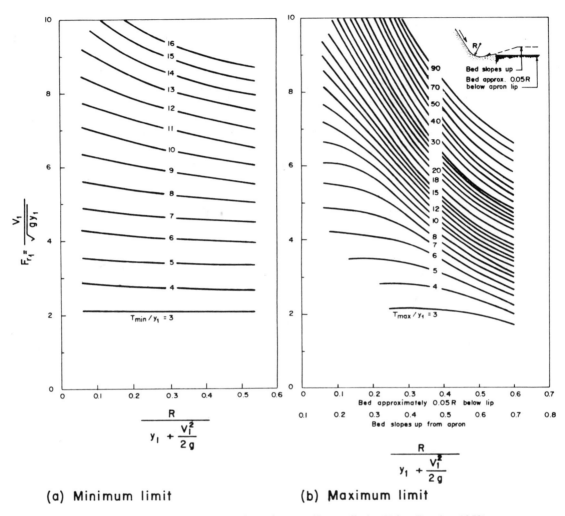

(a) Minimum limit   (b) Maximum limit

Figure 7-26   Minimum and maximum tailwater limits (After Peterka, 1958).

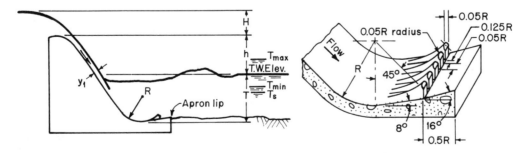

Figure 7-27   Definition sketch (After Peterka, 1958).

    i. By assuming a value for the head losses on the spillway (say, 5 to 10 percent), compute the flow velocity at the toe and hence determine $y_1 = Q/(BV_1)$ and the Froude number, F_r.

    ii. For F_r computed in step 1, determine the factor $R/(y_1 + V_1^2/2g)$ from Fig. 7-28 and hence the minimum bucket radius, R, and select a value for R.

    iii. For the computed values of F_r and $R/(y_1 + V_1^2/2g)$, determine T_{min}/y_1 from Fig. 7-26(a), and hence compute the minimum tailwater limit, T_{min}. Similarly, determine the maximum tailwater limit, T_{max}, from Fig. 7-26(b).

    iv. Set the bucket invert elevation so that the tailwater is between T_{min} and T_{max}. If possible, keep the apron lip and the bucket invert above the river bottom. For best performance, set the bucket invert so that tailwater depth $\simeq T_{min}$.

    v. Check the sweepout condition from Fig. 7-29.

    vi. Determine the tooth size, spacing, and other dimensions from Fig. 7-27.

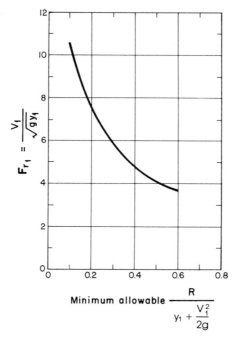

Figure 7-28 Minimum bucket radius (After Peterka, 1958).

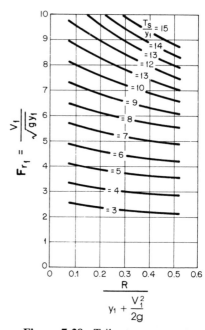

Figure 7-29 Tailwater sweepout depth (After Peterka, 1958).

7-9 Summary

In this chapter we discussed the differences between the analysis of rapidly and gradually varied flows. It was shown that it is necessary to know the pressure and velocity distributions in rapidly varied flows to apply the conservation laws of mass, momentum, and energy. The generation of shock waves in supercritical flow was discussed. Empirical relationships derived from the laboratory and field investigations on transitions, weirs, and hydraulic jump were presented. A brief discussion on spillways and energy dissipators was given.

Problems

7-1. If the flow depth at the entrance to a constant-width, rectangular transition with bottom slope S_o is equal to the critical depth, y_c, then prove that the flow depth y in the transition is given by the equation

$$S_o x = y + \frac{1}{2}\frac{y_c^3}{y^2} - \frac{3}{2}y_c$$

7-2. For the horizontal, frictionless channel transition shown in Fig. 7-30, prove that

$$\theta x = \frac{Q}{y\sqrt{2g(H_o - y)}} - \frac{Q}{\sqrt{g(2E/3)^3}}$$

in which y = flow depth at distance x, Q = rate of discharge, and H_o = total head. Assume the flow is critical at the entrance of the transition.

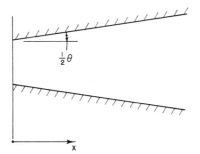

$\frac{1}{2}\theta$

x

Figure 7-30 Channel transition of Prob. 7-2.

7-3. Prove that for large values of $\mathbf{F}_{r1}$, the ratio of the sequent depths ($y_r = y_2/y_1$) and $\mathbf{F}_{r1}$ are linearly related. By using this relationship, show that

$$\frac{\Delta H}{H_1} = \left(1 - \frac{\sqrt{2}}{\mathbf{F}_{r1}}\right)^2$$

where $\Delta H = H_1 - H_2$, H = total head, and the subscripts 1 and 2 refer to the quantities upstream and downstream of the jump.

7-4. By applying the momentum principle, show that the head loss in a sudden expansion (Fig. 7-3) is $(V_1 - V_3)^2/(2g)$.

7-5. Prove that for a hydraulic jump, the ratio of the sequent depths, $y_r = y_2/y_1$, may be approximated as

$$y_r = \sqrt{2}\mathbf{F}_{r1} - \frac{1}{2} \qquad \text{for} \quad \mathbf{F}_{r1} > 2$$

7-6. Show that the head loss, h_l, in a classical hydraulic jump is

$$h_l = \frac{(y_2 - y_1)^3}{(4y_1 y_2)}$$

7-7. Determine the throw distance for a 20° flip bucket located at the end of a chute spillway. The bucket lip is 60 ft below the reservoir level. Assume the energy losses are 10 percent of the total available head.

7-8. Design a stilling basin for the spillway of Prob. 7-7. Assume a suitable tailrace rating curve.

7-9. Design a flip bucket, a roller bucket, and a stilling basin for an overflow spillway (six 50-ft-wide bays with 10-ft-wide piers) for a flow of 360,000 ft³/sec. The difference between the upstream and downstream water levels at this discharge is 140 ft.
  Which of these energy dissipators do you prefer for each of the following sites?

   i. Located in cold climate with a bridge downstream of the spillway
   ii. Excellent rock conditions
   iii. Large tailrace submergence

7-10. Compute and plot the rating curve for the spillway of Prob. 7-9.

7-11. Determine the rating curve for an overflow spillway having four 50-ft-wide bays. The piers are of type 2, the design head for the crest is 32 ft, the maximum reservoir level is at El. 948, and the crest is at El. 900. The river bottom upstream of the spillway is at El. 836.

7-12. An expression for the discharge Q through an overflow spillway can be written as

$$Q = CL_e H^{1.5}$$

In the customary English units (i.e., Q in ft³/sec and L_e and H in ft) $C = 3.8$ for a given spillway at a design head of 10 m. Will the value of C be the same in SI units (i.e., Q in m³/s, and L_e and H in m)? If not, what is the corresponding value in SI units? Determine the discharge for $L_e = 100$ m and $H = 10$ m.

7-13. Suppose you are the design engineer for the energy dissipators of three large projects. Each project has the following typical characteristics. Which type of energy dissipator would you select for each project? List, in point form, the other characteristics the project should have so that your selected energy dissipator will operate properly.

   i. Shallow water depth and good rock conditions exist in the tailrace area
   ii. High tailwater depth, rock quality is average
   iii. The variation of tailwater depth with respect to the spillway discharge is small and the river bottom is highly erodable.

7-14. For the reservoir levels at El 150 and 160 ft, determine the discharge through an overflow five-bay spillway having type 2 piers. Assume the abutment contraction coefficient to be zero. The crest is at El 120, river bottom at El 20, and the crest design head is 25 ft. Each bay is 60 ft wide.

7-15. Compute the spillway discharge for the upstream water levels at El 300 and 310 m. The spillway length is 20 m and there are no piers and no abutment contractions. The crest is at El 280 and is shaped for a design head of 18 m and the river bottom is at El. 244.

7-16. Determine the discharge through an overflow spillway having no piers for the upstream reservoir level at El 620 and El 648 ft. The abutments are shaped to prevent contractions. The design head is selected such that it is 0.75 of the maximum head. If the maximum permissible reservoir level for the design flow of 60,000 ft³/sec is El 652, select the appropriate spillway length and the crest level.

7-17. Determine the location of the hydraulic jump in the channel of Prob. 5-15.

7-18. The longitudinal section of a spillway is shown in Fig. 7-31. Its width remains constant at 100 ft, there are no piers, and the crest is shaped for a 24-ft design head. Compute:

   i. The spillway discharge for the maximum reservoir level of El. 165

   ii. The water surface profile in the chute

   iii. Required downstream water level for the formation of hydraulic jump in the stilling basin

   iv. The energy losses in the jump

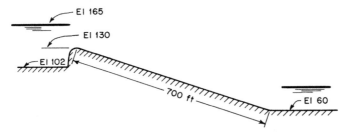

Figure 7-31 Spillway of Prob. 7-18.

7-19. If the downstream water level for the design discharge of 50,000 ft³/sec is at El. 72 ft, design an energy dissipator for the spillway of Prob. 7-18.

7-20. The water level in the reservoir upstream of a 50-m-wide spillway is at El. 200 m and water level in the downstream channel for a design flow of 2700 m³/s is at El. 50 m. If the stilling basin width is the same as that of the spillway, determine the floor level of the basin so that the jump is formed in the basin. Assume the losses in the spillway are negligible and no appurtenences are to be provided to stabilize the jump.

    What would be the basin level if you could use an end sill or one row of baffle blocks and an end sill?

REFERENCES

Abbott, D. E., and Kline, S. J. 1962. Experimental investigation of subsonic turbulent flow over single and double backward facing step, *Jour. Basic Engineering,* Amer. Soc. Mech. Engrs., 84:317–25.

Ackers, P.; White, W. R.; Perkins, J. A.; and Harrison, A. J. M.   1978. *Weirs and flumes for flow measurement,* John Wiley, New York, NY.

Advani, R. M. 1962. A new method for hydraulic jump in circular channels, *Water Power,* 16, no. 9:349–50.

Ali, K. H. M., and Ridgeway, A.   1977. Hydraulic jump in trapezoidal and triangular channels, *Proc.,* Institution Civil Engrs. (London), 63:203–14, 761–67.

Anderson, V. M.   1978. Undular hydraulic jump, *Jour. Hydraulics Div.,* Amer Soc. Civ. Engrs., 104, no. 8:1185–88 (Discussion, 1979, HY9: 1208–11; 1980, HY7: 1252–54).

Bakhmeteff, B. A., and Matzke, A. E. 1936. The hydraulic jump in terms of dynamic similarity, *Trans.*, Amer. Soc. Civil Engrs., 100:630–80.

――――. 1938. The hydraulic jump in sloped channels, *Trans.*, Amer. Soc. Mech. Engrs., HYD-60-1:111–18.

Bos, M. G. 1976. *Discharge measurement structures*, Report 4, Landbauwhogeschool, Laboratorium voor Hydraulica en Afvoerhydrologie, Wageningen, the Netherland.

Cassidy, J. J. 1970. Designing spillway crest for high head operations, *Jour. Hydraulics Division*, Amer. Soc. Civil Engrs., 96, HY3:745–53.

Chow, V. T. 1959. *Open-channel hydraulics,* McGraw-Hill, New York, N. Y.

Davis, C. V., and Sorensen, K. E. 1969. *Handbook of applied hydraulics*, 3rd ed., McGraw-Hill, New York, NY.

Elevatorski, E. A. 1958. Trajectory bucket-type energy dissipators, *Jour. Power Division*, Amer. Soc. Civil Engrs. (February):1553-1–17.

Falvey, H. T. 1980. Air-water flow in hydraulic structures, *Engineering Monograph no. 41*, U.S. Bureau of Reclamation, Denver, Colorado.

Formica, G. 1955. Esperienze preliminari sulle perdite di carico nei canali dovute a cambiamenti di sezione (Preliminary tests on head losses in channels due to cross-sectional changes), *L'Energia Elletrica,* Milan, 32, no. 7 (July):554.

Forster, J. W., and Skrinde, R. A. 1950. Control of the hydraulic jump by sills, *Trans.*, Amer. Soc. Civil Engrs., 115:973–1022.

Garg, S. P., and Sharma, H. R. 1971. Efficiency of hydraulic jump, *Jour. Hydraulics Division*, Amer. Soc. Civil Engrs., 97, HY3:409–420. (Discussions: vol. 97, HY9:1570–73; HY10:1790–95, HY11:1923; HY12:2107–10; vol. 98, HY1:278–84; vol. 99, HY3:527–29.)

Graber, S. D. 1982. Asymmetric flow in symmetric expansion, *Jour. Hydraulic Engineering,* Amer. Soc. Civil Engrs., 108, no 10:1082–1101.

Hager, W. H. 1986. Discharge measurement structures, Communication 1, Chaire de Constructions Hydrauliques, Ecole Polytechnique Federale de Lausanne, Lausanne, Switzerland.

――――. 1991a. *Energy dissipators and hydraulic jump,* Kluwer Academic, Dordrecht, the Netherlands.

――――. 1991b. Experiments on standard spillway flows, *Proc.*, Institution of Civil Engrs. (London) 91 (Sept):399–416.

Hager, W. H., and Bremen, R. 1989. Classical hydraulic jump: Sequent depths ratios, *Jour. Hydraulic Research*, International Assoc. for Hydraulic Research 27, no. 5:565–85.

Hager, W. H., and Sinniger, R. 1985. Flow characteristics in a stilling basin with an abrupt bottom rise, *Jour. Hydraulic Research*, International Assoc. for Hydraulic Research 23, no. 2:101–113 (Discussion: 24, no. 3:207–15.)

Hall, G. W. 1962. Discharge characteristics of broad-crested weirs using boundary layer theory, *Proc, Institution of Civil Engrs* (London), 22, June, 177.

Henderson, F. M. 1966. *Open channel flow*, MacMillan, New York, NY.

Ippen, A. T. 1950. Channel transitions and controls. In *Engineering Hydraulics,* edited by H. Rouse, Chap. 8, John Wiley, New York, NY.

Kabos, H. 1984. *Symposium on scale effects in modeling hydraulic structures,* Chap. 2. International Assoc. for Hydraulic Research, Delft, the Netherlands.

Kao, D. T. and Dean, P. S. 1976. Spatially varied flow in channel transitions, *Rivers 76,* Amer. Soc. Civ. Engrs., 1551–71.

Kindsvater, C. E. 1944. The hydraulic jump in sloping channels, *Trans.*, Amer. Soc. Civil Engrs., 109:1107–54.

Leliavsky, S. 1955, Irrigation and hydraulic design, Volume 1, Chapman and Hall, London.

Lenau, C. W., and Cassidy, J. J. 1969. Flow through spillway flip buckets, *Jour. Hydraulics Division*, Amer. Soc. Civil Engrs. HY2 (March):633–48.

Mason, P. J. 1982. The choice of hydraulic energy dissipator for dam outlet works based on a survey of prototype usage, *Proc.*, Institution of Civil Engrs. (London), 72, part 1:209–19. (Discussion: vol. 74 (1983):123–26).

Mason, P. J., and Arumugam, K. 1985. Free jet scour below dams and flip buckets, *Jour. Hydraulic Engineering*, Amer. Soc. Civil Engrs., 111, no. 2:220–35.

Mazumdar, S. K. 1967. Optimum length of transition in open channel expansive subcritical flow. *Jour. Institution of Engrs.*, (India) 48, no 3:463–76.

Mazumder, S. K., and Ahuja, K. C. 1978. Optimum length of contracting transition in open channel subcritical flow, *Jour. Inst. Engrs.* (India), 58, Pt. CI 5 (March):218–23.

McCorquodale, J. A. 1986. "Hydraulic jump and internal flows." In *Encyclopedia of fluid mechanics 1*, edited by N. P. Cheremisinoff, Chap. 6, Gulf Publishing Co., Houston, Texas.

McCorquodale, J. A., and Khalifa, A. 1983. Internal flow in hydraulic jumps, *Jour. Hydraulic Engineering*, Amer. Soc. Civil Engrs., 109, no. 5:684–701.

Mehrotra, S. C. 1976. Length of hydraulic jump, *Jour. Hydraulics Division*, Amer. Soc. Civil Engrs. 102, HY7:1027–33.

Mehta, P. R. 1979. Flow characteristics in two-dimensional expansion, *Jour. Hydraulics Division*, Amer. Soc. Civil Engrs. 105, no. 5:501–16.

Murphy, T. E. 1974. Spillway stilling basin, hydraulic jump type, *Memorandum*, Waterways Experiment Station (June 25), 11 pp.

Nashta, C. F., and Garde, R. J. 1988. Subcritical flow in rigid-bed open channel expansions, *Jour. Hydraulic Research*, Inter. Assoc. Hydraulic Research, 26, no. 1:49–65. (Discussion: 27, no. 4:556–58).

Ohtsu, I. 1976. Free hydraulic jump and submerged hydraulic jump in trapezoidal and rectangular channels, *Trans.*, Japanese Soc. Civil Engineering 8:122–25.

Peterka, A. J. 1958. *Hydraulic design of stilling basins and energy dissipators*, U.S. Bureau of Reclamation, Denver, Col (7th printing in 1983).

Rajaratnam, N. 1967. "Hydraulic jump." In *Advances in Hydrosciences*, edited by V. T. Chow, Academic Press, New York, 4:197-280.

Rajaratnam, N., and Subramanya, K. 1968. Profile of hydraulic jump, *Jour. Hydraulics Division*, Amer. Soc. Civil Engrs., 94, HY3:663–73.

Rajaratnam, N.; Subramanya, K.; and Muralidhar, D. 1968. Flow profile over sharp-crested weirs, *Jour. Hydraulics Division*, Amer. Soc. Civil Engrs. 94, HY3:843–47.

Rama-Murthy, A. S.; Basak, S.; and Rama, R. 1970. Open channel expansions fitted with local hump, *Jour. Hydraulics Division*, Amer. Soc. Civil Engrs., 96, no 5:1105–13.

Ranga Raju, K. G., 1981. Flow through open channels, Tata McGraw Hill, New Dehli, India.

Rao, B. V., and Seetharamiah, K. 1969. Separation control devices in diverging channels, *Proc.*, 13th Congress, International Assoc. Hydrology Research, Kyoto 1:113–21.

Rao, N. S. L. 1975. "Theory of weirs." In *Advances in Hydroscience*, Edited by V. T. Chow. Vol. 10, Academic Press, New York, NY.

Reese, A. J., and Maynard, S. T. 1987. Design of spillway crests, *Jour. Hydraulic Engineering*, Amer. Soc. Civil Engrs., 113, no. 4:476 – 90.

Roberson, J. A.; Cassidy, J. J.; and Chaudhry, M. H. 1988, *Hydraulic engineering,* Houghton Mifflin, Boston, MA.

Rouse, H.; Bhoota, B. V.; and Hsu, E. Y. 1951. Design of channel expansions, *Trans.,* Amer. Soc. Civil Engrs., 116:347

Skogerboe, G. V.; Austin, L. H.; and Bennel, R. S. 1971. Energy loss analysis for open channel expansion, *Jour. Hydraulics Division,* Amer. Soc. Civil Engrs., 97, no. 10:1719–36.

Smith, C. D., and Yu, J. N. 1966. Use of baffles in open channel transitions, *Jour. Hydraulics Division,* Amer. Soc. Civil Engrs., 92, no 2:1–16.

Sturm, T. W. 1985. Simplified design of contractions in supercritical flow, *Jour. Hydraulic Engineering* 111:871–75 (Discussion: 113, HY4:539–43).

U.S. Army Corps of Engineers. 1965. Hydraulic design of spillways, *Engineering Manual* EM1110-2-1603.

U.S. Army Corps of Engineers. 1977. *Hydraulic design criteria,* Waterways Experiment Station, Vicksburg, Miss.

U.S. Bureau of Reclamation. 1977 *Design of small dams,* 2nd ed., Denver, Co.

Vittal, N. 1978. Direct solution to problems of open channel transitions, *Jour. Hydraulic Division,* Amer. Soc. Civil E4, Bridge piers as channel obstructions, *Technical Bulletin no. 422,* U.S. Department of Agriculture. Nov.

Vittal, N., and Chiranjeevi, V. V. 1983. Open channel transitions: Rational method of design, *Jour. Hydraulic Engineering,* Amer. Soc. Civil Engrs., 109, no. 1:99–115.

White, W. R. 1977. Thin plate weirs, Proc. Inst. Civil Engrs. (London) 63, no. 2:255-69.

Yarnell, D. L. 1934. Bridge piers as channel obstructions, *Technical Bulletin* no. 422, U.S. Department of Agricultrue, Nov.

8 Computation of Rapidly Varied Flow

Aerial view of shock waves in the 55-m-wide spillway chute of Bennet Dam, British Columbia for a discharge of 570 m³/s (Courtesy of B. C. Hydro and Power Authority, Vancouver, B.C., Canada).

8-1 Introduction

Typical examples of natural and constructed open channels having rapidly varied flows are mountainous streams, rivers during periods of high floods, spillway chutes, conveyance channels, sewer systems, and outlet works. Unlike the case of gradually varied flows, a number of difficulties, such as the formation of roll waves, air entrainment, and cavitation, are encountered in the analysis of these flows. In addition, instabilities may develop if the Froude number exceeds a critical value, giving rise to roll waves or slug flow. Standing waves and large surface disturbances, commonly referred to as *shocks* or *standing waves*, are important aspects of rapidly varied flows and need to be considered in the analysis and design.

To compute supercritical flow in channel expansions, including the effects of bottom slope and friction, Liggett and Vasudev (1965) numerically integrated the steady, two-dimensional, shallow-water equations. However, these and many other procedures suitable for gradually varied flows cannot be used to compute flows with shocks or standing hydraulic jumps. By using shock-tracking techniques, Pandolfi (1975) analyzed flow around a blunted obstacle in a supercritical stream. Demuren (1979) computed the sub- and supercritical steady flows by using methods developed by Patankar and Spalding and compared the computed and experimental results. Although the agreement between computed and experimental results is fair, the ability of the numerical scheme to handle discontinuities is not clearly demonstrated. The method of characteristics was used for the analysis of two-dimensional supercritical flows by Bagge and Herbich (1967), Herbich and Walsh (1972), Villegas (1976), and Dakshinamoorthy (1977). Ellis and Pender (1982) used an implicit method of characteristics to compute high-velocity flows in channels of arbitrary alignment and slope. Like other characteristic-based procedures, this method cannot compute oblique jumps and requires many interpolations, which may seriously affect the accuracy of the solution. Jimenez and Chaudhry (1988), Bhallamudi and Chaudhry (1992), and Gharangik and Chaudhry (1991) utilized shock-capturing finite-difference methods to analyze rapidly varied flows; this chapter is based mainly on these investigations.

In this chapter,* we present finite-difference methods for the computation of rapidly varied flows. These are shock-capturing methods and do not require any special treatment if a shock develops in the solution. Three different formulations are discussed. The St. Venant equations, also referred to as the shallow-water equations, are assumed

---

*The material presented in this chapter will be easier to follow if the reader first becomes familiar with the unsteady flow equations and the finite-difference schemes of Chapters 11–15.

to describe these flows in the first two formulations, and Boussinesq terms are included in the third to account for nonhydrostatic pressure distribution. The validity of these computational procedures is verified by comparing the computed results with the analytical solutions and with the experimental measurements.

8-2 Governing Equations

The flow conditions are function of time in unsteady flows. If this function is a constant, then steady flow may be considered as a special case of unsteady flow. We may, therefore, solve the unsteady flow equations to analyze steady flows. This may be done by computing the flow conditions in the channel system for a sufficient length of time until steady-state conditions are obtained. This procedure offers a number of advantages for the solution of the governing equations; we discuss this in more detail later in this section.

The St. Venant equations (see Chapter 15 for their derivation and the simplifying assumptions upon which they are based) describing the two-dimensional unsteady flows may be written in a vector form as

$$\frac{\partial \mathbf{U}}{\partial t} + \frac{\partial \mathbf{E}}{\partial x} + \frac{\partial \mathbf{F}}{\partial y} + \mathbf{S} = 0 \tag{8-1}$$

in which

$$\mathbf{U} = \begin{pmatrix} h \\ uh \\ vh \end{pmatrix}; \qquad \mathbf{E} = \begin{pmatrix} uh \\ u^2h + \frac{1}{2}gh^2 \\ uvh \end{pmatrix}$$

$$\mathbf{F} = \begin{pmatrix} vh \\ uvh \\ v^2h + \frac{1}{2}gh^2 \end{pmatrix}; \qquad \mathbf{S} = \begin{pmatrix} 0 \\ -gh(S_{ox} - S_{fx}) \\ -gh(S_{oy} - S_{fy}) \end{pmatrix} \tag{8-2}$$

in which t = time; u = depth-averaged flow velocity in the x direction; v = depth-averaged flow velocity in the y direction; h = water depth measured vertically; g = acceleration due to gravity; $S_{o(x,y)} = \sin\alpha_{(x,y)}$ = channel bottom slope in the (x, y) directions; $\alpha_{(x,y)}$ = angles between the bottom of the channel and the (x, y) directions; $S_{f(x,y)}$ = friction slopes in the (x, y) directions; and the (x, y) coordinate system is as shown in Fig. 8-1. The friction slope S_f is calculated from the following steady-state formulas:

$$S_{fx} = \frac{n^2 u \sqrt{u^2 + v^2}}{C_o^2 h^{1.33}}; \qquad S_{fy} = \frac{n^2 v \sqrt{u^2 + v^2}}{C_o^2 h^{1.33}} \tag{8-3}$$

in which n = Manning roughness coefficient and C_o = correction factor for units ($C_o = 1$ in SI units and $C_o = 1.49$ in customary units).

Of all the simplifying assumptions made to derive the St. Venant equations, the hydrostatic pressure distribution is probably the weakest one for the present application. The pressure distribution is hydrostatic at all points except in the vicinity of a shock, such as a surge wave, hydraulic jump, etc. Although some details are lost in the vicinity

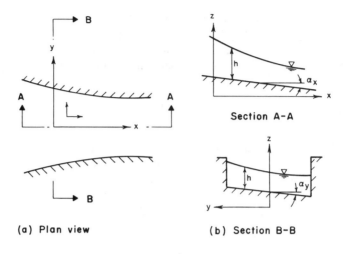

(a) Plan view   (b) Section B-B

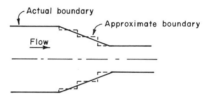

(c) Approximation of boundary

Figure 8-1 Notation.

of the shock if these equations are used for the analysis of rapidly varied flows, the overall results are adequate for engineering purposes (Cunge 1975). Liggett and Vasudev (1965) showed by using dimensional arguments that this assumption is valid as long as the "shallowness" parameter h_o/l_o is small, where h_o and l_o are water depth and characteristic length. Following the theory of Engelund and Munch-Petersen, Jimenez and Chaudhry (1988) showed that the shallow-water theory may reasonably represent smooth, steady, supercritical flow if the Froude number is not close to 1 and the depth-to-width ratio is of the order of 0.1 or less.

There are three independent variables in Eq. 8-1, x, y, and t. As we discussed in Chapter 1, the flow is steady if the variation of flow variables with respect to time is zero. Thus, we may deduce equations describing steady flow from Eq. 8-1 by dropping the time derivative term, $\partial \mathbf{U}/\partial t = 0$. In other words, the equation describing steady, two-dimensional flow in open channels is

$$\frac{\partial \mathbf{E}}{\partial x} + \frac{\partial \mathbf{F}}{\partial y} + \mathbf{S} = 0 \tag{8-4}$$

Characteristic Directions

According to the theory of characteristics, the characteristic directions, λ_i, of Eq. 8-4 are given by the eigenvalues of the matrix of coefficients of the nondivergent form of these equations (Jimenez and Chaudhry 1988), i.e.,

$$\lambda_1 = \frac{v}{u} \tag{8-5}$$

$$\lambda_{2,3} = \frac{uv \pm gh\sqrt{\mathbf{F}_r^2 - 1}}{u^2 - gh} \tag{8-6}$$

in which $\lambda_i = (dy/dx)_i$ are the slopes of the characteristics lines and $\mathbf{F}_r$ is the local Froude number, given by

$$\mathbf{F}_r = \frac{\sqrt{u^2 + v^2}}{\sqrt{gh}} = \frac{V}{\sqrt{gh}} \tag{8-7}$$

in which V is the magnitude of the velocity vector. It follows from Eqs. 8-6 that Eq. 8-4 is

- *Hyperbolic* if $\mathbf{F}_r > 1$;
- *Parabolic* if $\mathbf{F}_r = 1$; and
- *Elliptic* if $\mathbf{F}_r < 1$.

Note that Eq. 8-5 defines the direction of the streamlines. If θ is the angle between the velocity vector and the x-axis, then

$$\begin{aligned} u &= V\cos\theta \\ v &= V\sin\theta \end{aligned} \tag{8-8}$$

For an infinitely wide channel, the angular position of a small stationary wave (Engelund and Munch-Petersen 1953)

$$\mu = \sin^{-1}\frac{1}{\mathbf{F}_r} \tag{8-9}$$

By substituting Eqs. 8-8 and 8-9 into Eqs. 8-5 and 8-6, rearranging, and simplifying, we obtain the following expressions for the characteristic directions:

$$\begin{aligned} \left(\frac{dy}{dx}\right)_1 &= \tan\theta \\ \left(\frac{dy}{dx}\right)_{2,3} &= \tan(\theta \pm \mu) \end{aligned} \tag{8-10}$$

These equations permit a clear graphical interpretation. Figure 8-2 shows a streamline passing through point P, which makes an angle θ with the x-axis. It is clear from Eqs. 8-10 that, in addition to the streamline, two more characteristics pass through P: one at angle μ above the streamline (C^+) and the other at angle μ below the streamline (C^-). For the physical meaning of the Mach lines, it may be shown from more elementary considerations (see Henderson (1966, 239)) that they define the locus

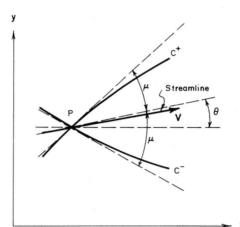

Figure 8-2 Characteristic directions.

of weak disturbances originating at point P. In other words, they bound the zone of influence of P.

Coordinate Transformations

Most real-life channels have irregular or curvilinear channel geometries. The inclusion of boundaries becomes a problem when analyzing these channels by the finite-difference methods. The grid points usually do not coincide with the boundaries, thus requiring interpolation procedures, which have proven to deteriorate the solution (Roache 1972). To avoid this problem, the coordinates may be transformed such that the coordinate axes coincide with the boundaries. For example, the following simple coordinate transformation yields good results in many cases.

Let us consider a symmetrical transition as shown in Fig. 8-3(a). Due to symmetry, only one-half of the channel may be analyzed. The following transformation of independent variables x and y converts the physical domain into a rectangular computational

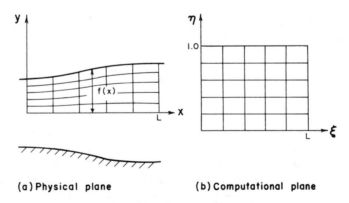

(a) Physical plane

(b) Computational plane

Figure 8-3 Coordinate transformation.

domain in coordinates ξ and η (Fig. 8-3b):

$$\xi = x$$
$$\eta = \frac{y}{f(x)} \tag{8-11}$$

in which $f(x)$ is the distance between the symmetry line and the left boundary at distance x (Fig. 8-3a). This transformation allows a uniformly spaced grid in the computational plane such that the boundaries now coincide with the lines $\eta = 0$ and $\eta = 1$. To do this, the governing equations are first transformed in terms of ξ and η by applying the chain rule of partial differentiation. Then, the resulting equations may be cast into conservation form by means of algebraic manipulations outlined by Anderson, (1984). This process gives the following divergent-free equation in the transformed coordinates:

$$\frac{\partial \mathbf{U}}{\partial t} + \frac{\partial \mathbf{G}}{\partial \xi} + \frac{\partial \mathbf{H}}{\partial \eta} + \widehat{\mathbf{S}} = 0 \tag{8-12}$$

in which

$$\mathbf{G} = f(\xi)\mathbf{E}$$
$$\mathbf{H} = \mathbf{G} - \eta\, f'(\xi)\mathbf{E} \tag{8-13}$$
$$\widehat{\mathbf{S}} = f(\xi)\mathbf{S}$$

and $f'(\xi) = df/d\xi$. Note that Eq. 8-12 is analogous to the original equation, Eq. 8-1. The former is in the x and y coordinates, whereas the latter is in the ξ and η.

Polar coordinates combined with transformations of the type given by Eq. 8-11 may be used to obtain a rectangular computational domain for the channels with general geometries.

8-3 Computation of Supercritical Flow

As we discussed in the last section, Eq. 8-4 is hyperbolic if $\mathbf{F}_r > 1$. Thus a channel having supercritical flow throughout its length may be analyzed by using the steady form of the shallow-water equations. In x-y coordinates, we obtained Eq. 8-4 from Eq. 8-1. A similar equation may be obtained from Eq. 8-12 in ξ-η coordinates. These equations are hyperbolic as long as the flow is supercritical. This offers special advantages for their numerical solution in the sense that we solve the equations directly to obtain the flow conditions and not in time until steady conditions are reached.

Finite-Difference Methods

The x-y plane (ξ-η plane for the transformed coordinates) is divided into a computational grid to solve Eq. 8-4 by the finite-difference methods. We will use the following notation to identify variables at different grid points. The superscript k and subscript j indicate nodes in the x and y directions, respectively. As discussed previously, when the local

Froude number is greater than 1, the system of equations describing steady flows (Eqs. 8-4) are hyperbolic (Abbott 1979). Therefore, a marching procedure may be used to integrate them. The solution is obtained by starting at the upstream end of the channel and advancing the computations first to $x_0 + \Delta x$, then to $x_0 + 2\Delta x$, and so on. In this case, the x direction is called the *marching direction*. This direction may be any one as long as the system is hyperbolic with respect to that particular marching direction (Kutler 1975). What this means is that the disturbances originating in the flow field should not travel opposite to the marching direction. According to Eqs. 8-9, Eq. 8-7 is hyperbolic with respect to the x direction if

$$u^2 - gh > 0 \tag{8-14}$$

Thus the coordinates are selected such that the marching direction is aligned with the predominant flow direction. Otherwise, the requirement given by Eq. 8-14 may not be fulfilled, even if the flow is supercritical.

It is desirable to use shock-capturing or through methods, since a complex oblique jump pattern may develop in many situations involving supercritical flow. The channels or structures having rapidly varied flow are usually short, and probably fewer than a few hundred marching steps are sufficient to compute the water surface profiles in them. Therefore, explicit methods suitable for hyperbolic systems may be utilized. Two explicit schemes—Lax and MacCormack schemes—were used by Jimenez and Chaudhry (1988). The Lax scheme is first-order accurate, and the MacCormack scheme is second-order accurate. They are probably the simplest of the available explicit, dissipative numerical schemes and have been widely used in fluid-flow applications (Anderson et al. 1984). We discuss only the application of the MacCormack scheme in this chapter; readers interested in the application of the Lax scheme should see Jimenez and Chaudhry (1988).

MacCormack scheme.   The MacCormack scheme is a two-step predictor-corrector scheme (Anderson et al., 1984). Considering x as the marching direction, application of the scheme to Eq. 8-4 yields the following equations:

Predictor

$$\mathbf{E}_j^* = \mathbf{E}_j^k - \frac{\Delta x}{\Delta y}(\mathbf{F}_{j+1}^k - \mathbf{F}_j^k) - \Delta x \mathbf{S}_j^k \tag{8-15}$$

Corrector

$$\mathbf{E}_j^{**} = \mathbf{E}_j^k - \frac{\Delta x}{\Delta y}(\mathbf{F}_j^* - \mathbf{F}_{j-1}^*) - \Delta x \mathbf{S}_j^* \tag{8-16}$$

$$\mathbf{E}_j^{k+1} = \frac{1}{2}(\mathbf{E}_j^* + \mathbf{E}_j^{**}) \tag{8-17}$$

An asterisk (*) indicates the values at the end of the predictor part and (**) refers to the values at the end of the corrector part. Another variation of the method is to use backward differences for the y derivative in the predictor part and forward differences in the corrector part. In some applications, alternation of both variations every other integration step is recommended (Roache 1972). Kutler (1975) showed that the shock resolution is best in problems involving discontinuities when the difference in the predictor part is in the direction of the propagation of discontinuity. The validity of this conclusion has been verified for the present application by Jimenez and Chaudhry (1988).

We consider the ξ coordinate as the marching direction if the equations are in the transformed coordinates and we use the MacCormack scheme to integrate the steady form of Eq. 8-12 (i.e., the equation obtained by substituting $\partial \mathbf{U}/\partial t = 0$ into Eq. 8-12). The stability requirement of the equations in the transformed coordinates is modified accordingly.

Stability.   The *Courant-Friedrichs-Lewy condition* (CFL for short) has to be satisfied for the preceding scheme to be stable. The CFL condition for Eq. 8-1 may be written as (Anderson, et al. 1984)

$$C_n = |\lambda_{\max}| \frac{\Delta x}{\Delta y} \leq 1 \tag{8-18}$$

in which $|\lambda_{\max}|$ is the maximum absolute value of the characteristic slopes, $|\lambda_i|$, and C_n is referred to as the *Courant number*. It follows from Eq. 8-9 that

$$|\lambda_{\max}| = \frac{|uv| + gh\sqrt{\mathbf{F}_r^2 - 1}}{u^2 - gh} \tag{8-19}$$

The truncation error in the MacCormack scheme is the smallest when the largest possible value of the Courant number, compatible with the preceding stability condition, is used (Anderson, et al. 1984). However, note that CFL is derived by neglecting the head-loss term and by using a linearized form of the governing equations.

Experience has shown that this stability condition is adequate for the analyses of systems having low head losses, although additional stability criteria may have to be satisfied for systems having large losses. Since the head losses are usually very small in a typical channel having rapidly varied flows, the previously mentioned CFL stability condition should be sufficient for the present application. The step size should be chosen so that Eq. 8-18 is satisfied at all points in the y direction since $|\lambda_{\max}|$ is a local function of h, u, and v. If this condition is not satisfied, then the step size should be reduced and the calculations repeated after each integration step to avoid instability.

Boundary Conditions

The preceding finite-difference equations are used at the interior grid points. To start the computations, we specify the initial conditions and we include the boundary conditions to simulate the boundaries of the channel and the inflow and outflow conditions at the ends of the channel.

Proper inclusion of the boundaries is very important for a successful application of any numerical technique, especially for hyperbolic systems, in which an error introduced at the boundaries is propagated and reflected throughout the grid. These errors may cause instability in many cases (Anderson, et al. 1984).

For the initial conditions, we specify all three variables (h, u, v) at all grid points.

It is sufficient to analyze one-half of a symmetrical system by means of a symmetrical boundary at the symmetry plane. In addition, we have to specify the boundary conditions for the channel boundaries.

For a *solid boundary* we enforce the condition that there is no mass flow through it. This may be done by the following equation, referred to as the *slip condition*:

$$\frac{v}{u} = \tan\theta \qquad (8\text{-}20)$$

in which θ is the angle between the wall and the x-axis. A *symmetry boundary* is similar to a solid boundary in that the normal velocity with respect to the symmetry plane should be zero. In addition, it is required that the normal gradients of all variables with respect to the symmetry plane vanish.

Several wall boundary techniques enforce in one way or another the basic requirement given by Eq. 8-20. The problem arises in applying this equation at the grid points along the wall. The values of all the variables are required for this purpose, and Eq. 8-20 does not provide all the needed information. Thus, these values are computed using information from the interior points plus the boundary condition. Abbett (1971) developed a technique that has proven to be successful in many supersonic flow computations. This procedure was adapted for the analysis of supercritical flows by Jimenez and Chaudhry (1988) and is discussed here. For simplicity, let us assume that the wall under consideration is aligned with the x-axis. Thus the boundary condition is given by $v = 0$.

The basic idea in the Abbett procedure (Abbett 1971; Kutler 1975) is to apply the numerical scheme up to the wall using one-sided differences as a first step. Then, to enforce the surface tangency requirement, a simple wave is superimposed on the solution to make the flow parallel to the wall. The details of the method are presented with reference to the MacCormack scheme.

Let us assume that the solution is being advanced from station k to station $k + 1$ (Fig. 8-4a). Since the first step of the MacCormack scheme as given by Eq. 8-15 involves a forward difference, it can be applied at the wall to yield the predicted

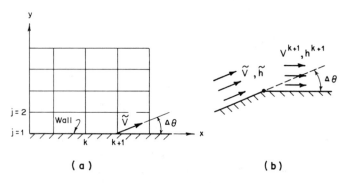

Figure 8-4 Abbett procedure for solid walls.

value $\mathbf{E}_1^\star$. It is followed by the corrector step (Eq. 8-16); however, the backward difference in the y direction is replaced by a forward difference to yield the first corrected value $\widetilde{\mathbf{E}}_1 = \frac{1}{2}(\mathbf{E}_1^\star + \mathbf{E}_1^{\star\star})$. We will use a tilde~on various variables to indicate values corresponding to $\widetilde{\mathbf{E}}$. From this value of $\widetilde{\mathbf{E}}_1$, we determine $\widetilde{h}$, $\widetilde{u}$, and $\widetilde{v}$ at station $k+1$. Generally the resulting velocity, $\widetilde{V}$, will not be tangent to the wall. Let this angle between $\widetilde{V}$ and the wall be $\Delta\theta$. Then,

$$\Delta\theta = \tan^{-1}\frac{\widetilde{v}}{\widetilde{u}} \qquad (8\text{-}21)$$

If $\Delta\theta$ is positive, an expansion wave is required to rotate $\widetilde{V}$ so that it becomes tangent to the wall; a contraction wave is necessary for the same purpose if $\Delta\theta$ is negative. Figure 8-4 shows a situation in which $\Delta\theta > 0$. This is equivalent to conditions produced by a wall turning away from the flow. The situation for $\Delta\theta < 0$ corresponds to a wall turning into flow.

A comparison of three available procedures for computing $\Delta\theta$ gave similar results for $|\Delta\theta| < 5°$ and $2 < \mathbf{F}_r < 8$ (Jimenez and Chaudhry 1988). Among these procedures, the following does not require an iterative solution, as do the other two, and is presented here.

Knapp (1951) found from experiments on curved channels that the following equation gives good results:

$$\frac{h^{k+1}}{\widetilde{h}} = \widetilde{\mathbf{F}}_r^2 \sin^2\left(\widetilde{\mu} - \frac{\Delta\theta}{2}\right) \qquad (8\text{-}22)$$

in which $\widetilde{\mu} = \sin^{-1}(1/\widetilde{\mathbf{F}}_r)$. This expression is obtained by assuming a constant magnitude of velocity, $\widetilde{V} = V^{k+1}$, through the cross wave.

The Abbett procedure may be applied to a curved wall, in which the angle $\Delta\theta$ includes the deviation of the wall because of its curvature.

Verification

The results for several cases computed by using the MacCormack scheme were compared with the analytical solutions and with the experimental results (Jimenez and Chaudhry 1988). Only two comparisons are presented here. The first is with the analytical solution and the second is with the experimental measurements.

Oblique hydraulic jump.   As we discussed in the last chapter, an oblique hydraulic jump or standing wave is produced when a vertical boundary is deflected inward into the flow, e.g., as in a channel contraction. This causes an abrupt depth increase, which is propagated from the point of deflection in the wall to the interior of the flow field at angle β with respect to the flow direction. Equation 7-17 is the analytical solution for the problem if the bottom friction and slope of the channel bottom are neglected.

Figure 8-5(a) shows the grid system used in the numerical computations. The x-axis is aligned with the wall downstream of the deflection point. The variables at

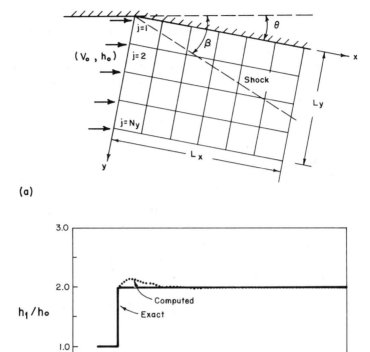

(a)

(b)

Figure 8-5 Oblique jump.

$x = 0$ are specified as $h = h_o$; $u = V_o \cos\theta$; and $v = -V_o \sin\theta$, where h_o and V_o are the approach flow depth and velocity and θ is the angle of wall deflection. Along the lower boundary $(y = L_y)$, a zero-order extrapolation from the immediate interior point is used, i.e., $\mathbf{E}_{N_y}^{k+1} = \mathbf{E}_{N_y-1}^{k+1}$. This is a good approximation, since the flow field remains undisturbed upstream of the wave front.

For the example shown here, $\mathbf{F}_{r0} = V_o/\sqrt{gh_o} = 4$ and $\theta = 12°$. The analytical solution for this is $h_1/h_o = 1.987$; $\beta = 25.505°$; and $V_1/V_o = 0.9282$, in which h_1 and V_1 are the downstream flow depth and velocity, respectively.

Figure 8-5(b) shows the water surface profiles along the wall computed by using $C_n = 0.98$. Since the Abbet procedure incorporates Eq. 7-17, the abrupt initial jump is computed exactly; it is then followed by a small overshoot, which is rapidly corrected. The same trend was observed for different Froude numbers and deflection angles.

A three-dimensional plot of the computed water surface for $C_n = 0.98$ is shown in Fig. 8-6. The analytical solution is indicated by a dashed line. The figure illustrates the shock-capturing capabilities of the scheme. The strength of the shock and its direction

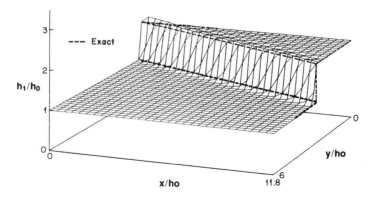

Figure 8-6   Computed water-surface profile.

are accurately predicted. The former is obtained exactly up to the third significant digit; the error in the latter is not easy to determine because of the spreading of the shock over two or three grid points. As mentioned earlier, the best resolution of the shock is obtained when the finite-difference approximation in the predictor part is in the direction of propagation of the discontinuity (Kutler 1975). For example, for the results shown in Fig. 8-6, the overshoot on the back side of the jump is approximately 4 percent of $h_1 - h_o$, as compared to the opposite alternative (using backward finite-differences in the predictor part for this example), for which the overshoot is about 25 percent.

Circular-arc contraction.   The flow in a contraction composed of circular arcs (Fig. 8-7) is analyzed. The results compared here are for an initial Froude number of 4 and an initial water depth of 0.030 m. The flow at the entrance of the contraction is uniform. This was the condition in the experiment for the assumed Manning n of 0.012 and for a bottom slope of $S_o = 0.072$. In the computations, a constant depth and uniform velocity distribution were assumed at the upstream section and 21 grid points in the y direction (for half-channel width) and a Courant number of 0.98 were used.

Figure 8-7(b) compares the computed water-surface profile at the wall with the experimental results reported in Ippen et al.'s (1951, Fig. 38) and Fig. 8-7(c) shows a three-dimensional plot of the computed water surface profile. The walls of the channel are not shown in this plot. The comparison of water depths in the length of the contraction, including the first peak, is good. Downstream of the transition, however, the disagreement between the experimental and computed results becomes large. This example shows that although a solution of the shallow-water equations simulates the general features of the flow, the prediction of the maximum water levels is unsatisfactory. This is because the disturbances as well as the depth-to-width ratio ($h/b \approx 0.2$ for the downstream channel) are large.

These comparisons show that this scheme gives satisfactory results if the assumption of hydrostatic pressure distribution is valid.

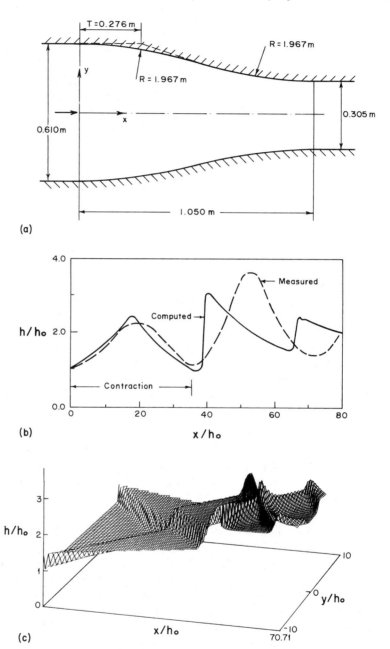

Figure 8-7   Circular-arc contraction.

8-4 Computation of Sub- and Supercritical Flows

In this section, two-dimensional, depth-averaged, *unsteady* flow equations in a transformed coordinate system (Eq. 8-12) are solved numerically to analyze flows in channel expansions and contractions. An unsteady flow model is used to obtain steady-flow solutions by treating the time variable as an iteration parameter and letting the solution converge to the steady state. Unlike the steady model of the last section, which can be used only for supercritical flows, the unsteady model is capable of simulating both sub- and supercritical flows. The approximation of the side wall boundaries of a physical system as shown by dotted lines in Fig. 8-1(c) may introduce large errors. It is better in such cases to convert the physical domain into a rectangular computational domain using simple transformations discussed in Sec. 8-2.

We first describe the details of the MacCormack scheme and procedures for including the initial and boundary conditions and then present results for its verification.

Numerical Solution

The MacCormack scheme (MacCormack 1969) is used to integrate numerically the transformed form of the governing equations (Eq. 8-12). Referring to the finite-difference grid shown in Fig. 8-8, these finite-difference approximations are as follows.

Predictor

$$\mathbf{U}_{i,j}^{\star} = \mathbf{U}_{i,j}^{k} - \frac{\Delta t}{\Delta \xi} \left(\mathbf{G}_{i+1,j}^{k} - \mathbf{G}_{i,j}^{k} \right) - \frac{\Delta t}{\Delta \eta} \left(\mathbf{H}_{i,j+1}^{k} - \mathbf{H}_{i,j}^{k} \right) - \mathbf{S}_{i,j}^{k} \Delta t \tag{8-23}$$

Corrector

$$\mathbf{U}_{i,j}^{\star\star} = \mathbf{U}_{i,j}^{k} - \frac{\Delta t}{\Delta \xi} \left(\mathbf{G}_{i,j}^{\star} - \mathbf{G}_{i-1,j}^{\star} \right) - \frac{\Delta t}{\Delta \eta} \left(\mathbf{H}_{i,j}^{\star} - \mathbf{H}_{i,j-1}^{\star} \right) - \mathbf{S}_{i,j}^{\star} \Delta t \tag{8-24}$$

in which the subscripts i and j refer to the grid points in the ξ and η directions, respectively. The superscript k refers to the variable at the known time level, $\star$ refers to the variables computed at the end of the predictor part, and $\star\star$ refers to the variables at the end of the corrector part.

Now, $\mathbf{U}$ at the unknown time level $k+1$ is determined from

$$\mathbf{U}_{i,j}^{k+1} = \frac{1}{2} \left(\mathbf{U}_{i,j}^{\star} + \mathbf{U}_{i,j}^{\star\star} \right) \tag{8-25}$$

As we discussed in Sec. 8-3, there are two other formulations of the MacCormack scheme in addition to that given here. These are to use backward finite differences in the predictor part and forward finite differences in the corrector part or to alternate the direction of differencing from one time step to the next. All three alternatives gave similar results for the steady-state solutions studied by Bhallamudi and Chaudhry (1992).

Initial and boundary conditions.   To start the unsteady state computations, the values of u, v, and h at time $t = 0$ are specified at all the grid points. In the present application, specification of their approximate values is sufficient, since these are needed only to start the computations that are continued until the solution converges to a steady state.

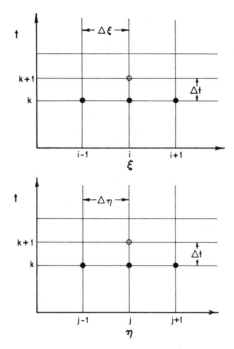

Fig..re 8-8 Computational grid.

Typical boundaries for a channel may be included in the analysis as follows.

Inflow and outflow boundaries.   The specification of inflow and outflow boundary conditions at the upstream and downstream ends depends on whether the flow is subcritical or supercritical (Stoker 1957; Verboom, et al. 1982). For two-dimensional supercritical flow, three boundary conditions have to be specified at the inflow boundary and none at the outflow boundary. For two-dimensional subcritical flow, however, two conditions are specified at the inflow boundary and one at the outflow boundary. The details of these open-boundary specifications and the approximations are discussed later (in the verification section) with reference to the particular problem solved.

Symmetry boundary.   A reflection procedure is used at a symmetry boundary (Roache 1972). In this procedure, the nonconservative flow variables u and h at the imaginary reflection points shown in Fig. 8-9(a) are specified as even functions with respect to the symmetry line. However, the normal velocity is specified as an odd function so that the average normal velocity at the boundary is zero. Note that the reflection procedure is exact for a symmetry line.

Solid-side-wall boundary.   A slip condition is used as the boundary condition for a side wall. Therefore, the resultant velocity at a solid wall is tangent to it. The reflection procedure used herein for the solid side wall is approximate and is not exact as was the case for a symmetry boundary procedure.

Referring to Fig. 8-9(b), the flow depth and the magnitude of resultant velocity at an imaginary reflection point are specified equal to their values at the corresponding

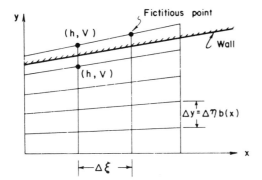

(a)

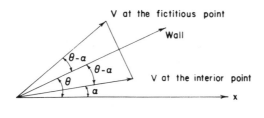

(b)

Figure 8-9 Imaginary reflection point.

interior grid point. The direction of the flow velocity, however, is determined such that the normal velocity at the wall is zero. If θ is the angle between the wall and the x-axis and α is the angle between the resultant velocity at the interior point and the x-axis (Fig. 8-9), then the velocity components u_o and v_o at the reflection point are given by

$$u_o = V \cos(2\theta - \alpha) \tag{8-26}$$

$$v_o = V \sin(2\theta - \alpha) \tag{8-27}$$

where V = resultant velocity at the interior point. Equations 8-26 and 8-27 are derived for a channel expansion; similar equations may be written for a channel contraction.

Stability.   The MacCormack scheme is stable if the Courant-Friedrichs-Lewy (CFL) condition is satisfied. This condition for two-dimensional flow in the transformed coordinates may be written as (Roache 1972)

$$C_n = \frac{(V + \sqrt{gh})\Delta t}{b(x)\Delta\xi\Delta\eta}\sqrt{\Delta\xi^2 + [b(x)\Delta\eta]^2} \leq 1 \tag{8-28}$$

where V = resultant velocity at the grid point. The numerical scheme is stable only if the above condition is satisfied at every grid point. Since the preceding condition is heuristic and is derived from a linearized form of the governing equations for one-dimensional flows, some numerical experimentation should be done before selecting an actual upper limit of the value of C_n.

Artificial viscosity.   The dispersive errors in the MacCormack scheme produce high-frequency oscillations near the steep gradients. To dampen these oscillations, (Anderson, et al. 1984) a procedure developed by Jameson et al. (1981) is used. This procedure smooths large gradients and leaves the smooth areas relatively undisturbed.

In this procedure, we first compute the following parameters using the computed values of h at the $k + 1$ time level.

$$\nu_{\xi_{i,j}} = \frac{|h_{i+1,j} - 2h_{i,j} + h_{i-1,j}|}{|h_{i+1,j}| + |2h_{i,j}| + |h_{i-1,j}|}$$

$$\nu_{\eta_{i,j}} = \frac{|h_{i,j+1} - 2h_{i,j} + h_{i,j-1}|}{|h_{i,j+1}| + |2h_{i,j}| + |h_{i,j-1}|} \tag{8-29a}$$

At the points where $h_{i,j-1}$ does not exist, we use

$$\nu_{\eta_{i,j}} = \frac{|h_{i,j+1} - h_{i,j}|}{|h_{i,j+1}| + |h_{i,j}|}$$

and where $h_{i,j+1}$ does not exist, we use

$$\nu_{\eta_{i,j}} = \frac{|h_{i,j-1} - h_{i,j}|}{|h_{i,j-1}| + |h_{i,j}|} \tag{8-29b}$$

Then, we determine from the following equations

$$\epsilon_{\xi_{i-1/2,j}} = \kappa \max \left(\nu_{\xi_{i-1,j}}, \nu_{\xi_{i,j}} \right)$$
$$\epsilon_{\eta_{i,j-1/2}} = \kappa \max \left(\nu_{\eta_{i,j-1}}, \nu_{\eta_{i,j}} \right) \tag{8-30}$$

where κ is a dissipation constant. The final values of the variable f at the new time step are computed from the following equation:

$$\begin{aligned} f_{i,j}^{k+1} = f_{i,j}^{k+1} &+ \left[\epsilon_{\xi_{i+1/2,j}} \left(f_{i+1,j}^{k+1} - f_{i,j}^{k+1} \right) - \epsilon_{\xi_{i-\frac{1}{2},j}} \left(f_{i,j}^{k+1} - f_{i-1,j}^{k+1} \right) \right] \\ &+ \left[\epsilon_{\eta_{i,j+1/2}} \left(f_{i,j+1}^{k+1} - f_{i,j}^{k+1} \right) - \epsilon_{\eta_{i,j-1/2}} \left(f_{i,j}^{k+1} - f_{i,j-1}^{k+1} \right) \right] \end{aligned} \tag{8-31}$$

in which f refers to u, v, and h. Equation 8-31 should be viewed as a FORTRAN replacement statement. The dissipation constant, κ, is used to regulate the amount of dissipation.

This procedure is equivalent to adding second-order dissipative terms to the original governing equations. The actual numerical eddy viscosity coefficient in the ξ direction is of the order of $\kappa \nu_\xi (\Delta \xi^2 / \Delta t)$. This indicates that the influence of κ on the results depends upon the gradients in the flow depth as well as on the grid size. As can be seen, its influence in the smooth regions is minimal since ν tends to be zero in such a case. A numerical grid is chosen such that the model gives convergent results and a finer grid does not improve the results significantly. The value of κ is selected such that it is as small as possible and at the same time smooths the high-frequency oscillations. A value of 0.3 for κ is recommended for an initial trial.

Verification

The computed results are compared with the laboratory test data for two cases; for other comparisons, see Bhallamudi and Chaudhry (1992).

Supercritical flow in symmetrical contraction.   In the laboratory tests reported by Ippen et al. (1951) on supercritical flow in the symmetrical, straight-wall contraction shown in Fig. 8-10, the upstream depth, h_o, was 0.0305 m and the upstream Froude number, $\mathbf{F}_{r_o}$, was 4.0.

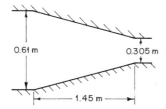

(a) Contraction

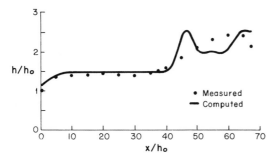

(b) Water level along wall

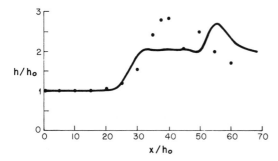

(c) Water level along centerline

Figure 8-10 Supercritical flow in a contraction.

The computations were done using a grid $\triangle \xi = 0.0483$ m and $\triangle \eta = 0.0476$ m. The dissipation coefficient, κ, was 0.8, the Courant number was 0.80, and the friction and bottom slopes were assumed to be zero. A depth of 0.0305 m, streamwise velocity of 2.188 m/s, and zero transverse velocity were specified at every grid point as the initial conditions. Starting with these initial values, the flow conditions were computed up to 3 s when the flow became steady. At the upstream boundary, h, u, and v were specified as 0.0305 m, 2.188 m/s, and zero, respectively. No condition was specified at the downstream boundary. The variables at the downstream end were, however, extrapolated from the interior points.

As shown in Fig. 8-10, the agreement between the computed water-surface profiles is good along the walls; and at the centerline, where the flows are smooth. However, this is not the case for the centerline water-surface profile in the vicinity of strong shocks. Although the computed maximum height of the shock is about the same as that in the experiment, the computed location differs significantly. Thus, the computed results may be used confidently for selecting the wall height; however, they are not accurate in the middle of the channel, which is more of an academic interest. The differences between the computed and measured results at the centerline of the transition may be due to the assumption of hydrostatic pressure distribution not being valid near steep gradients and the exclusion of the effects of air entrainment.

Hydraulic jump in a gradual expansion.

The computation of both sub- and supercritical flows is demonstrated by simulating a hydraulic jump in a gradual expansion. Figure 8-11 shows the general dimensions of the channel and the transition. The specified flow conditions in the channel were as follows: discharge = 0.007 m³/s; depth at the upstream end = 0.06 m; depth at the downstream end = 0.07 m; channel bottom slope, $S_{o_x} = 0.00017$; and Manning $n = 0.015$. For these conditions, the flow is supercritical at the inflow section (Froude number = 1.52) and subcritical at the outflow section (Froude number = 0.268).

In the computations, the flow depth $h = 0.07$ m, streamwise velocity $u = 0.222$ m/s and transverse velocity $v = 0$ m/s were specified as the initial conditions at all grid points. Since the flow is supercritical at the upstream end, three conditions have to be specified for the upstream boundary condition. For this purpose, the following values were used: $u = 1.167$ m/s, $v = 0.0$ m/s, and $h = 0.06$ m. The flow is subcritical at the downstream end. Therefore, only one condition, $u = 0.222$ m/s, was imposed as the downstream boundary, and the flow depth was determined from the positive characteristic equation. The length of the channel downstream of the transition was long enough so that the transverse velocity at the downstream end could be assumed as zero. A grid with $\triangle \xi = 0.15$ m and $\triangle \eta = 0.0476$ m was used. Computations were performed with a Courant numbers = 0.90 and $\kappa = 0.003$. Because of the presence of bottom slope and friction, the numerical oscillations were not significant. Therefore, a small amount of artificial viscosity was sufficient to obtain satisfactory results. Computations were done up to $t = 7s$ when the solution converged to a steady state. The computed water surface profile along the channel centerline is plotted in Fig. 8-11. Since the flow is supercritical at the inflow section and subcritical at the outflow section, a hydraulic

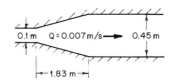

(a) Expansion

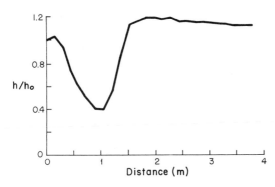

(b) Variation of water depth with distance

Figure 8-11 Hydraulic jump in gradual expansion.

jump is formed, as shown. This illustrates the ability of the model to handle mixed supercritical and subcritical flows.

8-5 Simulation of a Hydraulic Jump

To determine the jump location in a channel, Chow (1959) computed the water surface profiles for supercritical flow starting from the upstream end and the subcritical flow starting from the downstream end. The jump is formed at a location where the specific forces on both sides of the jump are equal. McCorquodale and Khalifa (1983) used the strip-integral method to compute the jump length, water surface profile, and pressures at the bed. To solve the St. Venant equations numerically, Abbott et al. (1969) used a finite-difference method and Katopodes (1984) used the finite-element method. In these simulations, the computations were continued until a steady state was reached. The location of the hydraulic jump is automatically computed as part of the solution. The water surface in rapidly varied flows, however, has steep gradients, and the assumption of hydrostatic pressure distribution may not be valid (Basco 1983). If we include the additional terms in the gradually varied flow equations to allow for nonhydrostatic pressure distribution, the resulting equations are referred to as the *Boussinesq equations.*

In this section, these equations are solved to compute the formation of hydraulic jump in a rectangular channel. The inclusion of initial and boundary conditions is discussed, and the importance of the Boussinesq terms is investigated.

Governing Equations

The Boussinesq equations* for one-dimensional flow in vector form may be written as

$$\frac{\partial \mathbf{U}}{\partial t} + \frac{\partial \mathbf{E}}{\partial x} = \mathbf{S} \tag{8-32}$$

in which

$$\mathbf{U} = \begin{bmatrix} h \\ uh \end{bmatrix}; \qquad \mathbf{E} = \begin{bmatrix} uh \\ u^2 h + \frac{1}{2} g h^2 - E \end{bmatrix}; \qquad \mathbf{S} = \begin{bmatrix} 0 \\ gh(S_o - S_f) \end{bmatrix}$$

and

$$E = \frac{1}{3} h^3 \left[\frac{\partial^2 u}{\partial x \, \partial t} + u \frac{\partial^2 u}{\partial x^2} - \left(\frac{\partial u}{\partial x} \right)^2 \right] \tag{8-33}$$

where E is called the Boussinesq term. It is introduced by the second-order term of pressure distribution along the water depth. It is clear that Eqs. 8-32 are reduced to the St. Venant equations if the Boussinesq term, E, is omitted from these equations.

Numerical Solution

The first- and second-order numerical schemes yield satisfactory results for the solution of St. Venant equations. However, the Boussinesq equations describing rapidly varied flow have third-order terms; and considerable effort must be expended to reduce the truncation errors while approximating these terms by finite differences (Abbott 1979). Therefore, it is necessary to employ third- or higher-order accurate methods to solve these equations numerically. For this reason, the two-four scheme developed by Gottlieb and Turkel (1976) is used here to solve these equations at the interior computational nodes.

The following finite-difference approximations are used in the two-four scheme.

Predictor

$$\mathbf{U}_i^* = \mathbf{U}_i^k + \frac{1}{6} \frac{\Delta t}{\Delta x} \left[\mathbf{E}_{i+2}^k - 8\mathbf{E}_{i+1}^k + 7\mathbf{E}_i^k \right] + \Delta t \mathbf{S}_i^k \tag{8-34}$$

Corrector

$$\mathbf{U}_i^{**} = \frac{1}{2} \left(\mathbf{U}_i^* + \mathbf{U}_i^k \right) + \frac{1}{12} \frac{\Delta t}{\Delta x} \left[-7\mathbf{E}_i^* + 8\mathbf{E}_{i-1}^* - \mathbf{E}_{i-2}^* \right] + \frac{1}{2} \Delta t \mathbf{S}_i^* \tag{8-35}$$

*For the derivation of these equations and the simplifying assumptions on which they are based, see Chapter 12.

The term $\partial^2 u / \partial x^2$ is approximated by using a three-point central finite-difference approximation in both the predictor and corrector parts. To approximate the term $(\partial u / \partial x)^2$, a forward finite-difference approximation in the predictor part and a backward finite-difference approximation in the corrector part are used.

To dampen the high-frequency oscillations near steep gradients, artificial viscosity (Jameson et al. 1981) is introduced as follows. A parameter ν_i is first computed using the computed flow depths at the $k + 1$ time level:

$$\nu_i = \frac{|h_{i+1} - 2h_i + h_{i-1}|}{|h_{i+1}| + 2|h_i| + |h_{i-1}|}$$

$$\epsilon_{i+1/2} = \kappa \max (\nu_{i+1}, \nu_i) \qquad (8\text{-}36)$$

in which κ is used to regulate the amount of dissipation. The computed variables u and h are then modified as

$$f_i^{k+1} = f_i^{k+1} + \epsilon_{i+1/2} \left(f_{i+1}^{k+1} - f_i^{k+1} \right) - \epsilon_{i-1/2} (f_i^{k+1} - f_{i-1}^{k+1}) \qquad (8\text{-}37)$$

in which f refers to both u and h; this equation should be viewed as a FORTRAN statement.

Initial and boundary conditions.   For the initial conditions, the flow at time $t = 0$ is assumed to be supercritical in the entire channel. By starting with the specified flow depth and velocity at section 1 (Fig. 8-12), the initial steady-state flow depth and flow velocity at all computational nodes are determined by numerically integrating the equation describing the gradually varied flow (Eq. 5-5). Since the computations are continued until steady conditions are reached, it is sufficient to specify only the approximate values of the initial flow depths and velocities.

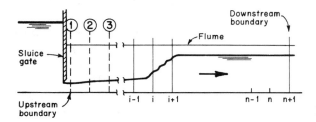

Figure 8-12 Definition sketch.

The flow conditions at the boundaries are computed as follows. At the *upstream boundary,* the flow depth, h, and velocity, u, are specified equal to their initial values, and they remain unchanged during the computations. At the downstream boundary, a constant flow depth is specified and the flow velocity is calculated from the characteristic form of Eq. 8-32 using a forward finite-difference approximation, i.e., the velocity, u_{i+1}^{k+1} at the unknown time level $k + 1$ is determined from the following equation (Chaudhry 1987):

$$u_{i+1}^{k+1} = u_i^k - \left(\frac{g}{c}\right)_i^k \left(h_{i+1}^{k+1} - h_i^k\right) + g\Delta t\left(S_o - S_f\right)_i \tag{8-38}$$

in which $c = \sqrt{gh}$ = celerity of a small gravity wave.

Stability conditions. The two-four scheme is stable if the following CFL condition is satisfied at each grid point:

$$\Delta t = C_n \frac{\Delta x}{|u| + \sqrt{gh}} \tag{8-39}$$

In this equation, C_n is the desired Courant number, which must be less than or equal to $\frac{2}{3}$ for the two-four scheme (Gottlieb and Turkel 1976).

Computational Procedure

The channel is divided into a number of equal-length reaches. Because the approximation of a second-order partial derivative requires values at the two neighboring nodes, it is not possible to calculate the variables at the computational nodes near the boundaries. Therefore, the flow equations at these nodes are first solved by neglecting the Boussinesq terms and by using the second-order MacCormack scheme for their solution. This should not significantly affect the overall accuracy of the solution in the region of interest, since the boundary nodes are located away from the jump location.

In the computations, the size of time step was restricted by the Courant stability condition and the spatial grid size. The Courant number was set equal to 0.65, since best results are obtained when it is approximately equal to $\frac{2}{3}$. To smooth high-frequency oscillations near the jump, the dissipation coefficient, κ, in the Jameson's formula (Eq. 8-37) was determined by a trial-and-error procedure. Trials values ranging from 0.01 to 0.05 indicated that a value of 0.03 provided the best results.

Several runs with different values of the upstream flow depth, velocity, Froude number, and downstream flow depth were made. The Manning n for the flume was determined by trial and error so that the computed water-surface profile matched with the measured water levels in the flume during the initial steady supercritical flow. The n values varied from 0.008 to 0.011, depending upon the flow depth, since the bottom of the flume is made up of metal and the sides are made up of glass. The initial steady-state depth and velocity at every computational node were first computed by assuming the flow to be supercritical throughout the flume. Then, the unsteady computations were started by increasing the downstream depth to the value measured during the experiment. The computations were continued until they converged to the final steady state for the specified end conditions.

Results

The size of the spatial grid, Δx, was varied from 0.15 m to 0.6 m. Simulations were also done by the second-order MacCormack scheme, neglecting the Boussinesq term.

No definite trend could be established that would indicate that reducing the value of Δx in the second-order method gave results tending toward those obtained by the fourth-order method. Thus, the computation of various nonlinear terms of the governing equations appears to play a more important role than the truncation errors introduced by the numerical scheme.

Figure 8-13 shows the water-surface profiles at different times following an increase in the downstream depth at time $t = 0$. The jump travels from the downstream end toward the upstream end and then moves back and forth until it stabilizes in one location.

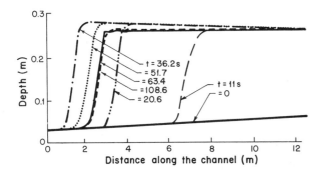

Figure 8-13 Water-surface profile at different times for $\mathbf{F}_r = 7$.

When the numerical solution converges to a steady state, the Boussinesq term is found to be small relative to the other spatial derivative terms in the vicinity of the hydraulic jump, and it is almost negligible in the regions away from the jump. The Boussinesq term at locations away from the jump virtually becomes zero, although the values of the other terms remain approximately the same. This is to be expected, since the flow surface in regions away from the jump is more or less smooth. The pressure distribution in such flows is hydrostatic and thus the Boussinesq term is negligible.

The computed results are compared with the measured results in Fig. 8-14. To conserve space, only the comparisons for $\mathbf{F}_r = 2.3$ and 7 are included herein; for similar comparisons for other values of $\mathbf{F}_r$, see Gharangik and Chaudhry (1991). The comparison of the computed and measured results generally shows that the fourth-order-accurate numerical models with or without Boussinesq terms give approximately the same results for all Froude numbers tested.

8-6 Summary

In this chapter we discussed the numerical modeling of rapidly varied flows. Three different formulations were presented. In the first, a steady form of the shallow-water equations was numerically integrated. This is valid only for supercritical flows. In the second formulation, the unsteady, gradually varied equations were solved with time as the iteration parameter. Since the pressure distribution in these two formulations is assumed hydrostatic, the computed results agree satisfactorily with the analytical

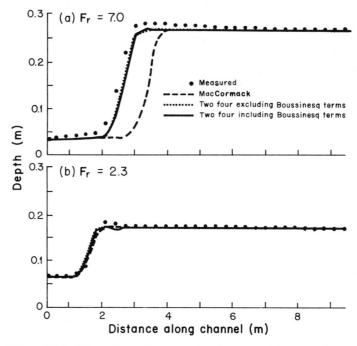

Figure 8-14 Comparison of computed and measured jump profiles.

solutions and with the experimental measurements in the regions where this assumption is valid. In the third formulation, the Boussinesq equations were solved by an explicit scheme that is second-order accurate in time and fourth-order accurate in space. The simulation of the formation of hydraulic jump in a rectangular channel was used for illustration purposes and for comparing the computed results with those measured in a laboratory flume.

REFERENCES

Abbett, M. 1971. Boundary conditions in computational procedures for inviscid, supersonic steady flow field calculations, Aerotherm Report 71-41.

Abbott, M. B. 1979. *Computational hydraulics; Elements of the theory of free surface flow,* Pitman, London.

Abbott, M. B.; Marshall, G.; and Rodenhuis, G. S. 1969. Amplitude-dissipative and phase-dissipative scheme for hydraulic jump simulation, *13th Congress IAHR* (Tokyo), 1 (August):313–29.

Anderson, D. A.; Tannehill, J. C.; and Pletcher, R. H. 1984. *Computational fluid mechanics and heat transfer,* McGraw-Hill, New York, NY.

Bagge, G. and Herbich, J. B. 1967. Transitions in Supercritical Open-Channel Flow, *Jour. Hydraulic Division,* Amer. Soc. Civil Engrs., 93, no. 5:23–41.

Basco, D. R.   1983. Introduction to rapidly varied unsteady, free-surface flow computation, *Water Resources Invention Report,* U.S. Geological Survey, Report No. 83-4284.

Bhallamudi, S. M., and Chaudhry, M. H.   1992. Computation of flows in open-channel transitions, *Jour. Hydraulic Research,* Inter. Assoc. Hyd. Research, no. 1:77-93.

Chaudhry, M. H.   1987. *Applied hydraulic transients,* 2nd ed., Van Nostrand Reinhold, New York, NY.

Chow, V. T.   1959. *Open channel hydraulics,* McGraw-Hill, New York, NY.

Cunge, J.   1975. "Rapidly varying flow in power and pumping canals." In *Unsteady Flow in Open Channels,* (Edited by K. Mahmood and V. Yevjevich) 539-86, Water Resources Publications. Littleton, CO.

Dakshinamoorthy, S.   1977. High velocity flow through expansions, *17th Congress, International Assoc. for Hydraulic Research* (Baden-Baden) 2:373–81.

Demuren, A. O.   1979. Prediction of steady surface-layer flows, Ph.D. diss., University of London.

Ellis, J. and Pender G.   1982. Chute spillway design calculations, *Proc.,* Inst. Civil Engrs., 73, Pt. 2 (June):299–312.

Engelund, F. and Munch-Petersen, J.   1953. Steady flow in contracted and expanded rectangular channels, *La Houille Blanche,* 8, no. 4 (August-September):464–74.

Fennema, R. J. and Chaudhry, M. H.   1986. Explicit numerical schemes for unsteady free–surface flows with shocks, *Water Resources Research,* 22, no. 13:1923–30.

Fennema, R. J. and Chaudhry, M. H.   1990. Numerical solution of two-dimensional transient free-surface flows, *Jour. of Hydraulic Engineering,* Amer. Soc. Civ. Engr., 116, no. 8 (August):1013–34.

Garcia, R., and Kahawita, R. A.   1986. Numerical solution of the St. Venant equations with the MacCormack finite-difference scheme, *International Jour. for Numerical Methods in Fluids,* 6:259–74.

Gharangik, A. and Chaudhry, M. H.   1991. Numerical simulation of hydraulic jump, *Jour. Hydraulic Engineering,* Amer. Soc. Civ. Engrs., 117, no. 9:1195–1211.

Gottlieb, D., and Turkel, E.   1976. Dissipative two-four methods for time-dependent problems, *Mathematics of Computation,* 30, no. 136 (October):703–23.

Henderson, F. M.   1966. *Open channel flow,* MacMillan, New York, NY.

Herbich, J. B. and Walsh, P.   1972. Supercritical flow in rectangular expansions, *Jour. Hydraulics Division,* Amer. Soc. Civil Engrs., 98, no. 9 (September):1691–1700.

Ippen, A. T. et al.   1951. *Proceedings of a symposium on high-velocity flow in open channels,* Trans., Amer. Soc. Civil Engrs., 116:265–400.

Jameson, A.; Schmidt, W.; and Turkel, E.   1981. Numerical solutions of the Euler equations by finite volume methods using Runge-Kutta time-stepping schemes, *AIAA 14th Fluid and Plasma Dynamics Conference,* Palo Alto, California, AIAA-81-1259.

Jimenez, O. F. and Chaudhry, M. H.   1988. Computation of supercritical free-surface flows, *Jour. of Hydraulic Engineering,* Amer. Soc. Civ. Engrs., 114, no. 4 (April):377–95.

Katopodes, N. D.   1984. A dissipative Galerkin scheme for open-channel flow, *Jour. Hydraulic Engineering,* Amer. Soc. Civil Engrs., 110, no. 4 (April):450–66.

Knapp, R. T.   1951. Design of channel curves for supercritical flow, *Trans.,* Amer. Soc. Civil Engrs., 116:296–325.

Kutler, P. 1975. "Computation of three-dimensional, inviscid supersonic flows." In *Progress in numerical fluid dynamics*, Lecture Notes in Physics No. 41, Springer-Verlag. New York:287–374.

Liggett, J. A. and Vasudev, S. U. 1965. Slope and friction effects in two-dimensional, high-speed flow, *11th Int. Congress IAHR*, Leningrad, vol. 1, paper 1.25.

MacCormack, R. W. 1969. The effect of viscosity in hypervelocity impact cratering, *Amer. Inst. Aeronautics and Astronautics,* Paper 69-354, Cincinnati, Ohio.

McCorquodale, J. A. and Khalifa, A. 1983. Internal flow in hydraulic jumps, *Jour. of Hydraulic Engineering,* 109, no. 5 (May):684–701.

McCowan, A. D. 1987. The range of application of Boussinesq-type numerical short wave models, *XXII Congress*, International Assoc. for Hydraulic Research, pp. 378–384.

Pandolfi, M. 1975. Numerical Experiments on Free Surface Water Motion with Bores, *Proc. 4th Int. Conf. on Numerical Methods in Fluid Dynamics*, Lecture Notes in Physics No. 35, Springer-Verlag, New York, NY:304–312.

Roache, P. J. 1972. *Computational fluid dynamics*, Hermosa Publishers, Albuquerque, NM.

Stoker, J. J. 1957. *Water waves,* Wiley Interscience Publishers, New York, NY.

Verboom, G. K.; Stelling, G. S.; and Officier, M. J. 1982. "Boundary conditions for the shallow water equations." In *Engineering applications of computational hydraulics,* edited by M. B. Abbott and J. A. Cunge Vol. 1, Pitman, Boston, MA.

Villegas, F. 1976. Design of the Punchiná spillway, *Water Power & Dam Construction* (November):32–34.

9 Channel Design

The California aqueduct: The canal is 715 km long, 33.5 m wide at bottom, 10.67 m deep and the maximum discharge is 104 m³/s. (Courtesy of the Metropolitan Water District of Southern California).

9-1 Introduction

The design of a channel involves the selection of channel alignment, shape, size, and bottom slope and deciding whether the channel should be lined to prevent the erosion of channel sides and bottom and to reduce seepage. Since a lined channel usually offers less resistance to flow than an unlined channel, the channel size required to convey a specified flow rate at a selected slope is smaller for a lined channel than if no lining were provided. Therefore, in some cases, a lined channel may be more economical than an unlined channel in spite of the cost of the lining.

Procedures are not presently available for selecting optimum channel parameters directly. Each site has unique features that require special considerations. Typically, the design of a channel is done by trial and error. Channel parameters are selected and an analysis is done to verify that the operational requirements are met with these parameters. A number of alternatives are considered, and their costs are compared. Then, the most economical alternative that gives satisfactory performance is selected. In this process, it is necessary to include maintenance costs while comparing different alternatives. Similarly, the costs of energy required if pumping is involved and the amount of revenues produced by hydropower generation must be included in the overall economic analysis.

The channel design may be divided into two categories, depending upon whether the channel boundary is erodible or non-erodible. For erodible channels, flow velocities are kept low so that the channel bottom and sides are not eroded. The minimum flow velocity in flows carrying sediment should be such that the material being transported is not deposited in the channel.

In this chapter, we first consider the design of rigid-boundary channels and then the design of erodible channels.

9-2 Rigid-Boundary Channels

In the design of a rigid-boundary channel, the channel cross section and size are selected such that the required discharge is carried through the channel for the available head with a suitable amount of freeboard. The *freeboard* is defined as the vertical distance between the design water surface and the top of the channel banks. Freeboard is provided to allow for unaccounted factors in design, uncertainty in the selected values of different parameters, disturbances on the water surface, etc.

The *channel alignment* is selected so that the channel length is as short as possible and at the same time meets other site restrictions and requirements, such as accessibility, right of way, and balancing of cut and fill amounts. The bottom slope is usually dictated by the site topography, whereas the selection of channel shape and dimensions take into consideration the amount of flow to be carried, the ease and economy of construction, and the hydraulic efficiency of the cross section. A triangular channel is used for small rates of discharge, and a trapezoidal cross section is used for large flows. For structural reasons, channels excavated through mountains or built underground usually have a circular or horseshoe shape. Normally, the Froude number is kept low (approximately up to 0.3) so that the flow surface does not become rough, especially downstream of obstructions and bends. Similarly, the flow velocity is selected such that the lining is not eroded and any sediment carried in flow is not deposited.

Normally, the channels are designed based on the assumption of uniform flow, although in some situations gradually varied flow calculations may be needed to assess the suitability of selected channel size for extreme events.

The maximum permissible velocity is not usually a consideration in the design of rigid-boundary channels if the flow does not carry large amounts of sediments. However, if the sediment load is large, then the flow velocities should not be too high to prevent erosion of the channel. The minimum flow velocity should be such that the sediment is not deposited, aquatic growth is inhibited, and sulfide formation does not occur. The lower limit for the minimum velocity depends upon the particle size and the specific gravity of sediments carried in flow. The channel size does not have significant affect on such a lower limit. Generally, the minimum velocity in a channel is about 0.6 to 0.9 m/s. Flow velocities of 12 m/s have been found to be acceptable in concrete channels for low sediment concentrations. The channel invert may be eroded at much lower velocities than this value if the flow carries sand or other gritty material.

The channel side slopes depend upon the type of soil in which the channel is constructed. Nearly vertical channel sides may be used in rocks and stiff clays, whereas side slopes of 1 vertical to 3 horizontal may be needed in sandy soils. For lined channels, U. S. Bureau of Reclamation recommends a value of 1 vertical to 1.5 horizontal.

To allow for waves and water-surface disturbances, a suitable amount of freeboard should be provided. It is not possible to specify a general formula for determining the freeboard under general conditions. As a rough estimate, the following formula, suggested by the U. S. Bureau of Reclamation, may be used:

$$F_b = \sqrt{ky} \qquad (9\text{-}1)$$

in which F_b = freeboard, in m, y = flow depth, in m, and k = coefficient varying from 0.8 for a flow capacity of about 0.5 m^3/s to 1.4 for a flow capacity exceeding 85 m^3/s. Table 9-1 lists suggested freeboards for canals based on the recommendations of the

Table 9-1 Suggested freeboard

Discharge (m^3/s)	< 1.5	1.5 to 85	> 85
Freeboard (m)	0.50	0.75	0.90

Central Board of Irrigation and Power, India (Raju 1983). These values are somewhat less than those given by Eq. 9-1.

Steps to design a rigid-boundary channel are as follows:

1. Select a value of roughness coefficient n for the flow surface and select bottom slope S_o based on topography and other considerations listed in the previous paragraphs.

2. Compute section factor from $AR^{2/3} = nQ/(C_o S_o^{1/2})$, in which $A =$ flow area, $R =$ hydraulic radius, $Q =$ design discharge, and $C_o = 1$ for SI units and $C_o = 1.49$ in customary units.

3. Determine the channel dimensions and the flow depth for which $AR^{2/3}$ is equal to the value determined in step 2. For example, for a trapezoidal section, select a value for the side slope s and compute several different ratios of bottom width B_o and flow depth y for which $AR^{2/3}$ is equal to that determined in step 2. Select a ratio B_o/y that gives a cross section near to the best hydraulic section, as discussed in the following section.

4. Check that the minimum velocity is not less than that required to carry the sediment to prevent silting.

5. Add a suitable amount of freeboard.

The following example illustrates this procedure.

Example 9-1

Design a trapezoidal channel to carry a discharge of 10 m³/s. The channel will be excavated through rock by blasting. The topography in the area is such that a bottom slope of 1 in 4000 will be suitable.

Given:
$Q = 10$ m³/s
Flow surface is blasted rock.
$S_o = 0.00025$

Determine:
$B_o = ?$
Total depth $= ?$

Solution
For the blasted rock surface, $n = 0.030$ and the side slopes may be almost vertical. Let us select a value for the side slope s as 1 horizontal to 4 vertical. The substitution of these values into the Manning equation yields

$$AR^{2/3} = \frac{nQ}{C_o S_o^{0.5}}$$
$$= \frac{0.030 \times 10}{(0.00025)^{1/2}}$$
$$= 18.97$$

Since the channel section is almost rectangular, let us select $B_o = 2y$. Then, $A = (B_o + \frac{1}{4}y)y = 2.25y^2$; $P = B_o + \frac{1}{2}\sqrt{17}y = 4.06y$; $R = (2.25y^2)/(4.06y) = 0.55y$. Hence,

$$AR^{\frac{2}{3}} = (2.25y^2)(0.55y)^{2/3}$$
$$= 1.518y^{2.67} = 18.97$$

Solving this equation for y, we get $y = 2.57$ m. Then, $B_o = 2 \times 2.57 = 5.14$ m. For ease of construction, let us use $B_o = 5$ m. Then, the corresponding value of y for which $AR^{\frac{2}{3}} = 18.97$ is determined by trial and error as 2.64 m.

$$\text{Freeboard} = \sqrt{0.8 \times 2.64} = 1.45\,\text{m}$$

As compared to this value, a freeboard of 0.75 m selected from Table 9-1 appears to be more appropriate. Then

$$\text{Total depth} = 2.64 + 0.75 = 3.39 \simeq 3.4\,\text{m}$$

The flow area for a flow depth of 2.64 m is 14.94 m^2. Therefore, the flow velocity = $10/14.94 = 0.67$ m/s. This is close to the minimum allowable flow velocity; thus, a bottom width of 5 m and a cross section depth of 3.4 m are satisfactory.

9-3 Most Efficient Hydraulic Section

A section that gives maximum discharge, Q, for a specified flow area, A, is called the *most efficient hydraulic section* or *best hydraulic section*. Since Q is proportional to $AR^{2/3}$ for a given channel (i.e., n and S_o are specifed) and $R = A/P$, we can say that the most efficient hydraulic section is the one that yields the minimum wetted perimeter, P for a given A.

Theoretically speaking, the most efficient hydraulic section yields the most economical channel. However, it must be kept in mind that the preceding formulation is oversimplified. For example, we did not take into consideration the possibility of scour and erosion, which may impose restrictions on the maximum flow velocity. And, for channel excavation we have to take into account the amount of overburden, viability of changing the bottom slope to suit the existing topographical conditions for minimizing the amount of excavation, ease of access, transportation of the excavated material to the disposal site, and the viability of the matching of the cut and fill volumes, etc. In addition, for a lined channel, the cost of lining as compared to the per unit cost of excavation has to be taken into consideration for an overall economical design.

The proportions for common cross sections so that they are the most efficient are derived in the following paragraphs.

Rectangular Section

For a rectangular channel, $A = By$ and $P = B + 2y$. For the best hydraulic section, we want to determine the ratio of B and y such that P is minimum for constant A. Now,

P can be written in terms of A and y as

$$P = \frac{A}{y} + 2y \tag{9-2}$$

Differentiating this expression for P with respect to y and then equating the resulting expression to zero, we obtain

$$\frac{dP}{dy} = \frac{-A}{y^2} + 2 = 0 \tag{9-3}$$

or

$$\frac{A}{y^2} = 2 \tag{9-4}$$

But $A = By$. Therefore,

$$\frac{By}{y^2} = 2 \tag{9-5}$$

or

$$y = \frac{1}{2}B \tag{9-6}$$

Thus, a rectangular cross section is the most efficient when the flow depth is one-half the channel width.

Triangular Section

Let us consider a symmetrical triangular section having side slope s horizontal to 1 vertical. Then

$$\begin{aligned} A &= sy^2 \\ P &= 2(\sqrt{1+s^2})y \end{aligned} \tag{9-7}$$

Substituting for y in terms of s and A, we obtain

$$P = 2\sqrt{1+s^2}\left(\frac{A}{s}\right)^{1/2} \tag{9-8}$$

Taking the square of both sides, this equation becomes

$$P^2 = 4\left(s + \frac{1}{s}\right)A \tag{9-9}$$

As we discussed previously, P should be minimum for a given A for the most efficient hydraulic section. For this condition, $dP/ds = 0$. By differentiating Eq. 9-9, we obtain

$$2P\frac{dP}{ds} = 4\left(1 - \frac{1}{s^2}\right)A = 0 \tag{9-10}$$

Hence, it follows from Eq. 9-10 that $s = 1$. Thus, a triangular section with the sides inclined at $45°$ is the most efficient triangular section.

Trapezoidal section

For a trapezoidal section (Fig. 9-1),

$$P = B_o + 2\sqrt{1 + s^2}\, y$$
$$A = (B_o + sy)y \tag{9-11}$$

The elimination of B_o from these two equations and the simplification of the resulting equation yield

$$P = \frac{A}{y} + y(2\sqrt{1 + s^2} - s) \tag{9-12}$$

If both A and y are constants and s is variable, then the condition for the most efficient section is $dP/ds = 0$. Hence, differentiating Eq. 9-12 with respect to s, equating the resulting equation to zero, and simplifying, we obtain

$$s = \frac{1}{\sqrt{3}}$$

or

$$\theta = 60^o \tag{9-13}$$

Now, let us consider A and s to be constants and y to be variable. Then, the condition for the most efficient section is $dP/dy = 0$. Differentiating Eq. 9-12 with respect to y, equating the resulting equation to zero, and simplifying, we obtain

$$B_o = 2(\sqrt{s^2 + 1} - s)y \tag{9-14}$$

Based on this equation, the top water-surface width is

$$B = B_o + 2sy = 2\sqrt{s^2 + 1}\, y \tag{9-15}$$

Thus, the top water-surface width is twice the length of the sloping side. In other words, these derivations show that the most efficient section is one-half of a hexagon.

Referring to triangle OCD of Fig. 9-1 and substituting expression for B from Eq. 9-15, we obtain

$$
\begin{aligned}
OC &= OD \sin \theta \\
&= \frac{1}{2} B \sin \theta \\
&= y
\end{aligned}
\tag{9-16}
$$

Thus a circle with radius y and with center at O is tangential to the channel bottom and sides.

9-4 Erodible Channels

If the channel bottom or sides are erodible, then the design requires that the channel size and bottom slope are selected so that the channel is not eroded. Two methods have

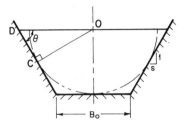

Figure 9-1 Trapezoidal section.

been used for the design of these channels: the permissible velocity method and the tractive force method. Both methods are discussed in this section.

Permissible Velocity Method

In the *permissible velocity method*, the channel size is selected such that the mean flow velocity for the design discharge under uniform flow conditions is less than the permissible flow velocity. The *permissible velocity* is defined as the mean velocity at or below which the channel bottom and sides are not eroded. This velocity depends primarily upon the type of soil and the size of particles, even though it has been recognized that it should depend upon the flow depth as well as whether the channel is straight or not. This is because, for the same value of mean velocity, the flow velocity at the channel bottom is higher for low depths than that at large depths. Similarly, a curved alignment induces secondary currents. These currents produce higher flow velocities near the channel sides, which may cause erosion.

A trapezoidal channel section is usually used for erodible channels. To design these channels, first an appropriate value for the side slope is selected so that the sides are stable under all conditions. Table 9-2, complied from data given by Fortier and Scobey (1926), lists recommended slopes for different materials.

Table 9-2 Suggested side slopes

Material	Side slope (sH:1V)
Rock	Nearly vertical
Stiff clay	$\frac{1}{2}$:1 to 1:1
Firm soil	1:1
Loose sandy soil	2:1
Sandy loam	3:1

The maximum permissible velocities for different materials are presented in Table 9-3. The values listed in this table are for a straight channel having a flow depth of about 1 m. As a rough estimate, Lane (1955) suggest reducing these values by 5 percent for slightly sinuous channels, 13 percent for moderately sinuous channels, and 22 percent

Table 9-3 Recommended permissible velocities*

Material	V (m/s)
Fine sand	0.6
Coarse sand	1.2
Earth	
Sandy silt	0.6
Silt clay	1.1
Clay	1.8
Grass-lined earth (slopes less than 5 percent)	
Bermuda grass	
Sandy silt	1.8
Silt clay	2.4
Kentucky blue grass	
Sandy silt	1.5
Silt clay	2.1
Poor rock (usually sedimentary)	
Soft sandstone	2.4
Soft shale	1.1
Good rock (usually igneous or hard metamorphic)	6.1

*Taken from U.S. Army Corps of Engineers (1970).

for very sinuous channels. For other flow depths, these velocities may be multiplied by a correction factor, k, to determine the permissible flow velocity (Mehrotra 1983). For very wide channels, $k = y^{1/6}$.

The steps for the design of a channel using permissible-velocity method are as follows:

1. For the specified material, select values of Manning n (from Table 4-1), side slope s (from Table 9-2), and the permissible velocity, V (from Table 9-3).
2. Determine the required hydraulic radius, R, from Manning formula and the required flow area, A, from the continuity equation, $A = Q/V$.
3. Compute the wetted perimeter, $P = A/R$.
4. Determine the channel bottom width, B_o, and the flow depth, y, for which the flow area A is equal to that computed in step 2 and the wetted perimeter P is equal to that computed in step 3.
5. Add a suitable value for the freeboard.

The following example illustrates this procedure.

Example 9-2

Design a channel to carry a flow of 6.91 m³/s. The channel will be excavated through stiff clay at a channel bottom slope of 0.00318.

Given:

$Q = 6.91$ m^3/s
$S_o = 0.00318$
Channel material is clay.

Determine:

$B_o = ?$
Depth $= ?$

Solution:

For stiff clay, $n = 0.025$, suggested side slope, $s = 1:1$ (from Table 9-2), and the permissible flow velocity (from Table 9-3) is 1.8 m/s. Hence,

$$A = 6.91/1.8 = 3.83 \text{ m}^2$$

Substituting values for V, n, and S_o into the Manning equation and solving for R, we get $R = 0.713$ m. Hence,

$$P = \frac{3.83}{0.713} = 5.37 \text{ m}$$

Substitution into expressions for P and A and equating them to these computed values, we obtain

$$B_o + 2.83y = 5.37$$

$$(B_o + y)y = 3.83$$

Elimination of B_o from these two equations yields

$$1.83y^2 - 5.37y + 3.83 = 0$$

Solution of this equation gives $y = 1.22$ m. The freeboard from Eq. 9-1 may be calculated as $\sqrt{0.8 \times 1.22} = 0.99$ m, whereas the suggested value in Table 9-1 is 0.75 m. Let us select a freeboard of 0.75 m. Then, the depth of the section is $1.22 + 0.75 = 1.97$ m. Select a depth of 2.0 m and bottom width of 1.9 m.

Tractive-Force Method

As compared to the permissible velocity, the scour and erosion process may be viewed in a more rational fashion by considering the forces acting on a particle lying on the channel bottom or on the channel sides. The channel is eroded if the resultant of forces tending to move the particle is greater than the resultant of forces resisting the motion; otherwise, it is stable. This concept, referred to as the tractive-force approach, was introduced by du Boys in 1879 and restated by Lane in 1955 (Chow, 1959).

The force exerted by the flowing water on the channel bottom and sides is called *tractive force*, or *drag force*. This is the force due to shear stress. In uniform flow, this force is equal to the component of weight of water acting in the direction of flow.

Let us consider a channel with bottom slope S_o. The weight of water in a reach of length L of this channel is γAL, in which $A =$ flow area. Now, the component of the weight of water in the downstream direction is γALS_o, in which $\gamma =$ specific weight of water. In uniform flow, this component of the weight of water is equal to the tractive force that acts over the wetted perimeter, P. Then, the unit tractive force or shear stress, $\tau_o = \gamma A L S_o/(PL) = \gamma R S_o$, in which R = hydraulic radius. In very wide channels, $R \simeq y$. Hence, $\tau_o = \gamma y S_o$.

The distribution of unit tractive force or shear stress over the channel perimeter is not uniform. Although many attempts have been made to determine this distribution, they have not been conclusive. As an approximation for a trapezoidal channel (Lane 1955), τ_o at the channel bottom may be assumed to be equal to $\gamma y S_o$, and at the channel sides, to be equal to $0.76\gamma y S_o$.

The shear stress at which the channel material just moves from a stationary condition is called *critical stress*, τ_c. The critical stress is a function of the material size and the sediment concentration. In addition, the critical stress at the channel sides is less than that at a level surface because the component of the weight along the side slope tends to roll the material down the slope, thereby causing instability.

Let us consider a particle lying on the channel side, as shown in Fig. 9-2. Let the slope of the side be θ, a = effective area, W_s = submerged weight of the particle, ϕ = angle of repose of the particle, and τ_s = shear stress on the channel sides. Two forces tending to move the particle are the tractive force, $a\tau_s$, due to flowing water and the component of the weight of particle along the side slope, $W_s \sin \theta$. The resultant of these two forces is

$$R = \sqrt{W_s^2 \sin^2 \theta + a^2 \tau_s^2} \qquad (9\text{-}17)$$

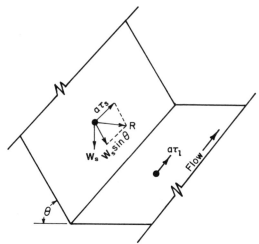

Figure 9-2 Forces acting on a particle.

The normal force, $W_s \cos \theta \tan \phi$, resists the particle motion. In this expression, ϕ = angle of repose of the side material. At the point of impending motion, the resultant of forces causing the motion is equal to the resultant of forces resisting the motion. Thus, for the impending motion,

$$W_s \cos \theta \tan \phi = \sqrt{W_s^2 \sin^2 \theta + a^2 \tau_s^2} \qquad (9\text{-}18)$$

It follows from this equation that

$$\tau_s = \frac{W_s}{a} \cos \theta \tan \phi \sqrt{1 - \frac{\tan^2 \theta}{\tan^2 \phi}} \tag{9-19}$$

For the impending motion of a particle on a level surface

$$W_s \tan \phi = a\tau_l \tag{9-20}$$

in which τ_l = shear stress at the impending motion of a particle on a level surface. Then

$$\tau_l = \frac{1}{a} W_s \tan \phi \tag{9-21}$$

It follows from Eqs. 9-19 and 9-21 that

$$K = \frac{\tau_s}{\tau_l} = \cos \theta \sqrt{1 - \frac{\tan^2 \theta}{\tan^2 \phi}} \tag{9-22}$$

which may be simplified as

$$K = \sqrt{1 - \frac{\sin^2 \theta}{\sin^2 \phi}} \tag{9-23}$$

This is the reduction factor for the critical stress on the channel sides.

The effect of the angle of repose should be considered only for coarse, noncohesive materials. For cohesive and fine, noncohesive materials, the gravity component causing the particle to roll down the side slope is much smaller than the cohesive forces and may thus be neglected. Figure 9-3 shows the curves prepared by the U. S. Bureau of Reclamation for the angle of repose for noncohesive material larger than 5 mm in diameter. The diameter in this figure is the diameter of a particle such that 25 percent of material by weight is larger than this diameter.

The critical shear stress for noncohesive material is shown in Fig. 9-4 and that for cohesive material is shown in Fig. 9-5. These values are for straight channels. Lane recommended reducing these values by 10 percent for slightly sinuous channels, 25 percent for moderately sinuous channels, and 40 percent for very sinuous channels.

The procedure for designing a channel by the tractive force approach involves the selection of a cross section such that the unit tractive force acting on the channel sides is equal to the permissible shear stress for the channel material. Then, we check that the unit tractive force on the channel bottom is less than the permissible stress.

The design steps are as follows:

1. For the channel material, select a side slope from Table 9-2, the angle of repose from Fig. 9-3, and the critical shear stress from Fig. 9-4 for noncohesive materials and from Fig. 9-5 for cohesive materials. Determine the permissible shear stress by taking into consideration whether the channel is straight or not.

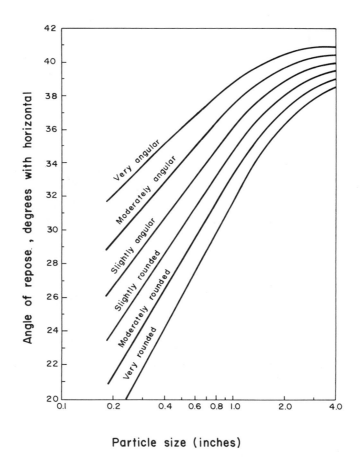

Particle size (inches)

Figure 9-3 Angles of repose for noncohesive material (After U.S. Bureau of Reclamation).

2. For the noncohesive material, compute the reduction factor, K, from Eq. 9-23. Then determine the permissible shear stress for the sides by multiplying by K the permissible stress determined in step 1.

3. Equate the permissible stress for the sides determined in step 2 to $0.76\gamma y S_o$ and determine y from the resulting equation.

4. For y determined in step 3 and for the selected values of the Manning n and the side slope, s, compute the bottom width, B_o, from the Manning equation for the design discharge.

5. Now, check that the shear stress on the bottom, $\gamma y S_o$, is less than the permissible shear stress of step 1.

The following example illustrates this procedure.

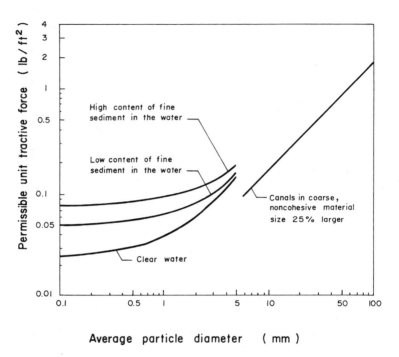

Figure 9-4 Permissible shear stress for noncohesive materials (After U.S. Bureau of Reclamation).

Example 9-3

Design a straight trapezoidal channel for a design discharge of 10 m³/s. The bottom slope is 0.00025 and the channel is excavated through fine gravel having particle size of 8 mm. Assume the particles are moderately rounded and the water carries fine sediment at low concentrations.

Given:

$$Q = 10 \text{ m}^3/s$$
$$S_o = 0.00025$$
Material: fine gravel, moderately rounded
Particle size = 8 mm

Determine:

$$B_o = ?$$
$$y = ?$$

Solution:

For fine gravel, $n = 0.024$ and $s = 3H : 1V$. Therefore, $\theta = \tan^{-1} \frac{1}{3} = 18.4°$.
From Fig. 9-3, $\phi = 24°$. Hence,

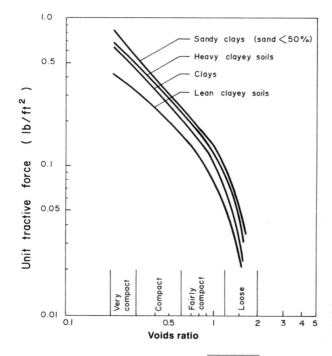

Figure 9-5 Permissible shear stress for cohesive materials (After Chow, 1959).

$$K = \sqrt{1 - \frac{\sin^2 \theta}{\sin^2 \phi}} = \sqrt{1 - \frac{0.1}{0.16}} = 0.63$$

From Fig. 9-4, the critical shear stress = 0.15 lb/ft² = 7.18 N/m². Since the channel is straight, we do not have to make a correction for the alignment. The permissible shear stress for the channel side is 7.18× 0.63 = 4.52 N/m².

Now, the unit tractive force on the side = $0.76\gamma y S_o = 0.76 \times 999 \times 9.81 y \times 0.00025$ = $1.862y$.

By equating the unit tractive force to the permissible stress, we obtain

$$1.862y = 4.52$$

or

$$y = 2.43 \text{ m}$$

The channel bottom width, B_o, needed to carry 10 m³/s may be determined from the Manning equation

$$\frac{1}{n}(B_o + sy)y \left(\frac{(B_o + sy)y}{B_o + 2\sqrt{1 + s^2}y} \right)^{2/3} \sqrt{S_o} = Q$$

By substituting $n = 0.024$, $s = 3$, $y = 2.43$, $S_o = 0.00025$, and $Q = 10$ m^3/s and solving for B_o, we obtain

$$B_o = 8.24 \text{ m}$$

For a selected freeboard of 0.75m, the depth of section $= 2.43 + 0.75 = 3.2$ m. For ease of construction, select a bottom width, $B_o = 8.25$ m.

9-5 Alluvial Channels

An alluvial channel is defined as a channel in which the flow transports sediment having the same characteristics as that of the material in the channel bottom. Such a channel is said to be *stable* if the sediment inflow into a channel reach is equal to the sediment outflow. Thus, the channel cross section and the bottom slope do not change due to erosion or deposition.

Two approaches have been used for the design of stable alluvial channels: (1) tractive force method and (2) regime theory. The tractive force approach is more rational, since it utilizes the laws governing sediment transport and resistance to flow. The regime theory is purely empirical in nature and was developed based on observations on a number of irrigation canals in the Indo-Pakistan subcontinent. Since the sediment concentration in these canals is usually less than 500 ppm by weight, the regime theory should be assumed to be applicable to channels carrying similar concentration of sediment load.

The tractive force approach was discussed in the previous section; a brief outline of the regime theory follows.

Regime Theory

Lacey (1930) defined a *regime* channel as a channel carrying a constant discharge under uniform flow in an unlimited incoherent alluvium having the same characteristics as that transported without changing the bottom slope, shape, or size of the cross section over a period of time.

Two types of equations have been extensively used for design in India and Pakistan. These are the Kennedy and the Lacey equations. The main limitation of the Kennedy equations is that they do not specify a stable width, thereby making an infinite number of depth-to-width ratios possible. However, experience shows that stability is possible only if the width does not vary over a wide range: Sides are scoured in a very narrow channel, whereas deposition occurs in a very wide channel. To take this factor into account, Lindley introduced a relationship between the nonsilting, nonscouring velocity and the bottom width. Lacey developed the following equations based on the analysis of a large amount of data collected on several irrigation canals in the Indian subcontinent:

$$P = 4.75\sqrt{Q}$$

$$f_s = 1.76 d^{1/2}$$

$$R = 0.47 \left(\frac{Q}{f_s}\right)^{1/3} \tag{9-24}$$

$$S = 3 \times 10^{-4} f_s^{5/3} Q^{1/6}$$

in which P = wetted perimeter in meters; R = hydraulic radius, in meters; Q = flow, in m^3/s; d = diameter of sediment, in mm; and f_s = silt factor, which takes into consideration the effect of sediment size on the channel dimensions. The particle size and silt factors for various materials are listed in Table 9-4.

Table 9-4 Particle size and silt factors for various materials*

Material	Size (mm)	Silt factor
Small boulders, cobbles, shingles	64–256	6.12 to 9.75
Coarse gravel	8–64	4.68
Fine gravel	4–8	2.0
Coarse sand	0.5–2.0	1.44–1.56
Medium sand	0.25–0.5	1.31
Fine sand	0.06–0.25	1.1–1.3
Silt (colloidal)		1.0
Fine silt (colloidal)		0.4–0.9

*Taken from Gupta (1989).

Combining these equations, the following resistance equation similar to the Manning equation is obtained:

$$V = 10.8 R^{2/3} S^{1/3} \tag{9-25}$$

Example 9-4

By using the regime approach, determine the cross section of an alluvial channel for a design flow of 8 m^3/s. The sediment carried by water is 0.4-mm sand.

Given:

   Q = 8 m^3/s
   d = 0.4 mm

Determine:

   B_0 = ?
   y = ?

Solution:

$$P = 4.75\sqrt{8} = 13.44\,\text{m}$$
$$f_s = 1.76(0.4)^{\frac{1}{2}} = 1.11$$
$$R = 0.47\left(\frac{8}{1.11}\right)^{0.333} = 0.907\,\text{m}$$

Let the side slopes be 0.5H : 1V. Then

$$A = (B_o + 0.5y)y = PR = 13.44 \times 0.907 = 12.18\,\text{m}$$
$$P = B_o + 2\sqrt{1 + (0.5)^2}\,y$$
$$= B_o + 2.24y = 13.44\,\text{m}$$

Substituting this expression for B_o into equation for A, we obtain

$$1.74y^2 - 13.44y + 12.18 = 0$$

Solution of this equation yields $y = 1.05$ m. Let us use a freeboard of 0.6 m. Then, the depth of the section is $1.05 + 0.6 = 1.65$ m and the width, $B_o = 13.44 - 2.24 \times 1.05 = 11.1$ m. Now,

$$S = 3 \times 10^{-4} \times (1.11)^{1.67}(8)^{0.167}$$
$$= 5.05 \times 10^{-4}$$

Problems

9-1. Design a power canal to carry a flow of 50 m³/s. The canal will be excavated through competent rock by blasting and will have a bottom slope of 0.0002.

9-2. Design an irrigation canal to irrigate 100 km² of farmland. The water demand is 0.1 m³/s/km² of land. The topography is flat and a bottom slope of 1 m per 2 km will be appropriate. The soil through which the channel is to be excavated is stiff clay.

9-3. A drainage channel has to be designed to carry runoff from 200 km². If the flow per square kilometer is 0.5 m³/s/km², determine the channel size using (a) the permissible velocity method, (2) the tractive force method, (3) regime theory. The material size is 2 mm and $S_o = 0.00002$.

9-4. Design a storm sewer for a new housing subdivision. The area of the subdivision is 4 km². The storm runoff may be taken as 0.15 m³/s/km². The topography is such that a channel bottom slope of 1 m in 2 km will be economical.

9-5. Design an irrigation canal for a design discharge of 1100 ft³/sec. The general slope in the area is 2 ft/mile and the soil is clay.

9-6. Design a flood control channel to carry a flow of 500 ft³/sec. A bottom slope of 0.003 will balance the cut and fill volumes. Assume the bottom and sides are paved with kiln-dried bricks.

9-7. A 2-km long horseshoe tunnel drilled through sound rock is to be used for river diversion during the construction of a dam. The bottom level at the tunnel inlet is at El. 100 and at the exit is El. 98.5. For a design flow of 100 m³/s, design the tunnel so that it will be free-flow. The downstream water level at design discharge is at El. 102. Plot the water surface profile in the tunnel.

For a flow of 150 m^3/s, the downstream water level is at El. 105. Determine the water surface profile in the tunnel for this flow.

9-8. Design a grass-lined channel to carry a flow of 100 ft^3/s. A 3 percent bottom slope is suitable for the terrain.

REFERENCES

American Society of Civil Engineers. 1979. *Design and construction of sanitary and storm sewers*, New York, NY.

Central Board of Irrigation and Power 1968. *Current practices in canal design in India*, New Delhi, India (June).

Chang, H. H. 1990. Hydraulic design of erodible bed channels, *Jour. Hydraulic Engineering*, Amer. Soc. Civil Engrs., 116, no. 1:87–101.

Chow, V. T. 1959. *Open channel hydraulics*, McGraw-Hill, New York, NY.

Fortier, S. and Scobey, F. C. 1926, Permissible canal velocities, Trans. Amer. Soc. Civil Engrs., 89, 940-956.

Gupta, R. S. 1989. *Hydrology and hydraulic systems*, Prentice Hall, Englewood Cliffs, NJ.

Lacey, G. 1930. Stable channels in alluvium, *Proc.*, Institute of Civil Engrs., London, 229, paper 4736.

Lane, E. W. 1955. Stable channel design, *Trans*, Amer. Soc. Civil Engrs., 120:1234-1260.

Mehrotra, S. C. 1983. Permissible velocity correction factors, *Jour. Hydraulic Engineering*, Amer. Soc. of Civil Engrs., 109, no. 2 (February): 305–308.

Raju, K.G. 1983. *Flow through open channels*, Tata McGraw-Hill, New Delhi, India.

U.S. Army Corps of Engineers. 1970. *Hydraulic design of flood control channels*, Report EM 1110-2-1601.

10 Special Topics

River Nile flood height measurement, at Hassan Pasha Palace, Rowdah Island, Cairo. Flood stage at this site has been recorded for 1150 years (Courtesy of Peter R. B. Ward, Ward and Assoc., Vancouver, BC., Canada).

10-1 Introduction

In this chapter, we present a brief discussion on the flows in a channel connecting two reservoirs, air entrainment, flow through culverts, and flow measurement. Most of this discussion involves application of the material presented in the previous chapters.

10-2 Flow in a Channel Connecting Two Reservoirs

Several different flow situations are possible, depending upon the system configuration and parameters. For example, the slope of the channel bottom may be mild or steep; the channel length may be short or long (a channel is considered short if the gradually varied flow profile extends to the end of the channel). The discussion in this section mainly deals with short channels, unless stated otherwise. We first consider a channel having mild bottom slope and then a channel with steep slope. Both qualitative discussion and procedures for flow computation are presented. In these discussions, we neglect the entrance and exit losses and the velocity head at the channel entrance and exit.

Mild Bottom Slope

The following three cases are possible:

1. Upstream reservoir level constant, downstream reservoir level variable
2. Downstream reservoir level constant, upstream reservoir level variable
3. Constant discharge, both upstream and downstream reservoir levels variable.

We discuss each of these cases one by one.

Upstream reservoir level constant, downstream reservoir level variable.

Figure 10-1 shows the channel system for this case. We are interested in plotting a curve between the channel discharge and the downstream reservoir level. This curve is referred to as the *delivery curve* for the channel.

There is no channel flow and the water surface in the channel is level (Line *ab* in Fig. 10-1) if the downstream reservoir level is at the same elevation as the upstream reservoir level. For the downstream reservoir levels above point *b*, water flows in the upstream direction. We refer to the flow from the downstream reservoir toward the upstream reservoir as negative.

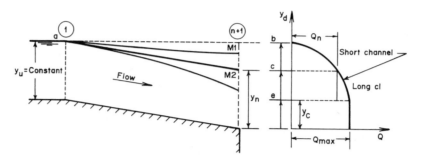

Figure 10-1 Upstream reservoir level constant, downstream reservoir level variable.

The channel discharge increases as we lower the water level in the downstream reservoir below point b. When the downstream level is at point c, the depth at the downstream end is the same as that at the upstream end, the flow surface in the channel is parallel to the channel bottom, and the flow in the channel is uniform. Let us call the discharge when the flow is uniform Q_n. As we lower the downstream water level further, two types of flow situations are possible, depending upon the channel length: If the channel is long (i.e., the water-surface profile from the downstream reservoir does not extend to the upstream reservoir), then the channel discharge does not increase above Q_n. However, if the channel is short, then the discharge increases as the downstream reservoir level is lowered below point c, as shown in Fig. 10-1. The channel discharge keeps on increasing until we reach point e, when the downstream flow depth corresponds to the critical depth. If we lower the water level further, it results in a free overfall and does not contribute to an increase in the channel discharge.

The preceding discussion has been mainly qualitative. Let us now discuss how to compute the channel discharge and the flow depth along the channel length.

As we discussed before, the channel discharge is zero if the water surfaces in the upstream and the downstream reservoirs are at the same level (i.e., at point b in Fig. 10-1).

The flow is uniform if the flow depths at the channel entrance and at the channel outlet are equal (i.e., at point c in Fig. 10-1). In this case, the flow depth at all channel sections is equal to the normal depth, y_n. The rate of discharge may be directly computed for this flow depth from the Manning equation or any other similar equation.

To determine the channel discharge and the water levels in the channel if the specified downstream reservoir level is between points c and b or below point c, we may use either of the following two procedures:

1. Trial-and-error approach
2. Simultaneous-solution approach

For the trial-and-error approach, we assume a value for the channel discharge and determine the value of y_c corresponding to this discharge. Then, we compute the water-surface profile in the channel, starting with a trial flow depth y_d at the downstream end. If the flow depth at the upstream end computed by this procedure is equal to the specified value y_u, then the assumed rate of discharge and the computed water levels are correct. Otherwise, we assume another value for the discharge and repeat this procedure.

In the simultaneous-solution procedure discussed in Chap. 6, we divide the channel into n reaches and call the section at the upstream end 1 and at the downstream end $n + 1$. The lengths of these reaches may not be equal. For the specified values of y_u and y_d (for plotting the delivery curve, we select several values of y_d and repeat this procedure for each depth one by one) and the channel parameters, we want to solve the resulting system of equations to determine the flow depths at $n + 1$ sections and the rate of discharge, Q. Thus, there are $n + 2$ unknowns and we need $n + 2$ equations. Two of these equations are provided by the upstream- and the downstream-end conditions, i.e., if the entrance and exit losses and the velocity head are neglected, then

$$y_1 = y_u \qquad (10\text{-}1)$$

$$y_{n+1} = y_d \qquad (10\text{-}2)$$

The remaining n equations are obtained by writing the energy equation between two consecutive channel sections, i.e.,

$$z_i + y_i + \frac{\alpha_i Q^2}{2g A_i^2} = z_{i+1} + y_{i+1} + \frac{\alpha_{i+1} Q^2}{2g A_{i+1}^2} + h_{f_i} \qquad (i = 1, 2, \ldots, n) \qquad (10\text{-}3)$$

in which $z =$ elevation of the channel bottom above a specified datum; $h_{f_i} =$ head losses between sections i and $i + 1$, and the subscripts i, $i + 1$, ... refer to quantities for i and $i + 1$ sections.

For water levels in the downstream reservoir at or below point e, the flow depth at the downstream end of the channel is equal to y_c. Therefore, we replace Eq. 10-2 by the following equation for the critical depth at a general cross section:

$$D_{n+1} = \frac{Q^2}{g A_{n+1}^2} \qquad (10\text{-}4)$$

in which $D_{n+1} =$ hydraulic depth at section $n + 1$.

Other than specifying the downstream flow depth by Eq. 10-4 instead of Eq. 10-2, we proceed as discussed in the previous paragraphs.

Downstream reservoir level constant, upstream reservoir level variable.

In this case, there is no flow in the channel if the water surface in the upstream reservoir is at the same level as that in the downstream reservoir (point a in Fig. 10-2). The channel flow is negative for the upstream reservoir level below this point.

The rate of discharge increases as we raise the upstream reservoir level above point a. At point c, the flow depth at the upstream end is the same as that at the downstream end. In such a situation, the flow is uniform and the flow depth at all channel sections is equal to the normal depth. With further increase in the upstream reservoir level, the channel discharge keeps on increasing until point d when the downstream flow depth, y_d, corresponds to the critical depth for the channel discharge. Any further increase

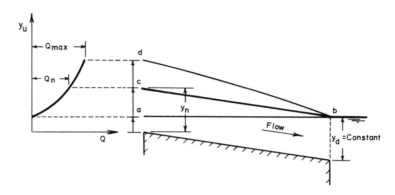

Figure 10-2 Downstream reservoir level constant, upstream reservoir level variable.

in the upstream reservoir level results in raising the water-surface profile in the entire channel.

To compute the rate of discharge and the water surface profile in the channel, we may use the simultaneous solution approach or compute them directly as follows. For the direct computations, we first determine Q_{max} assuming y_d as the critical depth. Then, for different values of discharge and of the downstream flow depth, we compute the water surface profiles starting at the downstream end with the selected value of y_d. This gives the upstream depth for the selected values of discharge and y_d. For the simultaneous-solution approach, two equations are provided by the upstream and the downstream end conditions, i.e., $y_1 = y_u$ and $y_{n+1} = y_d$ and the remaining n equations are obtained by writing the energy equation between two consecutive channel sections. The resulting system of equations is solved for the rate of discharge, Q, and the flow depths at $n + 1$ cross sections.

Variable upstream and downstream reservoir levels, constant rate of discharge.   In this case, y_u and y_d are both variable and the rate of discharge is specified.

Referring to Fig. 10-3, a line drawn at 45° and with an intercept of $S_o L$ on the y_d axis indicates that the upstream and downstream reservoirs have the same water levels. Therefore, all curves for different discharges approach this line asymptotically. A line drawn through the origin at 45° represents the uniform flow conditions since $y_u = y_d$.

The curve OC represents the critical flow conditions. This is for the case when $y_d = y_c$ for the specified discharge and y_1 is equal to the given value of y_u. If the channel is long, then the Q-contours extend to the left of the C-curve and they are parallel to the y_d axis.

To plot the curves shown in Fig. 10-3, we select a value for the rate of discharge, Q. Then, starting with flow depth y_d, we compute the water surface profile in the channel and determine the flow depth y_u. For this rate of discharge, we also compute the point on the C-curve by first determining the critical flow depth for the specified rate of discharge and then determining the corresponding upstream flow depth. A point

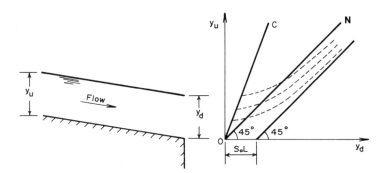

Figure 10-3 Both reservoir levels variable, constant discharge.

corresponding to the uniform flow will lie on the 45° line. By repeating this procedure for two or three values of downstream flow depths, we have sufficient number of points to draw the curve.

Steep Bottom Slope

In a channel with steep bottom slope the control is located at the upstream end. For the downstream reservoir level lower than the flow depth at the downstream end of the channel, the water levels in the channel do not depend upon the tailwater levels (Fig. 10- 4). However, as the tailwater level is raised above the flow depth at the downstream end, it starts to affect the water levels in the downstream portions of the channel, and a hydraulic jump may be formed. Upstream of the hydraulic jump, however, the water levels are not affected by the tailwater levels. As we raise the tailwater level further, the jump moves upstream until it reaches the upstream end. In such a case, the inlet becomes submerged and the channel discharge becomes dependent upon the tailwater level.

To compute the water-surface profile in such a channel, we first compute the rate of discharge using the critical flow conditions at the channel entrance. Then, starting with the critical depth at the upstream end, we compute the water-surface profile in the channel proceeding in the downstream direction. If the tailwater levels are high, then a hydraulic jump may be formed in the channel. In such a case, we compute the water-surface profile in the lower part of the channel by starting at the downstream end

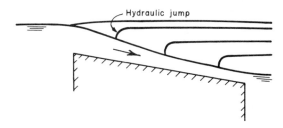

Figure 10-4 Channel with steep bottom slope.

with the specified flow depth and proceeding in the upstream direction. The location of the jump is determined by matching the specific forces upstream and downstream of the jump. Another procedure available for computing the water-surface profile and the location of the jump is to solve the unsteady flow equations. For this purpose, we solve the continuity and momentum equations subject to the end conditions and continue the computations until the solution converges to the steady-flow conditions. This was discussed in detail in Chap. 8.

10-3 Air Entrainment in High-Velocity Flow

As the boundary layer from the floor intersects the water surface (Falvey 1980; Wood 1991), air is entrained in high-velocity flow (Lane 1939) although no air is entrained in slow-moving water even when the boundary layer intersects the surface. Thus, some degree of turbulence must be exceeded for the air entrainment to begin. We refer to air entrainment as a process in which air enters the body of water (Falvey 1980; Wood 1991). The appearance of "white water" as such does not necessarily imply air entrainment, since it may be due to reflections coming from different angles. This conclusion has been confirmed by high-speed photography.

A designer needs to know the amount of air entrainment to select the height of side walls. It is also helpful in assessing the cavitation potential since air near the channel bottom reduces the possibility of cavitation. In addition, the friction losses are reduced by the presence of air next to the channel bottom and sides, as discussed later in this section.

The turbulent open-channel flow involving air entrainment may be divided into the following four vertical zones (Killen and Anderson 1969), as shown in Fig. 10-5:

1. Upper zone
2. Mixing zone
3. Underlying zone
4. Air-free zone

The upper zone comprises a small mass of flying water particles ejected from the mixing zone. This zone is not important for engineering applications. The mixing zone

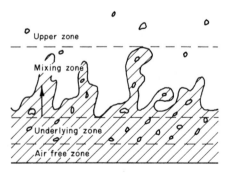

Figure 10-5 Vertical zones of aerated flow (After Killen and Anderson, 1969).

has surface waves of random amplitude and frequency. To prevent overtopping of the side walls, it is necessary to take the height of these waves into consideration. The knowledge of the characteristic of this zone is important, since all air entrained into flow or released from it passes through this zone. The surface waves do not penetrate the underlying zone, and the air concentration at any depth in this zone is determined by the number and the size of air bubbles. A number of correlations for the air-concentration distribution have been developed by using the turbulent-boundary-layer theory, although it must be pointed out that they are not totally reliable. The air-free zone exists only in that part of the channel where aeration is still developing. At the interface, the air concentration as well as the rate of change in concentration with depth is small.

The development of self-aerating flows in a wide channel may be divided into four regions (Wood 1991), as shown in Fig. 10-6:

1. Nonaerated flow region
2. Partially aerated flow region
3. Fully aerated flow region
4. Uniform fully developed aerated flow region

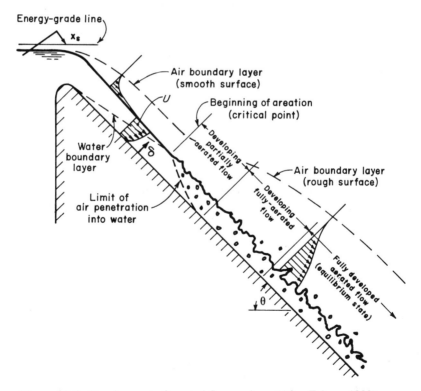

Figure 10-6 Development of aerated flow regimes (After Falvey, 1980).

In the nonaerated flow region, the turbulent boundary layer does not reach the flow surface and no air is entrained. In the partially and fully aerated flow regions, the air-concentration profiles vary with distance. However, the air does not reach the channel bottom in the partially aerated region, which is the case in the fully aerated region. In the uniform-flow region, the flow conditions reach an equilibrium state and are constant with distance.

Nonaerated Flow Region

As discussed earlier, the air entrainment starts at a point where the boundary layer from the bottom intersects the top water surface. This point is referred to as the *point of inception*. Several authors have presented empirical relationships for the location of this point based on experimental data. Keller and Rastogi (1975) computed the developing boundary layer by solving the time-averaged, two-dimensional Navier-Stokes equations. On the basis of these studies and other results for the values of different constants, Wood (1991) presented the following equation

$$\frac{\delta}{x_s} = 0.021 \left(\frac{x_s}{h_s}\right)^{0.11} \left(\frac{k_s}{x_s}\right)^{0.10} \tag{10-5}$$

in which δ = boundary layer thickness (defined as the perpendicular distance from the channel bottom to a depth where the velocity is 99 percent of the free streamline velocity); k_s = equivalent sand roughness; x_s = distance along the slope along which boundary layer grows (Fig. 10-6); and h_s = static head at the inception point ($\sin\theta = h_s/x_s$). This equation shows the relative importance of various parameters upon which the growth of boundary layer depends and is applicable for slopes of 5° to 70°.

Uniform Aerated Region

In this region, the mean-flow properties, such as the flow depth and the depth-averaged air concentration, $\overline{C}$, are functions of the discharge/unit width, q; channel bottom slope, S_o; bottom roughness; and the fluid properties.

By using the data obtained by Straub and Anderson on an artificially roughened 0.46-m-wide flume, Hager (1991) developed the following empirical expressions for this region.

$$\text{Average air concentration,} \quad \overline{C} = 0.75(\sin\theta)^{0.75} \tag{10-6}$$

This cross-sectional average was obtained by integrating the function $C(y)$ from the channel bottom to a depth y_r, where the air concentration is 90 percent. For the uniform aerated flow depth in a wide rectangular channel,

$$y_{99} = y_w + 1.35 y_w \left[\frac{\sin^3\theta\, y_w}{n^2 g^3}\right]^{1/4} \tag{10-7}$$

in which y_{99} = flow depth corresponding to 99 percent air concentration; y_w = depth corresponding to pure water; n = Manning constant; and g = acceleration due to gravity.

Friction factor.   The friction factor, f_e, in a uniform aerated flow is constant for the mean air concentration less than 30 percent (Wood 1991). As the air concentration increases, the value of f_e decreases rapidly. If f_w is the friction factor for flow with no entrained air, then f_e may be determined from the following equation (Hager 1991)

$$f_e = \frac{f_w}{1 + 10\overline{C}^4} \tag{10-8}$$

Velocity distribution.   For the mean air concentration up to 50 percent, the velocity distribution may be approximated as

$$\frac{V}{V_r} = \left(\frac{y}{y_r}\right)^{0.16} \tag{10-9}$$

in which y_r and V_r are the flow depth and flow velocity where the air concentration is 90 percent.

The flow velocity V_m of an air-water mixture may be related to the velocity of air and water phases as

$$V_m \rho_m = \rho_w V_w (1 - C) + \rho_a V_a C \tag{10-10}$$

in which the subscripts m, a, and w refer to the quantities for the air-water mixture, air, and water, respectively, and

$$\rho_m = \rho_w (1 - C) + \rho_a C \tag{10-11}$$

Flow calculations.   For a given channel bottom slope, the mean air concentration may be computed from Eq. 10-6. The friction factor for the aerated flow, f_e, may be determined from Eq. 10-8 if the friction factor for the flow of water, f_w, is available. The bulk flow depth, y_e, may then be computed from

$$y_e = \left(\frac{f_e q^2}{8 g S_o}\right)^{1/3} \tag{10-12}$$

The average water velocity, $V_w = q/y_e$ and the flow depth at 90 percent air concentration, $y_r = y_e/(1 - \overline{C})$. The data obtained on Aviemore dam shows that $V_r = 1.2V_w$. Now, the velocity distribution may be obtained from Eq. 10-9.

10-4 Flow Through Culverts

A short passage way for flow under a highway, railroad, or other embankment is referred to as a culvert. The culvert may be circular, rectangular, arch, or elliptical in shape. A rectangular culvert is referred to as a box culvert.

Although a culvert is a simple hydraulic structure, the computation of flow conditions through it may be somewhat complex. This is because several different flow conditions are possible and these conditions depend upon several parameters. The culvert may flow full, or partially full throughout its length, or in part of the length. The control may be at the upstream end (called *inlet control*), or it may be at the downstream end (called *outlet control*). Depending upon the head and tailwater levels, the control may shift from the inlet to the outlet and vice versa as these water levels change.

The flow through culvert has been classified into several types for analysis purposes (Chow 1959; Henderson 1966; Normann, et al., 1985). We may analyze these flows by utilizing the concepts we presented in the previous chapters. The discussion in the following paragraphs should be helpful for such analyses.

A curve relating the headwater level to the rate of discharge through the culvert is referred to as the *rating curve* or the *performance curve*. To determine the flow capacity, the curves for both inlet and outlet control should be plotted and the curve which gives lower discharge may be selected in those situations where it is not certain whether the control is at the inlet or at the outlet.

A culvert does not flow full even if the entrance is submerged if the head H at inlet is less than $1.5D$, where D is the height of the culvert at the entrance and H = the headpond water level − the culvert invert level. Similarly, a culvert having a square-edged entrance may not flow full even if the headwater is higher than the top of the culvert because of the flow contraction at the top.

Inlet Control

In the inlet control, the flow through the culvert mainly depends upon the inlet conditions, e.g., area, shape and configuration at the inlet. The flow in the culvert is supercritical and thus it is independent of the conditions in the culvert or in the tailwater area.

Figure 10-7 shows different possible flow conditions for the inlet control. The flow depth at the entrance is equal to the critical depth in case the entrance is not submerged. In such a case, the rate of discharge may be computed from the weir equation. For a submerged entrance, the flow springs clear from the top of the culvert if $H < 1.5D$. This limit may be even higher for a square-edged entrance. Then, the flow passes through the critical depth and the flow depth tends toward the normal depth. The flow in such a situation may be computed by using the orifice equation. A hydraulic jump may form inside the culvert depending upon the tailwater level. The downstream part of the culvert may be primed if the oulet is submerged.

The rate of discharge through a box culvert may be computed from the following equations (Henderson 1966).

Unsubmerged Entrance or H < 1.2D.

$$Q = \frac{2}{3}CBH\sqrt{\frac{2}{3}gH} \tag{10-13}$$

in which B = culvert width and the coefficient C accounts for the contraction on the sides. For square-edged sides, $C = 0.9$; and for slightly rounded sides, $C = 1$.

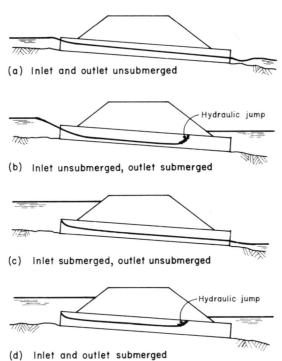

(a) Inlet and outlet unsubmerged

(b) Inlet unsubmerged, outlet submerged

(c) Inlet submerged, outlet unsubmerged

(d) Inlet and outlet submerged

Figure 10-7 Flow conditions for inlet control (After Normann, et al., 1985).

Submerged entrance and H > 1.2D.   In this case, the discharge may be computed from the orifice equation

$$Q = CBD\sqrt{2g(H - CD)} \tag{10-14}$$

in which C accounts for the contractions at the sides and the top. For a square-edge entrance, $C = 0.6$; for rounded edges, $C = 0.8$.

Outlet Control

In the outlet control, the culvert either flows full or partially full. In the later case, the flow is subcritical. The flow capacity depends upon the culvert area, shape, length, bottom slope, head losses in the culvert, and the headwater and tailwater levels.

Different flow conditions for the outlet control are shown in Fig. 10-8. The flow depth at the exit is critical if the tailwater level is at or below the critical depth. To compute the water-surface profile in a free-flow culvert, we start with the critical depth or with the tailwater level if it is higher than the critical depth. Then, we follow the procedure outlined in Sec. 10-2 for the case of constant downstream reservoir level. For a pressurized culvert, we determine the headwater level by applying the following energy equation between the tailwater and the headwater

$$Z_u + \frac{V_u^2}{2g} = H_l + Z_d + \frac{V_d^2}{2g} \tag{10-15}$$

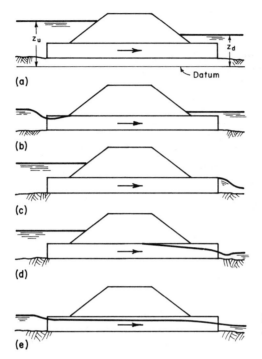

Figure 10-8 Flow conditions for outlet control (After Normann et al., 1985).

in which Z = elevation of the water level; H_l = sum of all the head losses between the headpond and the tailwater; and the subscripts u and d refer to the conditions on the upstream and on the downstream sides of the culvert. The head losses include the entrance loss, friction and form losses in the culvert and the exit losses. Normally, the flow velocities in the upstream and in the downstream areas are small and may be neglected. (Note that the velocity heads will cancel if they are equal.) If we express the total losses as $K Q^2$, then the preceding equation may be written as

$$Q = \frac{1}{K} \sqrt{Z_u - Z_d} \qquad (10\text{-}16)$$

Several modifications have been proposed to improve the entrance conditions (Harrison et al., 1972) to reduce losses, for vortex control, or for debris control (Reihsen and Harrison, 1971). A number of innovations for fish passage have also been proposed (Normann, et al., 1985).

10-5 Flow Measurement

Because the rate of discharge in a channel depends upon the flow depth as well as on the flow velocity, it is necessary to know each of them to determine the flow rate. For critical flow, however, only the flow depth is sufficient because of a unique relationship between the flow depth and the flow velocity. To measure the rate of flow in a channel, several different methods (Ackers et al. 1978; World Meteorological Organization, 1971; International Standards Organization 1977) are available—e.g., gauging structures, velocity-area methods, dilution techniques, and the slope-area method. To a limited extent, electromagnetic and ultrasonic methods have been employed on small channels with well-defined cross sections. In the laboratory or on a small channel, weirs or flumes and the velocity-area methods may be used. In the field or on a large channel, the dilution technique or the slope-area method may be utilized. A brief description of weirs and some other hydraulic structures was presented in Chapter 7 and flow through culverts was discussed in the previous section.

Velocity-Area Method

In the *velocity-area* method, we determine the depth-averaged velocity at a number of locations along a line perpendicular to the flow direction and then sum up the product of the flow velocities and the corresponding flow areas. A general procedure (Rantz et al. 1982] for determining the depth-averaged velocity is to average the velocity measured at $0.2d$ and $0.8d$ from the water surface, where d = total depth of flow. Alternatively, the velocity measured at $0.6d$ from the water surface has been used for this purpose. For a logarithmic velocity profile, Vanoni (1941) showed that the flow velocity measured at $0.632d$ from the water surface is equal to the theoretical logarithmic profile depth-averaged velocity. Walker (1988) proved that, for typical natural channels, the error introduced by averaging the flow velocities at $0.2d$ and $0.8d$ is less than 2 percent. By assuming logarithmic velocity distribution, he derived the following equation to determine the average flow velocity at a section:

$$\overline{V} = \frac{(1 + \ln d_2)V_1 - (1 + \ln d_1)V_2}{\ln(V_2/V_1)} \tag{10-17}$$

in which $\overline{V}$ = depth-averaged flow velocity; V_1 = velocity measured at depth d_1; and V_2 = velocity measured at depth d_2. This expression is useful for determining the depth-averaged flow velocity in situations where the flow velocities are measured at fixed locations, such as at automatic gauging stations, irrespective of the flow depth.

For natural channels, the flow section is subdivided into several vertical subsections. The flow velocity is measured at 0.2 and 0.6 of the flow depth in each subsection. The average of these two values is the depth-averaged flow velocity for the vertical section. The discharge through the subsection is computed by multiplying this average velocity and the flow area for the subsection. Then, the sum of the discharges through all the subsections is the total discharge in the channel.

Slope-Area Method

The *slope-area* method is based on the assumption of uniform flow in a channel reach. Therefore, caution must be exercised in applying this method to determine the rate of discharge during a flood. This is due to the fact that the assumption of uniform flow may be valid only during the periods when the flow is changing at a slow rate with respect to time.

A straight-channel reach of length L is selected (Benson and Dalrymple 1966), and the flow areas, A, and the conveyance factors, K, at the upstream and at the downstream ends of the reach are determined. A representative value for the Manning n for the channel is also selected. Let us refer to the quantities for the upstream and for the downstream ends by the subscripts u and d and the average value for the reach by a bar on the variable name. The expression for K for the Manning equation is $K = \left(\frac{C_o}{n}\right) A R^{2/3}$. If we assume that the velocity heads at the upstream and at the downstream ends of the selected reach are approximately the same, then the slope of the water surface is equal to the slope of the energy-grade line. The geometric mean of the conveyance factors at the upstream and at the downstream ends of the reach may be taken as the average conveyance factor for the reach, i.e., $\overline{K} = \sqrt{K_u K_d}$. Then, the discharge is computed from the following equation:

$$Q = \overline{K} S^{1/2} \tag{10-18}$$

in which the slope of the water surface, $S = (Z_u - Z_d)/L$; Z_u = the elevation of the water surface at the upstream end and Z_d = elevation of the water surface at the downstream end.

Flumes

The discharge measurement by means of flumes (Ackers et al. 1978) is based on the assumption that critical flow is produced by constricting the width, raising the bottom, or a combination of the two. Then, the measurement of a single flow depth is sufficient to determine the discharge. In flows carrying sediment or debris, a step rise in the channel bottom results in the deposition of sediment or the accumulation of debris on the upstream side of the raised part. To prevent this problem, a step rise is usually avoided. By selecting a suitable value for the throat width and for the bottom level at the throat section and in the downstream part, critical flow may be produced in the throat area followed by supercritical flow. A hydraulic jump is formed downstream of the supercritical flow part. Such flumes are referred to as the *standing wave flumes*. Parshall flume is a commonly used flume of this type since the twenties. It is available in various standard sizes for a range of discharges, and an extensive amount of experimental data is available for these designs. Figure 10-9 shows a typical layout and configuration of this flume.

Problems

10-1. A 10-m-wide and 1-km-long rectangular channel having a bottom slope of 0.0002 connects lakes B and C. Assume the channel is concrete-lined with $n = 0.013$ and channel bottom at Lake B entrance is El. 94.

   i. Plot the delivery curve for the channel if the water level in lake B is constant at El. 100 and the water level in lake C is variable.

   ii. Plot a curve between the discharge versus Lake B level if the water level in Lake C is constant at El. 98.

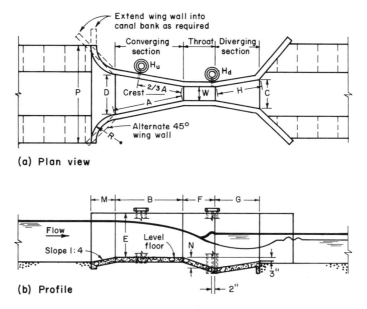

Figure 10-9 Parshall flume. (After Roberson, et al., 1988, *Hydraulic Engineering,* Houghton Mifflin).

iii. Plot a diagram between the water levels in both lakes and the channel discharge if the water levels in both lakes are variable.

10-2. An overflow spillway has a slope of 30°. For a head of 8 m above the spillway crest, compute the development of boundary-layer thickness along the spillway length.

10-3. In the uniform aerated flow region of the spillway of Prob. 10-2, determine the average air concentration, the flow depth corresponding to 99 percent air concentration, equivalent friction factor, and the bulk flow depth.

10-4. For a 2-m-wide, 4-m-high box culvert with a bottom slope of 0.0001, compute the rating curve for the culvert. Assume the downstream end of the culvert remains unsubmerged.

10-5. Plot the rating curve for the culvert of Prob. 10-4 if the bottom slope is 0.001. Assume the tailwater level remains below the culvert top at the downstream end.

10-6. If the culvert of Prob. 10-5 is 100 m long, compute and plot the water surface profile in the culvert if the tailwater level is 4.5 m above the culvert invert at the outlet.

10-7. Assuming a logarithmic velocity distribution, prove that the flow velocity at $0.368d$ ($d =$ flow depth) is the depth-averaged flow velocity.

10-8. Show that the average of the flow velocities at 0.2 and 0.8 depth gives a depth-averaged flow velocity with an error of 2 percent. Assume the velocity distribution is logarithmic.

10-9. The water level in the upstream lake of the channel system shown in Fig. 10-10 remains constant at El. 108 m; the water level in the downstream lake may vary between El. 98 and 108 m.

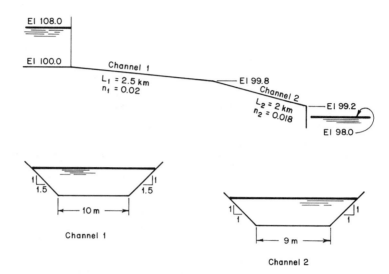

Figure 10-10   Channel system of Prob. 10-9.

    i. Compute the rates of discharge in the channel for different downstream lake levels and plot a curve between the discharge and the downstream lake level.

    ii. Compute and plot the water surface profile in the channel for the downstream levels of El. 98, 102, 104, and 106.

10-10. A branch channel (channel 3) takes off from the junction of channels 1 and 2 of Fig. 10-10. Channel 3 is 1 km long and has the same cross section as that of channel 2, and Manning n is 0.015. A lake with constant water level at El. 105 is located at the downstream end of the branch channel.

    i. Compute and plot the delivery curves for channels 2 and 3 for the different downstream lake levels downstream of channel 2.

    ii. Plot the water surface profiles in all three channels for the channel 2 lake levels of El. 98, 102, and 104.

REFERENCES

Ackers, P.; White, W. R.; Perkins, J. A.; and Harrison, A. J. M.   1978. *Weirs and flumes for flow measurement*, John Wiley, New York, NY.

Benson, M. A., and Dalrymple, T.   1966. General field and office procedures for indirect discharge measurements: U.S. Geological Survey techniques water-resources inv., Book 3, Chapter A1.

Bos, M. G. ed.   1976. *Discharge Measurement Structures,* Publication No. 20, International Institution for Land Reclamation and Improvement, Delft, The Netherlands.

Chow, V. T.   1959. *Open channel hydraulics,* McGraw-Hill, New York, NY.

Falvey, H. T.   1980. *Air-water flow in hydraulic structures,* Monograph No. 41, U.S. Bureau of Reclamation, Denver, Colo.

Hager, W. H.   1991. Uniform aerated chute flow, *Jour. Hydraulic Engineering,* Amer. Soc. Civil Engrs., 117, no. 4: 528-33.

Harris, J. D. 1982. Hydraulic design of culverts, Chapter D, Drainage Manual, Ontario Ministry of Transportation and Communication, Dawnsview, Ont.

Harrison, L. J.; Morris, J. L.; Normann J. M.; and Johnson, F. L. 1972. Hydraulic design of improved inlets for culverts, HEC No. 13, Bridge Division, Federal Highway Administration, Washington, DC.

Henderson, F. M. 1966. *Open channel flow,* Macmillan, New York, NY.

Hopping, P. L., and Hoopes, J. A. 1988. "Development of a numerical model to predict the behavior of air/water mixtures in open channels." In *Model Prototype Correlation of Hydraulic Structures,* edited by Philip Burgi, Amer. Soc. Civil Engrs., 419–28.

International Standards Organization. 1977. *Liquid flow measurement in open channels* International Standards Organization, Geneva, Switzerland.

Keller, R. J., and Rastogi, A. K. 1975. Prediction of flow development on spillways, *Jour. Hydraulics Div,* Amer. Soc. Civil Engrs., 101, no 9:1171–84.

Keller, R. J.; Lai, K. K.; and Wood, I. R. 1974. Developing region in self-aerating flows, *Jour. Hydraulics Div,* Amer. Soc. Civil Engrs., 100, no 4:553–68.

Killen, J. M., and Anderson, A. G. 1969. A study of the air-water interface in air entrained flow in open channels, 13*th Congress, International Association for Hydraulic Research* (Japan), 2:339–47.

Kulin, G., and Compton, P. R. 1975. *A guide to methods and standards for the measurement of water flow,* Special Publication No. 421, National Bureau of Standards, U.S. Department of Commerce, Washington, DC.

Lane, E. W. 1939. Entrainment on spillway faces, *Civil Engineering,* Amer. Soc. Civil Engrs., 9:89–96.

McClellan, T. J. 1971. Fish passage through highway culverts, PB 204 983, Federal Highway Administration, Region 8, Portland, Ore.

Normann, J. M., Houghtalen, R. J.; and Johnston, W. J. 1985. *Hydraulic design of highway culverts,* Report No. FHWA-IP-85-15, Federal Highway Administration, McLean, VA.

Parshall, R. L. 1926. The improved venture flume, Trans. Amer. Soc. Civil Engrs., 89:841.

Parshall, R. L. 1953. Parshall flumes of large size, *Bulletin, Colorado Agricultural Experiment Station,* no. 426A (March).

Rantz, S. E. et al. 1982. Measurement and computation of streamflow: Volume 1, Measurement of stage and discharge, *U.S. Geological Survey Water-Supply Paper 2175.*

Reihsen, G., and Harrison, L. J. 1971. Debris control structures, HEC No. 9, Bridge Div. Federal Highway Administration, Washington, DC.

Smith, A. G. 1974. Peak flows by the slope-area method, *Tech. Bull.* no. 79, Canadian Inland Waters Dir., Water Resources Branch, Ottawa, Ont., 31 pp.

Vanoni, V.A. 1941. Velocity distribution on open channels, *Civil Engineering,* Amer. Soc. Civil Engrs., 11(6), 356–57.

Walker, J. F. 1988. General two-point method for determining velocity in open channel, *Jour. Hydraulic Engineering,* Amer. Soc. Civil Engrs., 114, no. 7:801-805.

World Meteorological Organization. 1971. *Use of Weirs and Flumes in Stream Gauging,* WMO No. 280, Technical Note 117, Geneva, Switzerland.

Wood, I. R., ed. 1991. *Air entrainment in free-surface flows,* A.A. Balkema, Rotterdam, the Netherlands.

11 Unsteady Flow

Surge wave in Seton Canal produced by closing of turbine wicket gates at the downstream end of the canal (Courtesy of B.C. Hydro and Power Authority, Vancouver, B.C. Canada).

11-1 Introduction

In the previous chapters, we discussed steady flow in open channels. However, flows in the real world are usually unsteady, where the flow conditions vary with respect to time. The unsteadiness may be due to natural causes or due to human action. The analysis of unsteady flow is usually more complex than steady flow because unsteady-flow conditions are a function of space and time. Thus partial differential equations describe unsteady flows, and closed-form solutions are not available for these equations except in very simplified cases.

Unsteady flow is discussed in Chaps. 11 through 17. A brief introduction to this flow is presented in this chapter; governing equations are derived in the next chapter, and numerical methods for their solution are presented in Chaps. 13 and 14. Two-dimensional unsteady flow is discussed in Chap. 15, the finite-element method is outlined in Chap. 16, and a number of special topics related to unsteady flow are described in Chap. 17.

In this chapter, a number of commonly used terms are first defined. The causes of unsteady flow are then discussed and equations for the wave velocity are derived.

11-2 Definitions

A *wave* is defined as a temporal (i.e., with respect to time) or spatial (i.e., with respect to distance) variation of flow depth and rate of discharge. The *wave length, L,* is the distance between two adjacent wave crests or troughs and the *amplitude, z,* of a wave is the height between the maximum water level and the still-water level (Fig. 11-1).

By using different criteria, waves may be classified into several categories. A wave is called *oscillatory wave* if there is no mass transport in the direction of wave propagation, and it is called a *translatory wave* if there is net mass transport. For example, sea waves are oscillatory waves and flood waves are translatory. The translatory waves may be further classified as *solitary* or a *wave train*. A solitary wave has a rising and a falling limb and has a single peak. A wave train is, however, a group of waves in succession. A translatory wave having a steep front is called a *surge*.

A wave is called a *positive wave* if the water depth behind the wave is higher than the undisturbed flow depth, and it is called a *negative wave* if the flow depth behind the wave is lower than the undisturbed flow depth. A positive wave having a steep front is referred to as a *bore*, or a *shock*. The latter term is borrowed from gas dynamics.

As the wave passes a section, the entire flow depth is disturbed in a *shallow-water wave* while only the top layers, and not the entire section, are affected in a *deep-water*

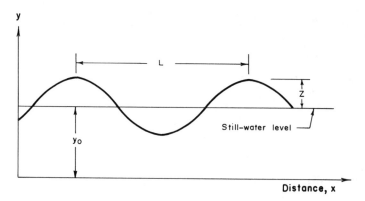

Figure 11-1 Wave length and amplitude.

wave. The ratio of wave length to the water depth is greater than 20 for shallow-water waves; this ratio is less than 20 for deep-water waves.

The *wave celerity* is defined as the relative velocity of a wave with respect to the fluid in which it is traveling, whereas *absolute wave velocity* is the velocity with respect to a fixed reference frame. Thus, the absolute wave velocity, $\mathbf{V_w}$, is equal to the vectorial sum of the flow velocity, $\mathbf{V}$, and the wave celerity, $\mathbf{c}$, i.e.,

$$\mathbf{V_w} = \mathbf{V} + \mathbf{c} \tag{11-1}$$

In a one-dimensional flow, the wave velocity is either in the direction of flow or in the opposite direction. Therefore,

$$V_w = V \pm c \tag{11-2}$$

The plus sign is used if the wave is traveling in the direction of flow and the negative sign is used if the wave is traveling opposite to the flow direction.

By neglecting the viscosity and surface tension, Airy derived the following expression (Henderson 1966) for the celerity of a small-amplitude wave:

$$c = \sqrt{\frac{gL}{2\pi} \tanh \frac{2\pi y_o}{L}} \tag{11-3}$$

in which y_o = undisturbed flow depth.

As we discussed in the previous paragraphs, y_o/L is large in deep-water waves. Thus, $\tanh 2\pi y_o/L \to 1$ and the celerity for the deep-water wave reduces to

$$c = \sqrt{\frac{gL}{2\pi}} \tag{11-4a}$$

The ratio y_o/L is very small for shallow-water waves; therefore, $\tanh 2\pi y/L \to 2\pi y_o/L$. Hence, the expression for the celerity of these waves becomes

$$c = \sqrt{gy_o} \tag{11-4b}$$

Note that Eq. 11-4b is valid only for small amplitude waves. We will derive a general expression for celerity in Sec. 11-4 and then deduce Eq. 11-4b from that expression.

11-3 Causes of Unsteady Flow

Transients in open channels are produced whenever flow conditions are disturbed by natural or artificial reasons. These reasons may be accidental or planned. The flow conditions include the flow depth and flow velocity. If the discharge is varied at a rapid rate or the channel has low friction, a *bore* may develop during the transient conditions.

Typical situations in which unsteady flows occur are as follows:

1. Surges in power canals or tunnels produced by starting or stopping of turbines or due to the opening or closing of the turbine gates to meet the load changes

2. Surges in upstream or downstream channels produced by starting or stopping of pumps and opening or closing of control gates

3. Waves in the navigation channels produced by the operation of navigation locks

4. Flood waves in streams, rivers, and drainage channels due to rainstorms and snowmelt

5. Tides in estuaries, bays, and inlets

6. Waves generated by landslides and avalanches in rivers, channels, reservoirs, and lakes

7. Waves produced by the failure of dams, dykes, or other control structures

8. Storm runoff in sewers and drainage channels

9. Circulation in lakes and reservoirs produced by wind or by temperature and density gradients

10. Waves in lakes, reservoirs, estuaries, bays, inlets, and oceans caused by wind storms, cyclones, and earthquakes

11-4 Height and Celerity of a Gravity Wave

In this section we will derive expressions for determining the celerity and the height of a gravity wave produced by a sudden change in discharge. These expressions are general and may be used for small- or large-amplitude waves. We will make the following simplifying assumptions in the derivations: (1) The channel is frictionless and the channel bottom is horizontal (same equations are obtained if the component of the weight of the liquid in the downstream direction is equal to the shear force acting on the channel sides and bottom); (2) the pressure distribution on both sides of the wave front is hydrostatic; (3) the velocity distribution is uniform on both sides of the wave front; (4) the wave is an abrupt discontinuity of negligible length; (5) the wave does not change in shape as it propagates in the channel; and (6) the water surface behind the wave is parallel to the initial water surface.

Let the flow be suddenly increased from Q_1 to Q_2 in a channel, which results in increasing the flow depth from y_1 to y_2, as shown in Fig. 11-2. Let this wave be traveling at absolute velocity V_w in the downstream direction. We consider this direction as positive. We are interested in determining the velocity of this wave and the wave height $y_2 - y_1$. These equations may be derived by using different methods.

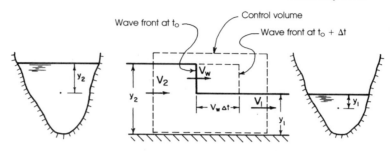

Figure 11-2 Definition sketch.

In the following paragraphs, we will apply the control-volume approach to the unsteady flow. However, we may convert the unsteady-flow situation to steady flow by applying velocity V_w on the entire system in the upstream direction. Then we may either use the control-volume approach as in this section, or we may apply the common continuity and momentum principles (as we did in Section 3-2) to derive these equations.

Let us consider a control volume, as shown in Fig. 11-2, in which the wave front has moved during time interval Δt, as shown. We will apply the Reynolds transport theorem (Roberson and Crowe 1990) in this section to derive the expressions for the wave height and the wave velocity.

Continuity Equation

Referring to Fig. 11-2,

$$\frac{d}{dt} \int_{cv} \rho d\mathcal{V} + \sum_{cs} \rho VA = 0 \tag{11-5}$$

in which ρ = mass density of water; $\mathcal{V}$ = volume of the control volume; V = flow velocity; and A = flow area. By using the subscripts 1 and 2 to denote quantities for sections 1 and 2, we may write

$$\sum_{cs} \rho VA = \rho V_1 A_1 - \rho V_2 A_2 \tag{11-6}$$

Since water may be assumed incompressible,

$$\frac{d}{dt} \int_{cv} \rho d\mathcal{V} = \rho \frac{d}{dt} \int_{cv} d\mathcal{V}$$

$$= \rho \times \text{rate of change of volume of liquid in the control volume}$$

$$= \rho \frac{\mathcal{V}_{t_o + \Delta t} - \mathcal{V}_{t_o}}{\Delta t} \tag{11-7}$$

$$= \rho \frac{\left[\mathcal{V}_{t_o} + V_w \Delta t (A_2 - A_1) \right] - \mathcal{V}_{t_o}}{\Delta t}$$

$$= \rho V_w (A_2 - A_1)$$

By substituting Eqs. 11-6 and 11-7 into Eq. 11-5 and simplifying, we obtain

$$A_1(V_1 - V_w) = A_2(V_2 - V_w) \tag{11-8}$$

Momentum Equation

The intensive property for the momentum equation is $\beta = V$. Therefore, according to the Reynolds transport theorem

$$\sum F = \frac{d}{dt} \int_{cv} \rho V d\mathcal{V} + \sum_{cs} V \rho V A \tag{11-9}$$

Referring to Fig. 11-2,

$$\sum F = \gamma \bar{y}_2 A_2 - \gamma \bar{y}_1 A_1 \tag{11-10}$$

and

$$\sum_{cs} \rho V^2 A = \rho A_1 V_1^2 - \rho A_2 V_2^2 \tag{11-11}$$

in which $\sum F$ = sum of the external forces on the control volume in the positive x direction; $\bar{y}$ = depth of the centroid of the flow section; and γ = specific weight of the liquid.

The time rate of change of momentum in the control volume is

$$\frac{d}{dt} \int_{cv} \rho V d\mathcal{V} = \rho V_w (A_2 V_2 - A_1 V_1) \tag{11-12}$$

Substituting Eqs. 11-10 to 11-12 into Eq. 11-9, utilizing Eq. 11-8, and simplifying the resulting equation, we obtain

$$\frac{A_1}{g}(V_1 - V_w)(V_1 - V_2) = \bar{y}_2 A_2 - \bar{y}_1 A_1 \tag{11-13}$$

Eliminating V_2 from Eqs. 11-8 and 11-13 and rearranging gives

$$(V_1 - V_w)^2 = \frac{g A_2}{A_1(A_2 - A_1)}(A_2 \bar{y}_2 - A_1 \bar{y}_1) \tag{11-14}$$

The velocity of the surge wave, V_w, must be greater than the undisturbed flow velocity, V_1, for the wave to propagate in the downstream direction. Therefore, it follows from the above equation that

$$V_w = V_1 + \sqrt{\frac{g A_2}{A_1(A_2 - A_1)}(A_2 \bar{y}_2 - A_1 \bar{y}_1)} \tag{11-15}$$

Transposing V_1 to the left-hand side,

$$V_w - V_1 = \sqrt{\frac{g A_2}{A_1(A_2 - A_1)}(A_2 \bar{y}_2 - A_1 \bar{y}_1)} \tag{11-16}$$

As we defined earlier, the celerity, c, of a wave is the velocity relative to the fluid in which it is traveling, i.e., $c = V_w - V_1$. Thus, the left-hand side of Eq. 11-16 is the celerity of the surge wave. Hence,

$$c = \sqrt{\frac{gA_2}{A_1(A_2 - A_1)}(A_2\bar{y}_2 - A_1\bar{y}_1)} \qquad (11\text{-}17)$$

The height of the surge wave, $y_2 - y_1$, produced by a sudden change in discharge may be determined from the following relationship between the flow depths and flow velocities at sections 1 and 2 (Fig. 11-2), derived by eliminating V_w from Eqs. 11-8 and 11-14:

$$A_2\bar{y}_2 - A_1\bar{y}_1 = \frac{A_1 A_2}{g(A_2 - A_1)}(V_1 - V_2)^2 \qquad (11\text{-}18)$$

Let us assume that we know the values of y_1 and V_1, or Q_1. Then, for a specified change in discharge from Q_1 to Q_2, we can determine the values of y_2 and V_2 from $Q_2 = V_2 A_2$ and Eq. 11-18 by trial and error. The value of the wave velocity, V_w, can then be determined from Eq. 11-15 for a wave traveling in the downstream direction. For a wave traveling in the upstream direction, use a negative sign with the radical term.

The preceding equations may be used for any cross section provided the simplifying assumptions we listed at the beginning of this section are valid.

For a *rectangular channel,* the preceding equations are simplified as follows. For a channel of width B, $A_1 = By_1$; $A_2 = By_2$; $\bar{y}_1 = \frac{1}{2}y_1$, and $\bar{y}_2 = \frac{1}{2}y_2$. Substituting these relationships into Eq. 11-17 and simplifying the resulting equation, we obtain

$$c = \sqrt{\frac{gy_2}{2y_1}(y_1 + y_2)} \qquad (11\text{-}19)$$

For waves of small height, $y_1 \simeq y_2 = y$ (say). Then Eq. 11-19 becomes

$$c = \sqrt{gy} \qquad (11\text{-}20)$$

Note that this is the same expression that we derived in Chapter 3 (Eq. 3-30).

Substituting $A_1 = By_1$ and $A_2 = By_2$ into Eq. 11-8, and simplifying gives

$$V_w = \frac{V_2 y_2 - V_1 y_1}{y_2 - y_1} \qquad (11\text{-}21)$$

Substitution of Eq. 11-19 into Eq. 11-2 for a wave traveling in the downstream direction and elimination of V_w from the resulting equation and Eq. 11-21 yields

$$(V_1 - V_2)^2 = \frac{g(y_1 - y_2)}{2y_1 y_2}(y_1^2 - y_2^2) \qquad (11\text{-}22)$$

Example 11-1

A 2.5-m-wide rectangular channel is carrying a flow of 5 m^3/s at a flow depth of 2 m. Determine the height of a surge wave and its velocity if the discharge is suddenly increased to 10 m^3/s at the upstream end.

Given:

$$Q_1 = 5 \text{ m}^3/\text{s}$$
$$Q_2 = 10 \text{ m}^3/\text{s}$$
$$y_1 = 2 \text{ m}$$
$$B = 2.5 \text{ m}$$

Determine:

$$y_2 = ?$$
$$V_w = ?$$

Solution:

$$V_1 = \frac{Q_1}{B y_1} = \frac{5}{2.5 \times 2} = 1 \text{ m/s}$$

Now,

$$Q_2 = B y_2 V_2$$

or

$$V_2 y_2 = \frac{10}{2.5} = 4$$

Substituting the values of y_1 and V_1 and $V_2 y_2 = 4$ into Eq. 11-22 and simplifying, we obtain

$$\left(1 - \frac{4}{y_2}\right)^2 = \frac{2.452(y_2 - 2)}{y_2}(y_2^2 - 4)$$

Solution of this equation by trial and error gives, $y_2 = 2.334$ m. Thus, the height of the surge = $2.334 - 2 = 0.334$ m.

Substituting values of y_1, y_2, V_1, and $V_2 y_2 = 4$ into Eq. 11-21, we obtain

$$
\begin{aligned}
V_w &= \frac{2 - 4}{2.0 - 2.334} \\
&= \frac{2}{0.334} \\
&= 5.99 \text{ m/s}
\end{aligned}
$$

11-5 Summary

In this chapter, a brief introduction to unsteady flow was presented. Commonly used terms were defined and causes that produce unsteady flow were discussed. Expressions were derived for determining the celerity and height of a gravity wave produced by a sudden change in discharge.

Problems

11-1. Consider a small wave of height Δy, which changes the flow velocity from V to $V + \Delta V$. By applying the continuity and momentum principles, derive Eq. 11-4b.

11-2. Determine the speed and height of a surge wave produced by instantaneously closing a downstream control gate in a 5-m-wide rectangular channel carrying a flow of 7.5 m³/s at a flow depth of 1.5 m.

11-3. If the width of the channel of Prob. 11-2 is reduced to 4 m at a distance of 500 m upstream of the control gate, determine the height of the wave in the constricted channel.

11-4. What will be the wave height if the width of the channel of Prob. 11-3 is increased to 7.5 m instead of reduced to 4 m?

11-5. By assuming that the shape of the wave does not change as it travels, the unsteady flow of Fig. 11-2 may be converted to steady flow by superimposing velocity V_w in the upstream direction on the entire system. Apply the continuity and momentum principles to the steady flow to derive Eqs. 11-8 and 11-17.

11-6. By utilizing the expression for the celerity of a small wave, show that a positive wave steepens and a negative wave flattens as they travel in a channel having negligible friction.

11-7. Derive expressions for the absolute wave velocities and heights of the reflected waves after two positive waves meet.

11-8. Derive the expressions for the wave velocities and the wave heights of the reflected and transmitted waves at a step rise in the channel bottom.

11-9. A monoclinal wave does not change shape as it propagates in a channel, and the flow is assumed to be uniform on the upstream and downstream sides of the wave. If the subscripts u and d refer to the variables for the upstream and downstream sides, prove that the velocity of this wave is

$$V_w = \frac{Q_u - Q_d}{A_u - A_d}$$

REFERENCES

Abbott, M. B.   1979. *Computational hydraulics: Elements of the theory of free surface flows,* Pitman, London.

Basco, D. R.   1983. *Computation of rapidly varied, unsteady free-surface flow,* U. S. Geological Survey Report WRI 83-4284.

Chow, V. T.   1959. *Open-channel hydraulics,* McGraw-Hill, New York, NY.

Cunge, J. A.; Holly, F. M., and Verwey, A.   1980. *Practical aspects of computational river hydraulics,* Pitman, London.

Dronkers, J. J.   1964. *Tidal computations in rivers and coastal waters,* North-Holland, Amsterdam, the Netherlands.

Henderson, F. M.   1966. *Open Channel Flow,* Macmillan, New York, NY.

Lai, C.   1986. Numerical modeling of unsteady open-channel flow, *Advances in Hydroscience,* 14:161–333.

Le Mehaste, B.   1976. *Hydrodynamics,* Springer Verlag, New York, NY.

Mahmood, K., and Yevjevich, V., eds.   1975, *Unsteady flow in open channels,* vol. 1, Water Resources Publications, Fort Collins, CO.

Roberson, J. A., and Crowe, C. T.   1990. *Engineering Fluid Mechanics,* 4th ed. Houghton Mifflin, Boston, MA.

Stoker, J. J.   1957. *Water waves,* Wiley Interscience, New York, NY.

12 Governing Equations for One-Dimensional Flow

Undular bore in a trapezoidal channel, $\mathbf{F}_r = 1.12$ (Courtesy of A. Treske, Hydraulic Laboratory, Obernach, Germany).

12-1 Introduction

Three conservation laws—mass, momentum, and energy—are used to describe open-channel flows. Two flow variables, such as the flow depth and velocity or the flow depth and the rate of discharge, are sufficient to define the flow conditions at a channel cross section. Therefore, two governing equations may be used to analyze a typical flow situation. The continuity equation and the momentum or energy equation are used for this purpose. Except for the velocity head coefficient, α, and the momentum coefficient, β, the momentum and energy equations are equivalent (Cunge, et al., 1980) provided the flow depth and velocity are continuous; i.e., there are no discontinuities, such as a jump or a bore. However, the momentum equation should be used if the flow has discontinuities, since, unlike the energy equation, it is not necessary to know the amount of losses in the discontinuities in the application of the momentum equation.

In this chapter, we will derive the continuity and momentum equations, usually referred to as de Saint Venant equations. Several investigators (Stoker 1957; Chow 1959; Dronkers 1964; Henderson 1966; Strelkoff 1969; Yen 1973; Liggett 1975; Cunge 1980; Lai 1986, and Abbott and Basco, 1990) derived these equations by using different procedures. For illustration purposes, we will use two different procedures in our derivations. We will use the Reynolds transport theorem for the prismatic channels having lateral inflows or outflows. The type of the governing equations is then discussed. The equations describing flows having nonhydrostatic pressure distribution are derived by integrating the continuity and momentum equations for two-dimensional flows. The chapter concludes by presenting integral forms of the governing equations.

12-2 St. Venant Equations

We will make the following assumptions in the derivation of the St. Venant equations.

1. The pressure distribution is hydrostatic. This is a valid assumption if the streamlines do not have sharp curvatures.
2. The channel bottom slope is small, so that the flow depths measured normal to the channel bottom and measured vertically are approximately the same.
3. The flow velocity over the entire channel cross section is uniform.
4. The channel is prismatic—i.e., the channel cross section and the channel bottom slope do not change with distance. The variations in the cross section or bottom

slope may be taken into consideration by approximating the channel into several prismatic reaches.

5. The head losses in unsteady flow may be simulated by using the steady-state resistance laws, such as the Manning or Chezy equation, i.e., head losses for a given flow velocity during unsteady flow are the same as that during steady flow.

Continuity Equation

The mass density of water is constant. Therefore, the continuity equation is the same as the law of conservation of mass.

Let us consider a control volume having fixed boundaries, as shown in Fig. 12-1. If the flow between section 1 and 2 is unsteady and nonuniform, then the rate of discharge, Q, flow velocity, V, and flow depth, y, are functions of distance x (measured positive in the downstream direction) and time, t. For applying the Reynolds transport theorem to the control volume, the extensive property, B = mass, M, and the intensive property, $\beta = \Delta m / \Delta m = 1$.

Referring to Fig. 12-1, substituting $B = M$ and $\beta = 1$ into Eq. 1-26, using subscripts 1 and 2 to indicate flow variables at sections 1 and 2, respectively, and noting that $dM/dt = 0$ (law of conservation of mass), we obtain

$$\frac{dM}{dt} = \frac{d}{dt} \int_{x_1}^{x_2} \rho A \, dx + \rho A_2 V_2 - \rho A_1 V_1 - \rho q_l (x_2 - x_1) = 0 \qquad (12\text{-}1)$$

in which A = flow area; ρ = mass density of water; and q_l = volumetric rate of lateral inflow or outflow per unit length of the channel between sections 1 and 2. The lateral inflow, q_l, into the channel is considered positive, whereas the outflow is considered negative. The lateral inflow or outflow may be due to infiltration, evaporation, or flows to or from the channel banks.

Since water may be assumed incompressible, mass density, ρ, is constant. Therefore, Eq. 12-1 may be written as

$$\frac{d}{dt} \int_{x_1}^{x_2} A \, dx + A_2 V_2 - A_1 V_1 - q_l (x_2 - x_1) = 0 \qquad (12\text{-}2)$$

We may derive the differential or integral form of the continuity equation from Eq. 12-2. The differential form requires that the flow variables are continuous, but there is no such restriction on the integral form. We derive the differential form in the following paragraphs and the integral form in Sec. 12-5.

Let us apply Leibnitz's rule* to the first term on the left-hand side of Eq. 12-1. To do this it is necessary that both A and $\partial A / \partial t$ are continuous with respect to both x and t. In other words, between x_1 and x_2, there are no abrupt discontinuities in the channel cross section (i.e., the channel width and the channel bottom do not have step changes)

*According to this rule

$$\frac{d}{dt} \int_{f_1(t)}^{f_2(t)} F(x, t) \, dx = \int_{f_1(t)}^{f_2(t)} \frac{\partial}{\partial t} F(x, t) \, dx + F(f_2(t), t) \frac{df_2}{dt} - F(f_1(t), t) \frac{df_1}{dt}$$

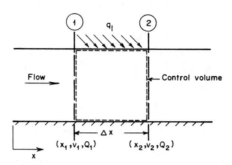

Figure 12-1 Definition sketch for continuity equation.

and in the flow depth (i.e., there is no bore or jump). By applying this rule, noting that $dx_1/dt = 0$ and $dx_2/dt = 0$, since the boundaries of the control volume are fixed, and $Q_1 = A_1 V_1$ and $Q_2 = A_2 V_2$, we obtain

$$\int_{x_1}^{x_2} \frac{\partial A}{\partial t}\, dx + Q_2 - Q_1 - q_l(x_2 - x_1) = 0 \tag{12-3}$$

Based on the mean value theorem* (to apply this theorem, it is necessary that both Q and $\partial Q/\partial x$ are continuous), Eq. 12-3 takes the form

$$\frac{\partial A}{\partial t} + \frac{\partial Q}{\partial x} = q_l \tag{12-4}$$

Equation 12-4 is referred to as the continuity equation in the *conservation* or *divergent form*. According to this equation, mass is conserved along any closed contour in the x-t plane if the right-hand side of this equation is zero. If the term on the right-hand side is not zero, then this term acts like a source or a sink, depending upon the sign of q_l.

For a channel having a regular cross section (i.e., the top water-surface width, B, is a continuous function of the flow depth y), the change in flow area, ΔA, for a small change in flow depth, Δy, may be approximated as $B \Delta y$. In the limit as $\Delta y \to 0$, $dA/dy = B$. Hence, Eq. 12-4 may be written as

$$B \frac{\partial y}{\partial t} + \frac{\partial Q}{\partial x} = q_l \tag{12-5}$$

Similarly, we may write $Q = VA$. Substituting this into Eq. 12-5, expanding the second term, and noting that $\partial A/\partial x = B \partial y/\partial x$ and the hydraulic depth, $D = A/B$, this equation becomes

$$\frac{\partial y}{\partial t} + D \frac{\partial V}{\partial x} + V \frac{\partial y}{\partial x} - \frac{q_l}{B} = 0 \tag{12-6}$$

*According to this theorem

$$\int_{x_1}^{x_2} F(x)\, dx = (x_2 - x_1) F(\zeta) \qquad x_1 < \zeta < x_2$$

Momentum Equation

For the momentum equation, the extensive property, $\mathcal{B}$ = momentum of water in the control volume = mV and the intensive property $\beta = V\,\Delta m/\Delta m = V$. In addition, according to the Newton's second law of motion, the rate of change of momentum is equal to the resultant force acting on the control volume, i.e., $\sum F = d\mathcal{B}/dt$. Substitution of these relationships into Eq. 1-26 yields

$$\sum F = \frac{d}{dt} \int_{x_1}^{x_2} V\rho A \ dx + V_2\rho A_2 V_2 - V_1\rho A_1 V_1 - V_x\rho q_l(x_2 - x_1) \qquad (12\text{-}7)$$

in which V_x is the component of the velocity of lateral inflow in the x-direction. Note that q_l is positive for lateral inflow and it is negative for lateral outflow.

By applying Leibnitz's rule and writing $Q = VA$, Eq. 12-7 becomes

$$\sum F = \int_{x_1}^{x_2} \rho \frac{\partial Q}{\partial t} \ dx + \rho Q_2 V_2 - \rho Q_1 V_1 - V_x\rho q_l(x_2 - x_1) \qquad (12\text{-}8)$$

By dividing throughout by $\rho(x_2 - x_1)$ and applying the mean value theorem, we obtain

$$\frac{\sum F}{\rho(x_2 - x_1)} = \frac{\partial Q}{\partial t} + \frac{\partial(QV)}{\partial x} - V_x q_l \qquad (12\text{-}9)$$

For simplicity, let us neglect the shear stresses on the flow surface due to wind and neglect the effects of the Coriolis acceleration. These are valid assumptions for typical hydraulic engineering applications. Since the channel is assumed to be prismatic, there are no external forces acting on the control volume due to changes in the channel cross section. Hence, referring to Fig. 12-2, the following forces are acting on the control volume:

$$\text{Pressure force, } F_1, \text{ acting on the upstream end } = \rho g A_1 \bar{y}_1 \qquad (12\text{-}10\text{a})$$

in which $\bar{y}_1$ = depth of the centroid of flow area A_1. Similarly,

$$\text{Pressure force, } F_2, \text{ acting on the downstream end } = \rho g A_2 \bar{y}_2 \qquad (12\text{-}10\text{b})$$

The component of weight of water in the control volume in the x-direction, F_3, may be written as

$$F_3 = \rho g \int_{x_1}^{x_2} A S_o \ dx \qquad (12\text{-}10\text{c})$$

in which S_o = channel bottom slope, considered positive sloping downward.

The frictional force, F_4, is due to shear between water and the channel sides and the channel bottom. This may be expressed in terms of the friction slope, S_f, or the energy gradient needed to overcome friction as

$$F_4 = \rho g \int_{x_1}^{x_2} A S_f \ dx \qquad (12\text{-}10\text{d})$$

For any exponential formula for friction losses, the expression for the friction slope may be written as

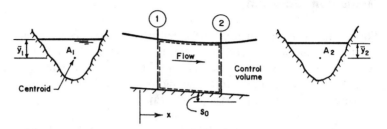

Figure 12-2 Definition sketch for momentum equation.

$$S_f = \frac{CV|V|^{m-1}}{R^p} \qquad (12\text{-}10\text{e})$$

in which C and p are the coefficients that depend upon the formula employed, $R =$ hydraulic radius, and m depends upon the flow type. For example, $m = 1$ for laminar flow; $m = 1.75$ for smooth, turbulent flow; and $m = 2$ for fully rough, turbulent flow. Hence, the resultant force acting on the control volume is

$$\sum F = F_1 - F_2 + F_3 - F_4 \qquad (12\text{-}11)$$

Substitution of expressions for F_1 through F_4 from Eq. 12-10 into Eq. 12-11 and division by $\rho(x_2 - x_1)$ yield

$$\frac{\sum F}{\rho(x_2 - x_1)} = \frac{g(A_1\bar{y}_1 - A_2\bar{y}_2)}{x_2 - x_1} + \frac{g}{x_2 - x_1}\int_{x_1}^{x_2} A(S_o - S_f)\,dx \qquad (12\text{-}12)$$

Equating the right-hand sides of Eqs. 12-9 and 12-12 and applying the mean value theorem, we obtain

$$\frac{\partial Q}{\partial t} + \frac{\partial(QV)}{\partial x} - V_x q_l = -g\frac{\partial}{\partial x}(A\bar{y}) + gA(S_o - S_f) \qquad (12\text{-}13)$$

This equation may be written as

$$\frac{\partial Q}{\partial t} + \frac{\partial}{\partial x}(QV + gA\bar{y}) = gA(S_o - S_f) + V_x q_l \qquad (12\text{-}14)$$

This equation is referred to as the momentum equation in the conservation form. This means that the momentum along any closed contour in the x-t plane is conserved if the right-hand side of Eq. 12-14 is zero (Cunge, et al., 1980). Nonzero terms on the right-hand side act as sources or sinks.

Now, $\Delta(A\bar{y}) = [A(\bar{y}+\Delta y) - \frac{1}{2}B(\Delta y)^2] - A\bar{y}$. Thus, by neglecting the higher-order terms and letting $\Delta y \rightarrow 0$, we obtain $(\partial/\partial y)(A\bar{y}) = A$. By utilizing this expression, we may write

$$\frac{\partial}{\partial x}(gA\bar{y}) = g\frac{\partial}{\partial y}(A\bar{y})\frac{\partial y}{\partial x} = gA\frac{\partial y}{\partial x}$$

Hence, Eq. 12-14 becomes

$$\frac{\partial Q}{\partial t} + \frac{\partial(QV)}{\partial x} + gA\frac{\partial y}{\partial x} = V_x q_l + gA(S_o - S_f) \qquad (12\text{-}15)$$

By expanding the first two terms on the left-hand side and rearranging, we get

$$
V\left[B\frac{\partial y}{\partial t} + A\frac{\partial V}{\partial x} + BV\frac{\partial y}{\partial x} - \frac{V_x}{V}q_l\right]
$$
$$
+ A\left(\frac{\partial V}{\partial t} + V\frac{\partial V}{\partial x} + g\frac{\partial y}{\partial x} + gS_f - gS_o\right) = 0 \tag{12-16}
$$

According to Eq. 12-6, the sum of the terms within the brackets is zero if $V_x = 0$ or if $V_x = V$. Hence

$$
\frac{\partial V}{\partial t} + g\frac{\partial}{\partial x}\left(\frac{V^2}{2g} + y\right) = g(S_o - S_f) \tag{12-17}
$$

This equation has been extensively used in the literature and has been called the momentum equation, equation of motion, and dynamic equation. Since it does not truly describe conservation of momentum, we refer to it as the *dynamic equation.*

We may rearrange the terms of Eq. 12-17 as

$$
S_f = S_o \left|-\frac{\partial}{\partial x}\left(\frac{V^2}{2g} + y\right)\right| - \frac{1}{g}\frac{\partial V}{\partial t}
$$

$\Big|$Steady, uniform

$\quad\quad\quad\quad\quad\quad$Steady, nonuniform

$\quad\quad\quad\quad\quad\quad\quad\quad$Unsteady, nonuniform

The significance of each term is clear from this equation. For steady-uniform flow, the slope of the energy-grade line is the same as the channel-bottom slope. The equation for steady, gradually varied flow is obtained by including the variation of the flow depth and the velocity head, i.e., by including the derivative with respect to distance x. The unsteadiness, or the local acceleration term, is needed to make the equation valid for unsteady-nonuniform flow.

12-3 General Remarks

The continuity and momentum equations form a set of nonlinear partial differential equations. A closed-form solution of these equations is not available except for very simplified cases. Therefore, numerical methods are used for their integration. To select a numerical scheme, it is helpful if we know the type of these equations, i.e., whether they are hyperbolic, parabolic, or elliptic. We may determine the type of these equations as follows.

Let us/write Eqs. 12-6 and 12-17 in the vector form as

$$
\frac{\partial \mathbf{U}}{\partial t} + \mathbf{A}\frac{\partial \mathbf{U}}{\partial x} = \mathbf{S} \tag{12-18}
$$

in which

$$\mathbf{U} = \begin{pmatrix} y \\ V \end{pmatrix}; \qquad \mathbf{A} = \begin{pmatrix} V & D \\ g & V \end{pmatrix}; \qquad \mathbf{S} = \begin{pmatrix} -\dfrac{q_l}{B} \\ g(S_o - S_f) \end{pmatrix} \qquad (12\text{-}19)$$

The eigenvalues of matrix $\mathbf{A}$ define the type of Eq. 12-18. These eigenvalues are determined from the equation

$$\begin{vmatrix} V - \lambda & D \\ g & V - \lambda \end{vmatrix} = 0 \qquad (12\text{-}20)$$

It follows from this equation that

$$(V - \lambda)^2 - gD = 0 \qquad (12\text{-}21)$$

Two roots of this equation are

$$\begin{aligned} \lambda_1 &= V + \sqrt{\frac{gA}{B}} \\ \lambda_2 &= V - \sqrt{\frac{gA}{B}} \end{aligned} \qquad (12\text{-}22)$$

The radical term of Eq. 12-22 is an expression for the wave celerity. Hence, the eigenvalues of $\mathbf{A}$ represent absolute wave velocities, $V + c$ and $V - c$.

The expressions for the eigenvalues, λ_1 and λ_2, show that both of them are real and distinct for sub- as well as supercritical flows. Therefore, Eqs. 12-18 are a set of *hyperbolic* partial differential equations. This type of equation represents the propagation of waves in different media. Computational procedures, referred to as the *marching procedures*, are suitable for the numerical integration of these equations.

12-4 Boussinesq Equations

If the streamlines in a flow have sharp curvatures, then the pressure distribution is not hydrostatic because of acceleration. In this section, we derive the equations to describe these flows by integrating the two-dimensional flow equations in the vertical direction and by utilizing the Boussinesq assumption. These equations are referred to as the *Boussinesq equations*. According to the Boussinesq assumption, the flow velocity, w, in the vertical direction varies from zero at the channel bottom to the maximum value at the flow surface.

In the derivation, we assume that the flow velocity in the lateral direction is zero; channel-bottom slope is small; flow velocity, u, in the x-direction is uniform over the flow depth, and the datum lies along the channel bottom (Fig. 12-3). In addition, we assume for simplicity that the channel is frictionless.

Continuity Equation

The continuity equation for a two-dimensional flow may be written as

$$\frac{\partial u}{\partial x} + \frac{\partial w}{\partial z} = 0 \qquad (12\text{-}23)$$

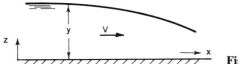

Figure 12-3 Notation.

in which $u =$ flow velocity in the x-direction and $w =$ flow velocity in the z-direction. By multiplying this equation throughout by dz and integrating it from the channel bottom ($z = 0$) to the free surface ($z = y$), we obtain

$$\int_0^y \frac{\partial u}{\partial x} dz + \int_0^y \frac{\partial w}{\partial z} dz = 0 \qquad (12\text{-}24)$$

Based on the Leibnitz rule, this equation becomes

$$\frac{\partial}{\partial x} \int_0^y u \, dz + \left(u \frac{\partial z}{\partial x} \right)\Big|_{z=0} - \left(u \frac{\partial z}{\partial x} \right)\Big|_{z=y} + w \Big|_{z=y} - w \Big|_{z=0} = 0 \qquad (12\text{-}25)$$

Let us refer to quantities for the channel bottom ($z = 0$) and for the water surface ($z = y$) by subscripts b and s respectively. Then

$$\frac{\partial}{\partial x} \int_0^y u \, dz + u_b \frac{\partial z_b}{\partial x} - u_s \frac{\partial z_s}{\partial x} + w_s - w_b = 0 \qquad (12\text{-}26)$$

For the velocity of the water surface, w_s, we may write

$$w_s = \frac{dy}{dt} = \frac{\partial y}{\partial t} + u_s \frac{\partial y}{\partial x} \qquad (12\text{-}27)$$

Since the datum lies along the channel bottom, $\partial z_b / \partial x = 0$. In addition, for a rigid channel bottom, $w_b = 0$. Hence, based on Eq. 12-27, Eq.12-26 may be written as

$$\frac{\partial y}{\partial t} + \frac{\partial (uy)}{\partial x} = 0 \qquad (12\text{-}28)$$

This equation may be expanded as

$$\frac{\partial y}{\partial t} + u \frac{\partial y}{\partial x} + y \frac{\partial u}{\partial x} = 0 \qquad (12\text{-}29)$$

or

$$\frac{\partial u}{\partial x} = -\frac{1}{y} \left(\frac{\partial y}{\partial t} + u \frac{\partial y}{\partial x} \right) \qquad (12\text{-}30)$$

Now, it follows from Eq. 12-23 that

$$\frac{\partial w}{\partial z} = -\frac{\partial u}{\partial x} \qquad (12\text{-}31)$$

Since u is assumed to be constant across the flow depth y, we can integrate Eq. 12-31 as

$$w = -\frac{\partial u}{\partial x} z \qquad (12\text{-}32)$$

Substitution of Eq. 12-30 into Eq. 12-32 yields

$$w = \frac{1}{y}\left(\frac{\partial y}{\partial t} + u\frac{\partial y}{\partial x}\right)z \tag{12-33}$$

Momentum Equation in the z-Direction

The momentum equation in the z-direction may be written as

$$\frac{\partial w}{\partial t} + u\frac{\partial w}{\partial x} + w\frac{\partial w}{\partial z} = -\frac{1}{\rho}\frac{\partial p}{\partial z} - g \tag{12-34}$$

By multiplying Eq. 12-34 by z and rearranging the terms, we obtain

$$\frac{p}{\rho} = \frac{\partial(zw)}{\partial t} + \frac{\partial(uzw)}{\partial x} + \frac{\partial(w^2z)}{\partial z} + gz + \frac{1}{\rho}\frac{\partial(pz)}{\partial z} - wz\left(\frac{\partial u}{\partial x} + \frac{\partial w}{\partial z}\right) - w^2 \tag{12-35}$$

Based on the continuity equation (Eq. 12-23), the sum of the terms in the parentheses on the right-hand side is zero. Hence, Eq. 12-35 becomes

$$\frac{p}{\rho} = \frac{\partial(zw)}{\partial t} + \frac{\partial(uzw)}{\partial x} + \frac{\partial(w^2z)}{\partial z} + gz + \frac{1}{\rho}\frac{\partial(pz)}{\partial z} - w^2 \tag{12-36}$$

Integration of this equation over the flow depth, y, yields

$$\int_0^y \frac{p}{\rho}\,dz = \int_0^y \frac{\partial(zw)}{\partial t}\,dz + \int_0^y \frac{\partial(uzw)}{\partial x}\,dz + \int_0^y \frac{\partial(w^2z)}{\partial z}\,dz$$
$$+ \int_0^y gz\,dz + \int_0^y \frac{1}{\rho}\frac{\partial(pz)}{\partial z}\,dz - \int_0^y w^2\,dz \tag{12-37}$$

Applying the Leibnitz rule and referring to the quantities for the free surface and for the channel bottom by the subscripts s and b respectively, this equation may be written as

$$\int_0^y \frac{p}{\rho}\,dz = \frac{\partial}{\partial t}\int_0^y wz\,dz - (wz)_s\frac{\partial y}{\partial t} + (wz)_b\frac{\partial z_b}{\partial t}$$
$$+ \frac{\partial}{\partial x}\int_0^y uwz\,dz - (uwz)_s\frac{\partial y}{\partial x}$$
$$+ (uwz)_b\frac{\partial z_b}{\partial x} + (w^2z)_s - (w^2z)_b$$
$$+ \frac{1}{2}gy^2 - \int_0^y w^2\,dz + \frac{1}{\rho}[(pz)_s - (pz)_b] \tag{12-38}$$

The flow conditions at the free surface and at the channel bottom are described by the following equations:

Free surface ($z = z_s$).

1. $w_s = \frac{\partial y}{\partial t} + u_s\frac{\partial y}{\partial x}$ (Eq. 12-27).
2. $p_s = 0$, since the pressure at the free surface is atmospheric.

Channel bottom $(\mathbf{z} = \mathbf{z_b})$.

1. $w_b = 0$, since $w = 0$ at the channel bottom.
2. $p_b z_b = 0$, since the channel bottom is used as datum.

Based on these conditions, Eq. 12-38 may be written as

$$\int_0^y \frac{p}{\rho}\,dz = \frac{\partial}{\partial t}\int_0^y wz\,dz + \frac{\partial}{\partial x}\int_0^y uwz\,dz + \frac{1}{2}gy^2 - \int_0^y w^2\,dz \qquad (12\text{-}39)$$

Momentum Equation in the x-Direction

The momentum equation in the x-direction may be written as

$$\frac{\partial u}{\partial t} + u\frac{\partial u}{\partial x} + w\frac{\partial u}{\partial z} = -\frac{1}{\rho}\frac{\partial p}{\partial x} \qquad (12\text{-}40)$$

Based on Eq. 12-24 and rearranging the terms, Eq. 12-40 may be written as

$$\frac{\partial u}{\partial t} + \frac{\partial u^2}{\partial x} + \frac{\partial(uw)}{\partial z} + \frac{1}{\rho}\frac{\partial p}{\partial x} = 0 \qquad (12\text{-}41)$$

By multiplying Eq. 12-41 throughout by dz, integrating from 0 to y, and applying the conditions at the free surface and at the bottom, we obtain

$$\frac{\partial(yu)}{\partial t} + \frac{\partial(yu^2)}{\partial x} + \int_0^y \frac{1}{\rho}\frac{\partial p}{\partial x}\,dz = 0 \qquad (12\text{-}42)$$

By applying the Leibnitz rule, this equation becomes

$$\frac{\partial(yu)}{\partial t} + \frac{\partial(yu^2)}{\partial x} + \frac{\partial}{\partial x}\int_0^y \frac{p}{\rho}\,dz - \left(\frac{p}{\rho}\right)_s \frac{\partial y}{\partial x} + \left(\frac{p}{\rho}\right)_b \frac{\partial z_b}{\partial x} = 0 \qquad (12\text{-}43)$$

As discussed in the previous paragraphs, $p_s = 0$ and $\partial z_b/\partial x = 0$. Substitution of these relationships and Eq. 12-39 into Eq. 12-43 and rearrangement of the terms of the resulting equation yield

$$\frac{\partial(yu)}{\partial t} + \frac{\partial}{\partial x}\left(yu^2 + \frac{1}{2}gy^2\right) + \frac{\partial}{\partial x}\left(\frac{\partial}{\partial t}\int_0^y wz\,dz + \frac{\partial}{\partial x}\int_0^y uwz\,dz\right.$$
$$\left. - \int_0^y w^2\,dz\right) = 0 \qquad (12\text{-}44)$$

Let us simplify the last term of this equation. By substituting Eq. 12-32 into the first expression of the last term, we obtain

$$\frac{\partial}{\partial t}\int_0^y wz\,dz = \frac{\partial}{\partial t}\int_0^y \left(-\frac{\partial u}{\partial x}z^2\right)dz$$

or

$$\frac{\partial}{\partial t}\int_0^y wz\,dz = \frac{\partial}{\partial t}\left(-\frac{\partial u}{\partial x}\frac{y^3}{3}\right) \qquad (12\text{-}45)$$

Similarly

$$\frac{\partial}{\partial x} \int_0^y uwz\, dz = \frac{\partial}{\partial x}\left(-u\frac{\partial u}{\partial x}\frac{y^3}{3}\right) \tag{12-46}$$

and

$$\int_0^y w^2\, dz = \left(\frac{\partial u}{\partial x}\right)^2 \frac{y^3}{3} \tag{12-47}$$

By substituting Eqs. 12-45 through 12-47 into the third term of Eq. 12-44 and expanding and simplifying the resulting expression, we get

$$\frac{\partial}{\partial x}\left(\frac{\partial}{\partial t}\int_0^y wz\, dz + \frac{\partial}{\partial x}\int_0^y uwz\, dz - \int_0^y w^2\, dz\right)$$
$$= -\frac{\partial}{\partial x}\left\{y^2\frac{\partial u}{\partial x}\left(\frac{\partial y}{\partial t}+u\frac{\partial y}{\partial x}\right)+\frac{y^3}{3}\left[\frac{\partial^2 u}{\partial x\partial t}+2\left(\frac{\partial u}{\partial x}\right)^2+u\frac{\partial^2 u}{\partial x^2}\right]\right\} \tag{12-48}$$

Based on Eq. 12-29, this equation may be written as

$$\frac{\partial}{\partial x}\left(\frac{\partial}{\partial t}\int_0^y wz\, dz + \frac{\partial}{\partial x}\int_0^y uwz\, dz - \int_0^y w^2\, dz\right)$$
$$= -\frac{1}{3}\frac{\partial}{\partial x}\left\{y^3\left[\frac{\partial^2 u}{\partial x\partial t}+u\frac{\partial^2 u}{\partial x^2}-\left(\frac{\partial u}{\partial x}\right)^2\right]\right\} \tag{12-49}$$

Hence, the momentum equation in the x-direction is

$$\frac{\partial yu}{\partial t}+\frac{\partial}{\partial x}\left\{yu^2+\frac{1}{2}gy^2-\frac{y^3}{3}\left[\frac{\partial^2 u}{\partial x\partial t}+u\frac{\partial^2 u}{\partial x^2}-\left(\frac{\partial u}{\partial x}\right)^2\right]\right\}=0 \tag{12-50}$$

By expanding various terms of this equation, we obtain

$$\frac{\partial yu}{\partial t}+\frac{\partial}{\partial x}\left(yu^2+\frac{1}{2}gy^2\right)+y^2\frac{\partial y}{\partial x}\left[\left(\frac{\partial u}{\partial x}\right)^2-u\frac{\partial^2 u}{\partial x^2}-\frac{\partial^2 u}{\partial x\partial t}\right]$$
$$+\frac{y^3}{3}\left(\frac{\partial u}{\partial x}\frac{\partial^2 u}{\partial x^2}-\frac{\partial^3 u}{\partial x^2\partial t}-u\frac{\partial^3 u}{\partial x^3}\right)=0 \tag{12-51}$$

Usually, the terms containing the product of derivatives and the third space derivatives are neglected (Basco 1983). Therefore, the preceding equation simplifies to

$$\frac{\partial yu}{\partial t}+\frac{\partial}{\partial x}\left(yu^2+\frac{1}{2}gy^2\right)-\frac{1}{3}\frac{\partial^3 u}{\partial x^2\partial t}=0 \tag{12-52}$$

Equations 12-28 and 12-52, referred to as the Boussinesq equations, describe flows if the pressure distribution is not hydrostatic. Note that if the additional terms accounting for the nonhydrostatic pressure distribution are neglected, these equations reduce to the Saint Venant equations.

12-5 Integral Forms

In Sec. 12-2, we developed differential form of the governing equations. For the derivation of these equations, we had to assume that the flow variables and the parameters of the channel section are continuous both in space and time. This was needed for the derivatives of various variables to exist. In this section, we will derive the integral form of the equations for which we do not have to make this assumption. Thus, we can use these equations to analyze flows having discontinuities such as bores and shocks.

By multiplying Eq. 12-2 by dt and integrating from t_1 to t_2, we obtain

$$\int_{x_1}^{x_2} (A_{t_2} - A_{t_1})\, dx + \int_{t_1}^{t_2} (Q_2 - Q_1)\, dt + \int_{t_1}^{t_2} \int_{x_1}^{x_2} q_l\, dx\, dt = 0 \qquad (12\text{-}53)$$

This is the continuity equation in the integral form.

Multiplying Eq. 12-7 by dt, integrating from t_1 to t_2, substituting expressions for $\sum F$ from Eq. 12-12, and noting that ρ is constant, we obtain

$$\int_{x_1}^{x_2} (Q_{t_2} - Q_{t_1})\, dx + \int_{t_1}^{t_2} Q_2 V_2\, dt - \int_{t_1}^{t_2} Q_1 V_1\, dt$$

$$- \int_{t_1}^{t_2} \int_{x_1}^{x_2} V_x q_l\, dx\, dt = g \int_{t_1}^{t_2} (A_1 \bar{y}_1 - A_2 \bar{y}_2)\, dt \qquad (12\text{-}54)$$

$$+ g \int_{t_1}^{t_2} \int_{x_1}^{x_2} A(S_o - S_f)\, dx\, dt$$

This is the momentum equation in the integral form.

12-6 Summary

In this chapter, we derived the continuity and momentum equations describing unsteady flow in open channels and presented different formulations of these equations. It was proved that these partial differential equations are hyperbolic. Then, these equations were derived for flows having nonhydrostatic pressure distribution based on the assumption that the vertical flow velocity varies from zero at the channel bottom to maximum at the water surface. The chapter concluded with the derivation of the integral form of the governing equations.

Problems

12-1. Figure 12-4 shows an infinitesimal length of a channel having a small bottom slope and no lateral inflow or outflow. By applying the law of conservation of mass between sections 1 and 2, derive the continuity equation for one-dimensional unsteady flow.

12-2. By using the Newton's second law of motion to the volume of water between sections 1 and 2 of Fig. 12-4, derive the momentum equation.

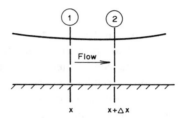

x   x+Δx   **Figure 12-4** Channel segment.

12-3. Deduce Eq. 5-5 from Eq. 12-17. Assume $\alpha = 1$.

12-4. If we use stage (the elevation of water surface above a specified datum), Z, instead of flow depth, y, show that the continuity and momentum equations for a prismatic channel become

$$\frac{\partial Z}{\partial t} + V\frac{\partial Z}{\partial x} + \frac{A}{B}\frac{\partial V}{\partial x} + VS_o = 0$$

$$\frac{\partial V}{\partial t} + V\frac{\partial V}{\partial x} + g\frac{\partial Z}{\partial x} + gS_f = 0$$

12-5. Starting with the momentum equation, show that the governing equation for steady, uniform flow is $S_f = S_o$.

12-6. If the wind stress on the flow surface is included, prove that the continuity and momentum equations become

$$B\frac{\partial y}{\partial t} + \frac{\partial Q}{\partial x} - q_l = 0$$

$$\frac{\partial Q}{\partial t} + \frac{Q}{A}\frac{\partial Q}{\partial x} + Q\frac{\partial}{\partial x}\left(\frac{Q}{A}\right) + gA\frac{\partial y}{\partial x}$$

$$-gA(S_o - S_f) - q_l u - k_w B V_w^2 \cos\theta = 0$$

in which $u = $ velocity component of the lateral flow in the positive x-direction, $V_w = $ wind velocity; and $k = $ dimensionless wind stress coefficient.

REFERENCES

Abbott, M. B. 1979. *Computational hydraulics: Elements of the theory of free surface flows,* Pitman, London.

Abbott, M. B., and Basco, D. R. 1990. Computational fluid dynamics, John Wiley, New York, NY.

Basco, D. R. 1983. *Computation of rapidly varied, unsteady free-surface flow,* U. S. Geological Survey Report WRI 83-4284.

Boussinesq, J. 1977. Essais sur la theorie des eaux courantes. *Memoires* presertes par divers, Savants a l'Academie des Sciences de l'Institut de France, 23:1-680; 24:1-64.

Chow, V. T. 1959. *Open-channel hydraulics,* McGraw-Hill, New York, NY.

Cunge, J. A., Holly, F. M.; and Verwey, A. 1980. *Practical aspects of computational river hydraulics,* Pitman, London.

de Saint-Venant, B.   1871. Theorie du mouvement non permanent des eaux, avec application aux crues de rivieras et a l'introduction des marces dans leur lit. *Comptes Rendus de l'Academic des Sciences,* 73: 147–54, 237–40.

Dronkers, J. J.   1964. *Tidal computations in rivers and coastal waters,* North-Holland, Amsterdam, Netherlands.

Henderson, F. M.   1966. *Open channel flow,* Macmillan, New York, NY.

Lai, C.   1986. Numerical modeling of unsteady open-channel flow, *Advances in Hydroscience* 14: 161–333.

Liggett, J. A.   1975. "Basic equations of unsteady flow." In *Unsteady flow in open channels,* edited by K. Mahmood and V. Yevjevich, vol. 1, Chap. 2, Water Resources Publications, Fort Collins, CO.

Stoker, J. J.   1957. *Water waves,* Wiley Interscience, New York, NY.

Strelkoff, T.   1969. One-dimensional equations of open-channel flow, *Jour. Hydraulics Division,* Amer. Soc. Civil Engrs., 95, HY3:861–66.

Yen, B. C.   1973. Open-channel flow equations revisited, *Jour. Engineering Mechanics Division,* Amer. Soc. Civil Engrs., 99, EM 5:979–1009.

13 Numerical Methods

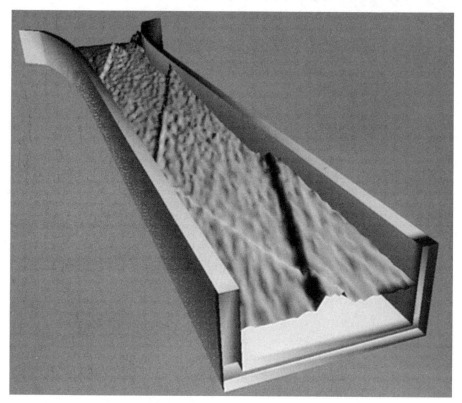

Computed shock waves in a fan-shaped contraction by a two-dimensional finite-difference mathematical model $F_r = 4$, $b_1/b_2 = 2$ (Courtesy of P. Rutschmann, VAW, Swiss Federal Institute of Tech., Zurich, Switzerland).

13-1 Introduction

In Sec. 12-3, we showed that the unsteady flow in open channels is described by a set of hyperbolic partial differential equations. These equations describe the conservation of mass and momentum in terms of the partial derivatives of dependent variables: flow velocity, V, and flow depth, y. However, for practical applications, we need to know the values of these variables instead of the values of their derivatives. Therefore, we integrate the governing equations. Because of the presence of nonlinear terms, a closed-form solution of these equations is not available, except for very simplified cases. Therefore, they are integrated numerically for which several numerical methods are available.

In this chapter, we introduce the method of characteristics and discuss necessary boundary and initial conditions for the numerical solution of governing equations. Various available numerical methods are listed, and their advantages and disadvantages are briefly discussed.

13-2 Method of Characteristics

Monge developed a graphical procedure in 1789 for the integration of partial differential equations. He called this procedure the *method of characteristics*. It was used by Massau (1889) and Craya (1946) for analyzing surges in open channels and it was utilized to investigate the propagation of flood waves (Isaacson et al., 1954) and other unsteady flow problems. Although it has become a standard method for the analysis of transients in closed conduits, its applications to open channels has become almost negligible. The concept of characteristic curves is helpful in understanding the propagation of waves and the development of boundary conditions for explicit finite-difference methods. We present a brief development of the method; readers interested in details should see Stoker (1957); Abbott (1966, 1979); and Lai (1986).

Let us rewrite Eqs. 12-6 and 12-17 for prismatic channels having no lateral inflow or outflow:

$$\frac{\partial y}{\partial t} + D_h \frac{\partial V}{\partial x} + V \frac{\partial y}{\partial x} = 0 \tag{13-1}$$

$$\frac{\partial V}{\partial t} + V \frac{\partial V}{\partial x} + g \frac{\partial y}{\partial x} = g(S_o - S_f) \tag{13-2}$$

in which V = flow velocity; y = flow depth; $D_h = A/B$ = hydraulic depth;* A = flow area; B = top water-surface width; S_o = channel-bottom slope; S_f = slope of the energy-grade line; x = distance along the channel length; t = time; and g = acceleration due to gravity.

By multiplying Eq. 13-1 by an unknown multiplier, λ, adding it to Eq. 13-2, and rearranging the terms of the resulting equation, we obtain

$$\left[\frac{\partial V}{\partial t} + (V + \lambda D_h)\frac{\partial V}{\partial x}\right] + \lambda\left[\frac{\partial y}{\partial t} + (V + \frac{g}{\lambda})\frac{\partial y}{\partial x}\right] = g(S_o - S_f) \tag{13-3}$$

Since $V = V(x,t)$ and $y = y(x,t)$, we may write the following expressions for the total derivatives

$$\frac{DV}{Dt} = \frac{\partial V}{\partial t} + \frac{\partial V}{\partial x}\frac{dx}{dt}$$
$$\frac{Dy}{Dt} = \frac{\partial y}{\partial t} + \frac{\partial y}{\partial x}\frac{dx}{dt} \tag{13-4}$$

A comparison of Eqs. 13-3 and 13-4 shows that we can write the terms inside the brackets as total derivatives if we define the unknown multiplier λ such that

$$V + \lambda D_h = \frac{dx}{dt} = V + \frac{g}{\lambda} \tag{13-5}$$

or

$$\lambda_{1,2} = \pm\sqrt{\frac{g}{D_h}} = \pm\sqrt{\frac{gB}{A}} \tag{13-6}$$

In Chapter 12, we derived the expression for the *celerity* of a gravity wave as $c = \sqrt{gA/B}$. Thus, by defining $\lambda_1 = g/c$, we can write Eq. 13-3 as

$$\frac{DV}{Dt} + \frac{g}{c}\frac{Dy}{Dt} = g(S_o - S_f) \tag{13-7}$$

if

$$\frac{dx}{dt} = V + c \tag{13-8}$$

Similarly, by defining $\lambda_2 = -g/c$, we can write Eq. 13-3 as

$$\frac{DV}{Dt} - \frac{g}{c}\frac{Dy}{Dt} = g(S_o - S_f) \tag{13-9}$$

if

$$\frac{dx}{dt} = V - c \tag{13-10}$$

Note that Eq. 13-7 is valid if Eq. 13-8 is satisfied and Eq. 13-9 is valid if Eq. 13-10 is satisfied. Equation 13-8 plots as a curve in the x-t plane (Fig. 13-1). This curve is referred to as the *positive characteristic*, C^+. Similarly, Eq. 13-10 plots as the *negative characteristic, C^-*. In other words, Eq. 13-7 is valid along the positive characteristic AP and Eq. 13-9 is valid along the negative characteristic BP. Equations

*Only in this section, we will use D_h for the hydraulic depth instead of D in order to differentiate between the symbols for the hydraulic depth and for the total derivatives.

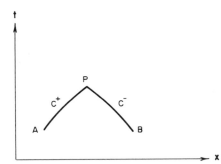

Figure 13-1 Positive and negative characteristics.

13-7 and 13-9 are called the *compatibility equations*. Thus, by means of these simple algebraic manipulations, we have eliminated the space variable, x, from the governing equations and converted the equations into ordinary differential equations. However, we had to pay a price in transforming the partial differential equations into the ordinary differential equations—i.e., Eqs. 13-1 and 13-2 are valid for any values of x and t; the transformed equations are, however, valid only along the characteristics.

By multiplying Eqs. 13-7 and 13-9 by dt and integrating along the characteristics AP and BP, we obtain

$$\int_A^P dV + \int_A^P \frac{g}{c} \, dy = g \int_A^P (S_o - S_f) \, dt \tag{13-11}$$

and

$$\int_B^P dV - \int_B^P \frac{g}{c} \, dy = g \int_B^P (S_o - S_f) \, dt \tag{13-12}$$

To evaluate these integrals, c and S_f along the characteristics should be known, i.e., y and V along the characteristics should be known, since c and S_f are functions of y and V. However, y and V are the unknowns we want to determine. Therefore, we have to make some approximation to evaluate these integrals. For this purpose, let us assume that the values of c and S_f computed by using V and y at A and B are valid along the entire characteristics AP and BP, respectively. Based on this approximation, Eqs. 13-11 and 13-12 may be written as

$$V_P - V_A + \left(\frac{g}{c}\right)_A (y_P - y_A) = g(S_o - S_f)_A(t_P - t_A) \tag{13-13}$$

and

$$V_P - V_B - \left(\frac{g}{c}\right)_B (y_P - y_B) = g(S_o - S_f)_B(t_P - t_B) \tag{13-14}$$

in which the subscripts A, B, and P refer to the quantities at points A, B, and P, respectively.

If we know the values of V and y at points A and B, then their values at point P may be determined by solving Eqs. 13-13 and 13-14 simultaneously. We may write these equations in a slightly different form by combining the known and unknown quantities as follows:

$$V_P = C_p - C_A y_P \tag{13-15}$$

and

$$V_P = C_n + C_B y_P \tag{13-16}$$

in which we have combined the known quantities in the two constants, C_p and C_n,

$$
\begin{aligned}
C_p &= V_A + C_A y_A + g(S_o - S_f)_A (t_P - t_A) \\
C_n &= V_B - C_B y_B + g(S_o - S_f)_B (t_P - t_B) \\
C &= \frac{g}{c}
\end{aligned}
\tag{13-17}
$$

Note that C_p and C_n are constants during the time interval $t_P - t_A$ and $t_P - t_B$, respectively; although they may vary from one time interval to another.

Characteristics

In the previous pargraphs, we defined the characteristics as the curves that are plots of $dx/dt = V \pm c$. In this section, we discuss their physical significance.

As we mentioned in Chap. 3, a flow disturbance (depth and/or velocity) propagates in two directions if the flow is subcritical and only in the downstream direction if the flow is supercritical. The absolute velocity at which this disturbance travels is $V \pm c$. If we plot this propagation in the x-t plane assuming the disturbance is at point C at time $t = 0$ and distance $x = x_o$, then its influence will be felt in the shaded region shown in Fig. 13-2a. This region is referred to as the *zone of influence*. The curve defining the propagation in the downstream direction is called as C^+ and that in the upstream direction as C^-. Any point outside the shaded region is not affected at all by the propagation of the disturbance.

Now, let us discuss the conditions that affect the flow conditions at point P. To understand this, let us draw the characteristics in the backward direction through point P (Fig. 13-2b). Then, only the conditions within the shaded region affect the flow conditions at point P. In other words, we may say that the flow conditions at point

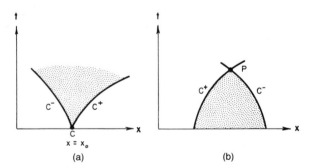

Figure 13-2 Zones of influence and dependence.

P depend upon the flow conditions in the shaded region. This is called the *zone of dependence.*

Depending upon the relative magnitude of the flow velocity and the celerity, a disturbance may or may not travel in the upstream direction. If the flow is subcritical, then the characteristic directions are both positive and negative. For critical flow, one of the characteristic directions is zero, whereas the second is positive. In supercritical flows, both characteristic directions are positive. Figure 13-3 shows the characteristics for different types of flow.

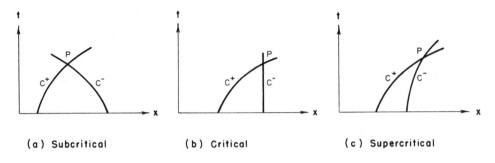

(a) Subcritical   (b) Critical   (c) Supercritical

Figure 13-3   Characteristics for subcritical, critical, and supercritical flow.

Mathematically speaking (Cunge, et al., 1980), the discontinuities in the first and higher derivatives of the dependent variables and the physical parameters that appear in the governing equations propagate along the characteristics. Thus the discontinuities in the slope of the water surface $\partial y/\partial x$ or in the velocity gradient $\partial V/\partial x$ propagate along the characteristics at the velocity of the shallow-water waves. However, this is not the case for the discontinuities in the flow variables themselves, i.e., a bore does not propagate at the velocity of the shallow-water waves.

Let us now reformulate Eqs. 13-7 and 13-9. By assuming B to be constant, we can write

$$
\begin{aligned}
\frac{dc}{dt} &= \frac{d}{dt}\left(\frac{gA}{B}\right)^{1/2} \\
&= \frac{1}{2}\frac{g}{c}\frac{dy}{dt}
\end{aligned}
\tag{13-18}
$$

Thus, Eqs. 13-7 and 13-9 may be written as

$$
\frac{d}{dt}(V + 2c) = S
\tag{13-19}
$$

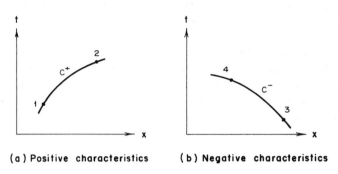

(a) Positive characteristics   (b) Negative characteristics

Figure 13-4 Positive and negative characteristics.

in which $S = g(S_o - S_f)$. This equation is valid along the positive characteristic, $dx/dt = V + c$. Similarly, Eq. 13-9, which is valid along the negative characteristic, may be written as

$$\frac{d}{dt}(V - 2c) = S \tag{13-20}$$

The term on the right-hand sides of Eqs. 13-19 and 13-20 is referred to as the *source term*. The source term, S, is zero if the channel is frictionless and the channel bottom slope is zero, or if the flow is uniform (i.e., $S_f = S_o$). If $S = 0$, it follows from Eqs. 13-19 and 13-20 that

$$V + 2c = J^+ \tag{13-21}$$

and

$$V - 2c = J^- \tag{13-22}$$

The constants J^+ and J^- are called *Riemann invariants*. Note that their values may vary from one characteristic to another.

For a sloping prismatic channel, including the friction losses, integration of Eqs. 13-19 and 13-20 along the characteristics (Fig. 13-4) yields

$$\left[V + 2c\right]_1^2 = \int_{t_1}^{t_2} S \, dt \tag{13-23}$$

and

$$\left[V - 2c\right]_3^4 = \int_{t_3}^{t_4} S \, dt \tag{13-24}$$

If the right-hand sides of Eqs. 13-23 and 13-24 are nonzero but are sufficiently small, then the left-hand sides are called *Riemann quasi-invariants*. By a proper selection of t_1 and t_2, the right-hand sides may be made small (Abbott, 1979). Riemann invariants in an analogous form cannot be derived for nonprismatic channels (Cunge, et al., 1980) because of the additional terms introduced in the source term S.

13-3 Initial and Boundary Conditions

In the analysis of unsteady flow in open channels, we usually start our calculations at a specified time. The flow conditions at this starting time are referred to as the *initial conditions*. Since the boundaries of all physical systems are located at finite distances, we have to specify in our calculations some particular conditions at the limits or boundaries of the physical system. These conditions are called the *boundary conditions*. In this section, we closely follow Cunge, et al. (1980) as we discuss how to specify these conditions.

Figure 13-5 shows the computational domain for a one-dimensional system. The computations are to start at time $t = t_o$; the upstream end of the system is at $x = x_o$; and the downstream end is at $x = x_1$. We call the upstream and downstream flow directions with reference to the initial steady-state flow directions. To facilitate understanding, let us consider the channel to be horizontal and frictionless. Thus the Riemann invariants are given by Eqs. 13-21 and 13-22.

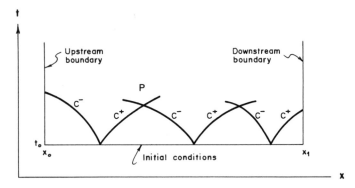

Figure 13-5 Computational domain for a one-dimensional system.

If we specify the initial steady-state flow depth and flow velocity at $t = t_0$, then the values of Riemann invariants are defined along the characteristics, since c is a function of the flow depth. We may independently specify either V or y anywhere on this characteristic; the other is then determined from the Riemann invariant. At any other point in the computational domain where the positive and negative characteristics intersect, V and y are determined from the values of J^+ and J^- for the characteristics passing through point P. Thus, we specify the *flow depth* and the *flow velocity* at the computational nodes for the initial conditions. At the boundaries, however, we specify the flow depth or the flow velocity or a relationship between the flow velocity and flow depth (e.g., a rating curve). The latter condition should be independent of the compatibility conditions.

Figure 13-6 shows the characteristics at the limits of the computational domain for the subcritical and supercritical flows. The following rule (Cunge, et al., 1980) may be followed for specifying the conditions at the boundaries (the initial conditions may

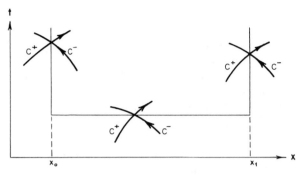

a) Subcritical flow

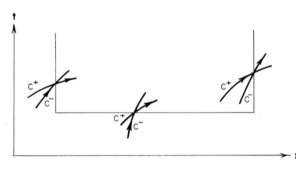

b) Supercritical flow

Figure 13-6 Characteristics at the limits of computational domain (After Cunge, et al. 1980).

be considered as a special case of the boundary condition in time): One condition has to be specified for each characteristic entering the computational domain at its limits. For example, two characteristics enter the computational domain at time $t = t_0$. Thus, two conditions have to be given as the initial conditions. At the upstream boundary, one condition is needed for subcritical flows and two conditions are needed for supercritical flows. At the downstream boundary, we have to specify one condition for the subcritical flow and none for the supercritical flows.

These additional conditions at the boundaries should be independent of the governing equations and of the Riemann invariants. In addition, if two conditions are specified, they must be independent of one another. For example, we may not specify y and $\partial y / \partial x$ independently as the initial conditions. Similarly, we may not specify the flow variables y and V independently such that the continuity or momentum equations are not satisfied.

In the previous paragraphs, we discussed the zone of dependence and the zone of influence. By referring to them, we can see how the initial and boundary conditions affect the solution. For channels having low friction losses, the effect of initial conditions may persist for a long time. In such situations, these conditions should be specified accurately. Similarly, the effect of boundary conditions are carried through for some time. In certain situations, it may be difficult to define proper initial conditions— e. g., tidal flow in an estuary. In such cases, we start the computations with assumed initial conditions and compute the flow conditions for two or three tidal periods. Usually, after two tidal cycles, conditions become periodic. However, to investigate a new system, the flow conditions between any two consecutive tidal cycles should be compared; the computations should be carried out for one more cycle if they differ significantly from one another. However, they may be assumed correct if they are close.

13-4 Characteristic Grid Method

In Section 13-2, we derived equations for the characteristic method in which the time and space interval are specified. The method is, therefore, referred to as the *method of specified intervals* (The details of the method are presented in the next section). To use specified intervals, interpolations become necessary either in time or in space, since all characteristics do not pass through the grid points. In this section, we integrate the characteristic form of the equations so that interpolations are not necessary, since we integrate along the characteristics and we do not use specified intervals for time and distance. Instead, the locations of the computational nodes after the initial conditions depend upon the computed values of V and y and are determined by the intersection of the characteristics.

Let us rewrite Eqs. 13-19 and 13-20 as well as the equations of characteristics along which they are valid. The equation

$$\frac{d}{dt}(V + 2c) = g(S_o - S_f) \tag{13-25}$$

is valid if

$$\frac{dx}{dt} = V + c \tag{13-26}$$

Similarly, the equation

$$\frac{d}{dt}(V - 2c) = g(S_o - S_f) \tag{13-27}$$

is valid if

$$\frac{dx}{dt} = V - c \tag{13-28}$$

Let us assume that the flow depth (and hence c) and the flow velocity are known at points L and R (Fig. 13-7). Note that L and R may not be at the same time level. The characteristics passing through L and R intersect at point P. The location of this point is not known a priori but is determined from Eqs. 13-26 and 13-28.

By multiplying Eqs. 13-25 through 13-28 by dt, integrating, and applying the limits, we get

$$(V + 2c)_P = (V + 2c)_L + g \int_L^P (S_o - S_f) \, dt \tag{13-29}$$

$$x_P = x_L + \int_L^P (V + c) \, dt \tag{13-30}$$

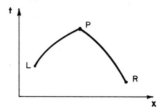

Figure 13-7 Definition sketch.

$$(V - 2c)_P = (V - 2c)_R + g \int_R^P (S_o - S_f)\, dt \tag{13-31}$$

$$x_P = x_R + \int_R^P (V - c)\, dt \tag{13-32}$$

Note that we have not made any approximation in the integration of Eqs. 13-29 through 13-32. However, approximations become necessary while evaluating the integrals, since we do not know the values of V, c, and S_f along the characteristics. By using the trapezoidal rule to evaluate the integrals, Eqs. 13-29 through 13-32 may be written as

$$x_P = x_L + (t_P - t_L)\left[\frac{1}{2}(V_P + c_P) + \frac{1}{2}(V_L + c_L)\right] \tag{13-33}$$

$$V_P = V_L + 2(c_L - c_P)$$
$$\qquad + \frac{1}{2}g(t_P - t_L)\left[(S_o - S_f)_P + (S_o - S_f)_L\right] \tag{13-34}$$

$$x_P = x_R + (t_P - t_R)\left[\frac{1}{2}(V_P - c_P) + \frac{1}{2}(V_R - c_R)\right] \tag{13-35}$$

$$V_P = V_R + 2(c_P - c_R)$$
$$\qquad + \frac{1}{2}g(t_P - t_R)\left[(S_o - S_f)_P + (S_o - S_f)_R\right] \tag{13-36}$$

Equations 13-33 through 13-36 are four nonlinear algebraic equations in four unknowns, namely, V_P, y_P, x_P, and t_P; the corresponding values of these variables and the coefficients at points L and R are known. These equations may be solved simultaneously to determine the values of these unknowns. By proceeding similarly, we determine V and y at later times.

At time $t = 0$, we compute the initial conditions at discrete grid points along the channel length; the distance between these points may not be equal. The distance and time where the flow conditions after the initial conditions will be computed are determined from Eqs. 13-33 through 13-36.

Note that we can use Eqs. 13-33 through 13-36 only at the interior points. At the upstream end, we cannot use Eqs. 13-33 and 13-34, since there is no grid point upstream of the upstream end. However, the distance at the upstream end is fixed, say $x = 0$. Still we need one more equation to have a unique solution for V, y, and t. This equation is provided by the condition imposed by the boundary. This condition might be in the form of specifying the flow depth, flow velocity, or a relationship between these two variables that is independent of the characteristic equations.

Similarly, at the downstream end, we cannot write Eqs. 13-35 and 13-36, since there is no grid point downstream of the downstream boundary. We may again specify the distance x (for example, $x = L$). The additional equation needed for a unique solution is again provided by specifying the flow depth, flow velocity, or a relationship between V and y.

By proceeding in a similar manner, we can compute the flow conditions in the channel for any desired length of time. If the conditions are needed at a specified location, they may be determined by interpolation from the values computed at the neighboring points.

For continuous flows, the procedure gives satisfactory results, although it may be necessary to interpolate the channel parameters or the computed values. In addition, once a bore is formed in the solution, it is possible to have multiple values for the flow variables, and the procedure fails. If we plot the characteristics on the x-t plane as the computations progress, then the formation of the bore or steep-fronted wave is indicated by the convergence of the characteristics.

13-5 Method of Specified Intervals

In the *method of specific intervals*, we specify the size of the spatial and time grids. (Lister, 1969; Wylie and Streeter, 1983; Lai, 1986) If we draw characteristics through a grid point, then they do not pass through the neighboring grid point. For example, the characteristics through point P do not pass through points A and B, instead they intersect at points R and S (Fig. 13-8). To compute conditions at point P, the conditions at points R and S should be known. These may be determined by interpolation from the known values at points A, B, and C as follows. We are using linear interpolations in the following derivation; higher-order interpolations may be utilized, if necessary.

Referring to Fig. 13-8, we may write

$$\frac{V_C - V_R}{V_C - V_A} = \frac{x_C - x_R}{x_C - x_A} = \frac{x_P - x_R}{x_C - x_A} = \frac{(V_R + c_R)\Delta t}{\Delta x} \tag{13-37}$$

Similarly, we may write

$$\frac{c_C - c_R}{c_C - c_A} = \frac{(V_R + c_R)\Delta t}{\Delta x} \tag{13-38}$$

Elimination of c_R from these two equations yields

$$V_R = \frac{V_C - \frac{\Delta t}{\Delta x}(c_A V_C - c_C V_A)}{1 + (V_C - V_A + c_C - c_A)\frac{\Delta t}{\Delta x}} \tag{13-39}$$

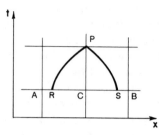

Figure 13-8 Interpolation.

Now, c_R and y_R may be determined from

$$c_R = \frac{c_C - V_R \frac{\Delta t}{\Delta x}(c_C - c_A)}{1 + \frac{\Delta t}{\Delta x}(c_C - c_A)}$$

$$y_R = y_C - \frac{\Delta t}{\Delta x}(V_R + c_R)(y_C - y_A) \tag{13-40}$$

For subcritical flow, point S lies between C and B. Proceeding as before

$$V_S = \frac{V_C - \frac{\Delta t}{\Delta x}(c_B V_C - c_C V_B)}{1 - \frac{\Delta t}{\Delta x}(V_C - V_B - c_C + c_B)}$$

$$c_S = \frac{c_C + V_S \frac{\Delta t}{\Delta x}(c_C - c_B)}{1 + \frac{\Delta t}{\Delta x}(c_C - c_B)} \tag{13-41}$$

$$y_S = y_C + \frac{\Delta t}{\Delta x}(V_S - c_S)(y_C - y_B)$$

By using these interpolated values, y_P and V_P may be determined from Eqs. 13-15 and 13-16, where

$$C_p = V_R + \frac{g}{c_R}y_R + g(S_o - S_f)_R \Delta t$$

$$C_n = V_S - \frac{g}{c_S}y_S + g(S_o - S_f)_S \Delta t \tag{13-42}$$

13-6 Other Numerical Methods

In addition to the characteristic method, the following methods have been used for the numerical integration of hyperbolic partial differential equations:

1. Finite-difference methods

2. Finite-element methods

3. Spectral method

4. Boundary-element method

The characteristic method, finite-difference methods, and finite-element methods have been employed for the analysis of unsteady open-channel flow. We presented the details of the method of characteristics in the previous two sections. The finite-difference methods are discussed in the next two chapters. The finite-element method (Baker 1983), presented in Chapter 16, has been used only to a limited extent for open-channel analysis (Katopodes 1984). It does not offer any significant advantage as compared with the other methods for one-dimensional flow problems, and several difficulties have to be overcome if a shock or bore is formed in the solution.

A number of problems arise in the application of the spectral method if the boundary conditions are nonperiodic (Canuto et al. 1988). As compared with the other methods, the boundary-element method (Liggett, 1984; Brebbia and Dominguez, 1989) has not proven to be very successful for time-dependent problems. We will not discuss these two methods.

13-7 Summary

In this chapter, we described the concept of characteristics and discussed the inclusion of initial and boundary conditions in the numerical computations. The characteristic method was introduced, and its two different formulations were presented. These formulations are the method of specified intervals and the characteristic grid method. A brief introduction to the other available numerical methods for the integration of shallow-water equations was presented.

Problems

13-1. A trapezoidal channel having a bottom width of 6.1 m and side slopes of $1.5H : 1V$ is carrying a flow of 126 m³/s at a flow depth of 5.79 m. The bottom slope is 0.00008, Manning n = 0.013 and the channel length is 5 km. There is a constant-level reservoir at the upstream end of the channel. A sluice gate at the downstream end is suddenly closed at time $t = 0$.
a. Compute the transient conditions until $t = 2000$ s by using the
    i. Grid of characteristics
    ii. Method of specified intervals.

  b. Plot the computed flow depth in the channel at $t = 0$, 500, 1000, 1500, and 2000 s.
  c. Plot the variation of the flow depth with time at 1.5, 2.5, 3 and 5 km from the reservoir.
13-2. By using different values of $\Delta t / \Delta x$, investigate the effect of interpolation error on the computed wave height and wave shape. Use data of Prob. 13-1.
13-3. Show that the method of characteristic grid fails when a shock or bore is formed. Use data for Prob. 13-1 except let the gate close in 10 s and let the wave propagate in the channel.

REFERENCES

Abbott, M. B. 1966. *An introduction to the method of characteristics*, Thames and Hudson, London, and American Elsevier, New York, NY.

———— 1975. "Method of characteristics," Chap. 3 and "Weak Solution of the equations of open channel flow," Chap. 7. In *Unsteady open channel flow*, edited by K. Mahmood, and V. Yevjevich. Water Resources Publications, Fort Collins, CO.

————, 1979. *Computational Hydraulics; Elements of the Theory of Free Surface Flows*, Pitman, London.

Abbott, M. B., and Verwey, A. 1970. Four-point method of characteristics, *Jour., Hydraulics Division*, Amer. Soc. Civil Engrs., 96, 12: 2549–64.

Amein, M., and Fang, C. S. 1970. Implicit flood routing in natural channels, *Jour., Hydraulics Division*, Amer. Soc. of Civil Engrs., 96, 12: 2481–2500.

Anderson, D. A.; Tannehill, J. C.; and Pletcher, R. H. 1984. *Computational fluid mechanics and heat transfer*, McGraw-Hill, New York, NY.

Baker, J. A. 1983. *Finite-element computational fluid dynamics*, McGraw-Hill, New York, NY.

Brebbia, C. A., and Dominguez, J. 1989. *Boundary elements, an introductory course*, Computational Mechanics Publications, London, England.

Canuto, C.; Hussaini, M. Y.; Quarteroni, A.; and Zang, T. A. 1988. *Spectral methods in fluid dynamics*, Springer-Verlag, New York, NY.

Chaudhry, M. H. 1987. *Applied hydraulic transients*, 2nd ed. Van Nostrand Reinhold, New York, NY.

Craya, A. 1946. Calcul graphique des regimes variables dans les canaux, *La Houille Blanche*, no. 1 November 1945–January 1946:79–138; no. 2 (March 1946):117–30.

Cunge, J.; Holly, F. M.; and Verwey, A. 1980. *Practical aspects of computational river hydraulics*, Pitman, London.

Fread, D. L., and Harbaugh, T. E. 1973. Transient simulation of breached earth dams, *Jour. Hydraulics Division*, Amer. Soc. Civil Engrs., 99, no. 1:139–54.

Isaacson, E.; Stoker, J. J.; and Troesch, B. A. 1954. Numerical solution of flood prediction and river regulation problems (Ohio-Mississippi Floods), *Report II*, Inst. Math. Sci. Rept. IMM-NYU-205, New York University, New York, NY.

Katopodes, N. 1984. A dissipative Galerkin scheme for open-channel flow, *Jour. Hydraulic Engineering*, Amer. Soc. Civil Engrs., 110, 4:450–66.

Lai, C. 1986. "Numerical modeling of unsteady open-channel flow". In *Advances in Hydroscience*, Vol. 14, Academic Press, New York, NY, 161-333.

———— 1988. Comprehensive method of characteristics models for flow simulation, *Jour. Hydraulic Engineering*, Amer. Soc. Civil Engrs., 114, no. 9:1074–97.

Lax, P. D. 1954. Weak solutions of nonlinear hyperbolic partial differential equations and their numerical computation, *Communications on Pure and Applied Mathematics*, 7:159–63.

Leendertse, J. J. 1967. Aspects of a computational model for long-period water-wave propagation, *Memo* RM-5294-PR, Rand Corporation (May).

Liggett, J. A. 1984. "The boundary element method—some fluid applications". In *Multidimensional fluid transients*, edited by M. H. Chaudhry and C. S. Martin, Amer. Soc. Mech. Engrs. New York, NY, pp. 1–8.

Lister, M. 1961. "Method of Characteristics," *Chapter in Numerical Methods for Computers,* edited by Fox, Wiley, New York, NY.

Massau, J. 1889. L'integration graphique and appendice au memoire sur l'integration graphique, *Assoc. des Ingenieurs sortis des Ecoles Speciales de Gand,* Belgium, Annales, 12:185–444.

Price, R. K. 1974. Comparison of four numerical flood routing methods, *Jour. Hydraulics Division,* Amer. Soc. Civil Engrs., 100, no. 7:879–99.

Richtmyer, R. D., and Morton, K. W. 1967. *Difference methods for initial value problems,* 2nd. ed., Wiley, New York, NY.

Stoker, J. J. 1957. *Water waves,* Wiley, Interscience, New York, NY.

Strelkoff, T. 1970. Numerical solution of St. Venant equations, *Jour. Hydraulics Division,* Amer. Soc. Civil Engrs., 96, no. 1:223–52.

Terzidis, G., and Strelkoff, T. 1970. Computation of open channel surges and shocks, *Jour. Hydraulics Division,* Amer. Soc. Civil Engrs., 96, no. 12:2581–2610.

Wylie, E. B., and Streeter, V. L. 1983. *Fluid transients,* FEB Press, Ann Arbor, Michigan.

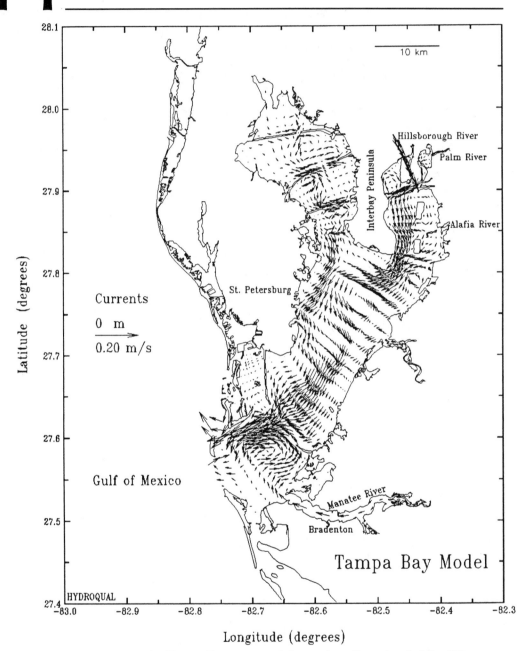

Surface current patterns in Tampa Bay computed by a three-dimensional finite-difference mathematical model (Courtesy of A. F. Blumberg, HydroQual, Inc., Mahwah, NJ. After Galperin et al., 1992).

14-1 Introduction

We discussed in Chapter 12 that de Saint Venant equations are nonlinear partial differential equations for which a closed-form solution is not available except for very simplified cases. In Chapter 13, we briefly presented several numerical methods that may be used for their integration. Of these methods, the finite-difference methods have been utilized very extensively; details of some of these methods are outlined in this chapter. Either a conservation or nonconservation form of the governing equations may be used in some methods, whereas only one of these forms may be used in others. A conservation form should be preferred, since it conserves various quantities better and it simulates the celerity of wave propagation more accurately (Cunge et al., 1980; Miller and Chaudhry 1989) than the nonconservation form.

We first discuss a number of commonly used terms. Then, a number of explicit and implicit finite-difference methods are presented, and the inclusion of boundary conditions in these methods is outlined. The consistency of a numerical scheme is briefly discussed and the stability conditions are then derived. The results computed by different schemes for a typical problem are compared.

14-2 Terminology

In this section, we briefly introduce a number of terms commonly used in finite difference applications.

Finite-Difference Approximations

To simplify presentation, let us first consider a function $f(x)$ of one independent variable x. Let us assume that the value of this function, $f(x_o)$, at x_o is known. Then, by using a Taylor series expansion, the function $f(x_o + \Delta x)$ may be written as

$$f(x_o + \Delta x) = f(x_o) + \Delta x f'(x_o) + \frac{(\Delta x)^2}{2!} f''(x_o) + O(\Delta x)^3 \qquad (14\text{-}1)$$

in which a prime refers to derivative with respect to x, e.g., $f'(x_o) = dy/dx$ evaluated at $x = x_o$, and $O(\Delta x)^3$ indicates terms of third- or higher-order of Δx.

Similarly $f(x_o - \Delta x)$ may be expanded as

$$f(x_o - \Delta x) = f(x_o) - \Delta x f'(x_o) + \frac{(\Delta x)^2}{2!} f''(x_o) + O(\Delta x)^3 \qquad (14\text{-}2)$$

Equation 14-1 may be written as

$$f(x_o + \Delta x) = f(x_o) + \Delta x f'(x_o) + O(\Delta x)^2 \qquad (14\text{-}3)$$

It follows from Eq. 14-3 that

$$\left.\frac{df}{dx}\right|_{x=x_o} = \frac{f(x_o + \Delta x) - f(x_o)}{\Delta x} + O(\Delta x) \qquad (14\text{-}4)$$

Similarly, it follows from Eq. 14-2 that

$$\left.\frac{df}{dx}\right|_{x=x_o} = \frac{f(x_o) - f(x_o - \Delta x)}{\Delta x} + O(\Delta x) \qquad (14\text{-}5)$$

Neglecting the $O(\Delta x)$ terms in Eqs. 14-4 and 14-5, we obtain

$$\left.\frac{df}{dx}\right|_{x=x_o} = \frac{f(x_o + \Delta x) - f(x_o)}{\Delta x} \qquad (14\text{-}6)$$

and

$$\left.\frac{df}{dx}\right|_{x=x_o} = \frac{f(x_o) - f(x_o - \Delta x)}{\Delta x} \qquad (14\text{-}7)$$

The finite-difference approximation of Eq. 14-6 is referred to as the *forward finite difference* and that of Eq. 14-7 is called the *backward finite difference*. Note that in both of these cases, the discarded terms are of the first order of Δx. Therefore, both forward and backward finite-difference approximations are referred to as *first-order accurate*.

Now, let us subtract Eq. 14-2 from Eq. 14-1, rearrange the terms, and divide by Δx.

$$\left.\frac{df}{dx}\right|_{x=x_o} = \frac{f(x_o + \Delta x) - f(x_o - \Delta x)}{2\Delta x} + O(\Delta x)^2 \qquad (14\text{-}8)$$

By neglecting the last term, we obtain

$$\left.\frac{df}{dx}\right|_{x=x_o} = \frac{f(x_o + \Delta x) - f(x_o - \Delta x)}{2\Delta x} \qquad (14\text{-}9)$$

This approximation is referred to as the *central finite-difference approximation*. Note that the neglected term in this case is of the order of $(\Delta x)^2$; therefore, it is referred to as *second-order accurate*.

Figure 14-1 shows a geometrical representation of the forward, backward, and central finite-difference approximations. The forward finite-difference approximation replaces the slope of the tangent to the curve at B by the slope of line BC, the backward finite-difference approximation replaces this slope by the slope of line AB, and the central finite-difference approximation replaces it by the slope of the chord line AC. It is clear from these figures that the central finite-difference approximation is more accurate than the forward or backward finite-difference approximations.

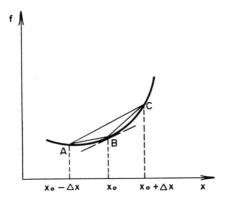

Figure 14-1 Finite-difference approximations.

Now, let us consider finite-difference approximations for a partial derivative. Let us consider a function $f(x, t)$. We have two independent variables: x and t. We may divide the x-t plane into a grid as shown in Fig. 14-2. The grid interval along the x-axis is Δx and the grid interval along the t-axis is Δt. In this figure, we are assuming the grid size is uniform along each axis, although it is not necessary to do so. For brevity, we will call the $i\Delta x$ grid point i and the $(i + 1)\Delta x$ grid point $i + 1$. For the time axis, we will use k for $k\Delta t$ grid point and $k + 1$ for the $(k + 1)\Delta t$ grid point. To refer to different variables at these grid points, we will use the number of the spatial grid as a subscript and that of the time grid as a superscript. For example, the flow depth, y, at the ith spatial grid point and kth time grid point will be denoted by y_i^k. We will denote the known time level by superscript k and the unknown time level by $k + 1$. By the known time level, we mean that the values of different dependent variables are known at this time and we want to compute their values at the unknown time level. The known conditions may be the values specified as the *initial conditions* or they may have been computed during the previous time step.

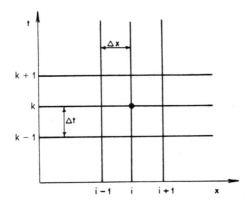

Figure 14-2 Finite-difference grid.

If the computations progress from one step to the next, then the procedure is referred to as a *marching procedure*. Most of the phenomena described by hyperbolic partial differential equations are solved by using marching procedures. The conditions specified at time $t = 0$ are referred to as the *initial conditions*. The conditions specified at the channel ends are called the *end*, or *boundary, conditions*.

There are several possibilities for approximating the partial derivatives. The spatial partial derivatives replaced in terms of the variables at the known time level are referred to as the *explicit* finite differences, whereas those in terms of the variables at the unknown time level are called *implicit* finite differences. Thus, referring to Fig. 14-2, if k is the known time level and $k + 1$ is the unknown time level, then some typical finite-difference approximations for the spatial partial derivative, $\partial f / \partial x$, at the grid point (i, k) are as follows:

Explicit finite differences

   Backward:

$$\frac{\partial f}{\partial x} = \frac{f_i^k - f_{i-1}^k}{\Delta x}$$

   Forward:

$$\frac{\partial f}{\partial x} = \frac{f_{i+1}^k - f_i^k}{\Delta x} \qquad (14\text{-}10)$$

   Central:

$$\frac{\partial f}{\partial x} = \frac{f_{i+1}^k - f_{i-1}^k}{2\Delta x}$$

Implicit finite differences

   Backward:

$$\frac{\partial f}{\partial x} = \frac{f_i^{k+1} - f_{i-1}^{k+1}}{\Delta x}$$

   Forward:

$$\frac{\partial f}{\partial x} = \frac{f_{i+1}^{k+1} - f_i^{k+1}}{\Delta x} \qquad (14\text{-}11)$$

   Central:

$$\frac{\partial f}{\partial x} = \frac{f_{i+1}^{k+1} - f_{i-1}^{k+1}}{2\Delta x}$$

Many other finite-difference approximations are possible if the values at three or more grid points, instead of just two grid points, are used as in Eqs. 14-10 and 14-11.

14-3 Explicit Finite-Difference Schemes

Several explicit finite-difference schemes have been proposed for the solution of hyperbolic partial differential equations. In this section, we will present a number of typical schemes that have been employed in hydraulic engineering.

Unstable Scheme

To solve de Saint Venant equations, we may select the following finite-difference approximations:

$$\frac{\partial f}{\partial x} = \frac{f_{i+1}^k - f_{i-1}^k}{2\Delta x}$$

$$\frac{\partial f}{\partial t} = \frac{f_i^{k+1} - f_i^k}{\Delta t} \tag{14-12}$$

in which, for brevity, f refers to both dependent variables y and V.

This finite-difference scheme is inherently unstable; i.e., computations become unstable irrespective of the size of grid spacing. This may be proved analytically following the stability analysis procedure presented in Sec. 14-6.

French (1985) used this scheme to solve shallow-water equations. Since the friction-losses were increased artificially by a large amount to make the scheme stable, the accuracy of the computed results is questionable.

Diffusive Scheme

Lax (1954) presented a slight variation of the unstable scheme discussed in the previous section. This scheme is one of the simplest to program and yields satisfactory results (Chaudhry 1987) for typical hydraulic engineering applications.

General formulation.   In this scheme the partial derivatives and other variables are approximated as follows:

$$\frac{\partial f}{\partial x} = \frac{f_{i+1}^k - f_{i-1}^k}{2\Delta x}$$

$$\frac{\partial f}{\partial t} = \frac{f_i^{k+1} - f^*}{\Delta t}$$

$$f^* = \frac{1}{2}(f_{i-1}^k + f_{i+1}^k) \tag{14-13}$$

$$D^* = \frac{1}{2}(D_{i-1}^k + D_{i+1}^k)$$

$$S_f^* = \frac{1}{2}(S_{f_{i-1}}^k + S_{f_{i+1}}^k)$$

in which, for brevity, we are again using f for both dependent variables y and V. We use these finite-difference approximations in the conservation and nonconservation forms of the governing equations as follows.

Nonconservation form.   Substitution of these expressions for the partial derivatives and for the coefficients of Eqs. 12-6 and 12-17 and simplification of the resulting equations yield

$$
\begin{aligned}
y_i^{k+1} &= \frac{1}{2}(y_{i-1}^k + y_{i+1}^k) - \frac{1}{2}\frac{\Delta t}{\Delta x}D_i^*(V_{i+1}^k - V_{i-1}^k) \\
&\quad - \frac{1}{2}\frac{\Delta t}{\Delta x}V_i^*(y_{i+1}^k - y_{i-1}^k) \\
V_i^{k+1} &= \frac{1}{2}(V_{i-1}^k + V_{i+1}^k) - \frac{1}{2}\frac{\Delta t}{\Delta x}g(y_{i+1}^k - y_{i-1}^k) \\
&\quad - \frac{1}{2}\frac{\Delta t}{\Delta x}V_i^*(V_{i+1}^k - V_{i-1}^k) + g\Delta t(S_o - S_f^*)
\end{aligned}
\tag{14-14}
$$

Conservation form.   The governing equations in the conservation form may be written in matrix form as

$$
\mathbf{U}_t + \mathbf{F}_x + \mathbf{S} = \mathbf{0}
\tag{14-15}
$$

in which

$$
\mathbf{U} = \begin{pmatrix} A \\ VA \end{pmatrix}; \quad
\mathbf{F} = \begin{pmatrix} VA \\ V^2A + gA\bar{y} \end{pmatrix}; \quad
\mathbf{S} = \begin{pmatrix} 0 \\ -gA(S_o - S_f) \end{pmatrix}
\tag{14-16}
$$

and $A\bar{y}$ = moment of flow area about the free surface. This moment may be computed from $\int_0^{y(x)} [y(x) - \eta]\sigma(\eta)\,d\eta$ in which σ is the water-surface width at depth η.

Substitution of the finite-difference approximations of Eq. 14-13 into Eq. 14-15 yields

$$
\mathbf{U}_i^{k+1} = \frac{1}{2}(\mathbf{U}_{i+1}^k + \mathbf{U}_{i-1}^k) - \frac{1}{2}\frac{\Delta t}{\Delta x}(\mathbf{F}_{i+1}^k - \mathbf{F}_{i-1}^k) - \mathbf{S}^*\Delta t
\tag{14-17}
$$

Once the values of A and VA have been determined at the $(k+1)$ time level, we determine the values of variables of interest, y and V, and then proceed to the next time step.

Boundary conditions.   Equations 14-14 or 14-17 may be used at the interior grid points to compute the unsteady flow depth and flow velocity. At the boundaries, however, we cannot use these equations, since there are no grid points outside the flow domain. Therefore, different procedures have been proposed to include the boundaries in the computations (for details, see Liggett and Cunge 1975). For one-dimensional flows, the specified-interval approach gives acceptable results and is employed herein.

In this procedure, we solve the positive characteristic equation (Eq. 13-15) simultaneously with the condition imposed by the boundary for the downstream-end condition and the negative characteristic equation (Eq. 13-16) with the upstream-end condition for the upstream boundary. The end condition may specify time variation of flow depth, flow velocity (or discharge), or a combination of these variables. For example, the flow depth remains constant for a constant-level reservoir; flow velocity is always zero for a dead-end or completely closed gate; and a relationship between the flow depth and discharge is specified for a rating curve. Similarly, discharge through a partially open gate is a function of the flow depth upstream of the gate.

For an intermediate boundary, e.g., a junction of two channels, we solve the positive and negative characteristic equations simultaneously with the continuity and energy equations for the junction. For example, for a junction of channels i and $i + 1$ (Fig. 14-3), we have two computational nodes at the junction. One of these nodes is the last node on channel i and the second junction node is the first node on channel $i + 1$.

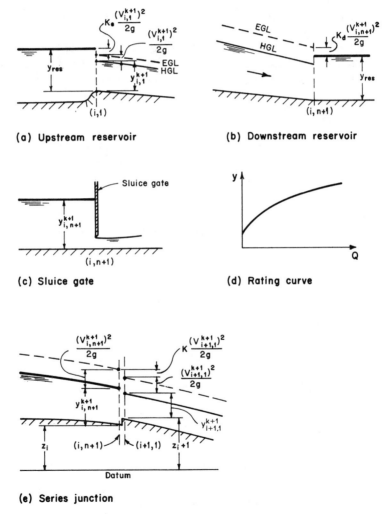

(a) Upstream reservoir

(b) Downstream reservoir

(c) Sluice gate

(d) Rating curve

(e) Series junction

Figure 14-3 Typical boundaries.

We consider the nodes to be close to each other, so that the channel length between them may be neglected. However, they are separate nodes in the sense that they may have different flow depths and flow velocities, and the head losses at the junction are assumed to be concentrated between these two nodes. For such a boundary, we need four equations, since we have two unknowns (flow depth and flow velocity) per computational node. We use the positive characteristic equation for the last node on channel i and the negative characteristic equation for the first node on channel $i+1$. The remaining two equations are the continuity and energy equations between the two nodes at the channel junction. For a junction of more than two channels, we write the positive characteristic equations for the channels upstream of the junction and the negative characteristic equations for the channel downstream of the junction. The additional equations are provided by the continuity and energy equations between channel nodes at the junction.

Referring to Fig. 14-3, the following equations describe the conditions imposed by different boundaries. These equations are solved simultaneously with the positive and/or negative characteristic equations to determine the flow conditions at the boundary nodes.

Upstream reservoir:

$$y_{i,1}^{k+1} = y_{\text{res}} - (1+k)\frac{(V_{i,1}^{k+1})^2}{2g}$$

Downstream reservoir:

$$y_{i,n+1}^{k+1} = y_{\text{res}} - (1-k)\frac{(V_{i,n+1}^{k+1})^2}{2g}$$

Sluice gate:

$$Q_{i,n+1}^{k+1} = C_d A_g \sqrt{2g y_{i,n+1}^{k+1}} \tag{14-18}$$

Rating curve:

$$Q = f(y_{i,n+1}^{k+1})$$

Series junction:

$$V_{i,n+1}^{k+1} A_{i,n+1}^{k+1} = V_{i+1,1}^{k+1} A_{i+1,1}^{k+1}$$

$$z_i + y_{i,n+1}^{k+1} + \frac{(V_{i,n+1}^{k+1})^2}{2g} = z_{i+1} + y_{i+1,1}^{k+1} + \frac{(V_{i+1,1}^{k+1})^2}{2g}$$

in which k = loss coefficient (entrance, exit, or junction); y_{res} = reservoir depth; Q = rate of discharge; C_d = discharge coefficient; A_g = area of the gate opening; and z = elevation of the channel bottom above a specified datum.

For completeness we have included the velocity head and the head-loss terms in Eq. 14-18. These terms are usually very small and may be neglected. In that case, these conditions are simplified considerably, i.e., the flow depth in the reservoir is equal to that in the channel at the channel entrance and exit, and the flow depths in the channels at the junction are equal to each other. In addition, this simplifies the algorithm if there is a possibility of flow reversal.

Stability.   For the stability* of the scheme, it is necessary that the Courant number, C_n, is less than or equal to 1, where

$$C_n = \frac{\text{actual wave velocity}}{\text{numerical wave velocity}} = \frac{|V| \pm c}{\Delta x / \Delta t} \tag{14-19}$$

Thus, the computational time interval depends upon the spatial grid spacing, flow velocity, and celerity, which is a function of the flow depth. Since the flow depth and the flow velocity may significantly change during the computations, it may be necessary to reduce the size of computational time interval for stability. The time interval should be such that C_n is as close to 1 as possible. If it is substantially less than unity, then the interval size should be increased to improve accuracy and to prevent the smearing of bores and steep waves.

MacCormack Scheme

The *MacCormack scheme* is an explicit, two-step predictor-corrector scheme (MacCormack 1969; Anderson, et al., 1984) that is second-order accurate both in space and time and is capable of capturing the shocks without isolating them. This scheme has been applied for analyzing one-dimensional, unsteady, open-channel flows by Fennema and Chaudhry (1986; 1987) and Dammuller, et al., (1989).

General formulation.   For one-dimensional flow two alternatives of this scheme are possible. In one alternative, backward finite differences are used to approximate the spatial partial derivatives in the predictor part and forward finite differences are utilized in the corrector part. The values of variables determined during the predictor part are used during the corrector part. In the second alternative, forward finite differences are used in the predictor part and backward finite difference are used in the corrector part. A general recommended procedure is to alternate the direction of differencing from one time step to the next; i.e., use alternative 1 during one time step, use alternative 2 during the next time step, and alternate this sequence thereafter. Recent investigations show that better results are produced if the direction of differencing in the predictor step is the same as that of the movement of the wave front.

The finite-difference approximations for the *first alternative* of this scheme are given in the following paragraphs; equations for the second alternative of the scheme may be written similarly by reversing the direction of the spatial finite-difference approximations as discussed in the previous paragraphs.

*We present a detailed discussion in Sec. 14-6.

Predictor

$$\frac{\partial \mathbf{U}}{\partial t} = \frac{\mathbf{U}_i^* - \mathbf{U}_i^k}{\Delta t}$$

$$\frac{\partial \mathbf{F}}{\partial x} = \frac{\mathbf{F}_i^k - \mathbf{F}_{i-1}^k}{\Delta x}$$

(14-20)

in which the notation of Fig. 14-2 is used and the superscript * refers to variables computed during the predictor part.

Substitution of these finite differences into Eq. 14-15 and simplification of the resulting equation yield

$$\mathbf{U}_i^* = \mathbf{U}_i^k - \frac{\Delta t}{\Delta x}\left(\mathbf{F}_i^k - \mathbf{F}_{i-1}^k\right) - \mathbf{S}_i^k \Delta t$$

(14-21)

The computed value of $\mathbf{U}_i^*$ gives A^* and Q^*, from which we determine the values of V^* and y^*. We compute these for all the computational nodes. These values are then used in the corrector part to compute $\mathbf{F}^*$ and $\mathbf{S}^*$.

Corrector

$$\frac{\partial \mathbf{U}}{\partial t} = \frac{\mathbf{U}_i^{**} - \mathbf{U}_i^k}{\Delta t}$$

$$\frac{\partial \mathbf{F}}{\partial x} = \frac{\mathbf{F}_{i+1}^* - \mathbf{F}_i^*}{\Delta x}$$

(14-22)

Substituting these finite differences and $\mathbf{S} = \mathbf{S}_i^*$ into Eq. 14-15, we obtain

$$\mathbf{U}_i^{**} = \mathbf{U}_i^k - \frac{\Delta t}{\Delta x}(\mathbf{F}_{i+1}^* - \mathbf{F}_i^*) - \mathbf{S}_i^* \Delta t$$

(14-23)

in which the superscript ** refers to the values of the variables after the corrector step. The value of $\mathbf{U}_i$ at the unknown time level $k + 1$ is given by

$$\mathbf{U}_i^{k+1} = \frac{1}{2}(\mathbf{U}_i^* + \mathbf{U}_i^{**})$$

(14-24)

Boundary conditions. The preceding equations are for the interior points. The boundary grid points may be included in the analysis in the same manner as we discussed for the Lax scheme. The simulation of the boundary points is, therefore, first-order accurate. Whether this first-order simulation of boundaries in an otherwise second-order scheme affects the overall accuracy of computed results is controversial. MacCormack heuristically showed that if the order of accuracy of the end conditions is one less than that of the interior points, then the overall accuracy of the computed results is not impaired. However, others question the validity of this statement.

Stability.   The MacCormack scheme is stable if $C_n \leq 1$. For the computations to be stable, this condition must be satisfied at each grid point during every computational interval.

Artificial viscosity.   The solution obtained by a finite-difference scheme has dissipative errors if the leading term of the truncation error in the scheme has even derivatives, and the solution has dispersive errors if the leading term has odd derivatives. The dispersive errors usually produce oscillations in the computed results in the vicinity of steep wave fronts. These oscillations are purely due to numerical errors and have nothing to do with the physical phenomenon being simulated. To smooth these oscillations, artificial viscosity is added to the scheme, for which several procedures have been reported. In this section we summarize a procedure presented by Jameson, et al. (1981). This procedure smoothes the oscillations where large gradients are present; however, it leaves the relatively smooth areas undisturbed.

To apply this procedure to open-channel flows, we first compute the following parameter based on the normalized form of the computed water-surface gradients:

$$\nu_i = \frac{|y_{i+1} - 2y_i + y_{i-1}|}{|y_{i+1}| + 2|y_i| + |y_{i-1}|}$$ (14-25a)

$$\epsilon_{i+1/2} = \kappa \max(\nu_{i+1}, \nu_i)$$

in which κ is used to regulate the amount of artificial viscosity. At the grid points where y_{i+1} and y_{i-1} do not exist, we use the following expressions instead of that of Eq. 14-25(a):

$$\nu_i = \frac{|y_i - y_{i-1}|}{|y_i| + |y_{i-1}|}$$

$$\nu_i = \frac{|y_{i+1} - y_i|}{|y_{i+1}| + |y_i|}$$ (14-25b)

The computed dependent variable f is then modified as

$$f_i^{k+1} = f_i^{k+1} + \epsilon_{i+1/2}(f_{i+1}^{k+1} - f_i^{k+1}) - \epsilon_{i-1/2}(f_i^{k+1} - f_{i-1}^{k+1})$$ (14-25c)

in which f refers to both y and V; this equation should be considered as a FORTRAN replacement statement.

Lambda Scheme

In the *Lambda scheme*, we first transform the governing equations into λ-form and then discretize them according to the sign of the characteristic directions, thereby enforcing the correct signal direction. This allows analysis of flows having sub- and super-critical flows. This scheme was proposed by Moretti (1979) and has been used for the analysis of unsteady open-channel flow by Fennema and Chaudhry (1986).

General formulation.   By multiplying Eq. 13-2 by c/g, adding it to Eq. 13-1, and rearranging the terms of the resulting equation, we obtain

$$\left[\frac{\partial y}{\partial t} + (V + c)\frac{\partial y}{\partial x}\right] + \frac{c}{g}\left[\frac{\partial V}{\partial t} + (V + c)\frac{\partial V}{\partial x}\right] = c(S_o - S_f) \tag{14-26}$$

Let $V + c = \lambda^+$ and let us use superscript + to indicate the space derivative in the λ^+ direction in order to enforce proper signal direction, as discussed in the following paragraph. Then the above equation may be written as

$$\frac{\partial y}{\partial t} + \lambda^+\frac{\partial y}{\partial x} + \frac{c}{g}\left(\frac{\partial V}{\partial t} + \lambda^+\frac{\partial V}{\partial x}\right) = c(S_o - S_f) \tag{14-27}$$

By multiplying Eq. 13-2 by c/g, subtracting from Eq. 13-1, and letting $V - c = \lambda^-$, we obtain

$$\frac{\partial y}{\partial t} + \lambda^-\frac{\partial y}{\partial x} - \frac{c}{g}\left(\frac{\partial V}{\partial t} + \lambda^-\frac{\partial V}{\partial x}\right) = -c(S_o - S_f) \tag{14-28}$$

Addition of Eqs. 14-27 and 14-28 yields

$$\frac{\partial y}{\partial t} + \frac{1}{2}\left[\lambda^+\left(\frac{\partial y}{\partial x}\right)^+ + \lambda^-\left(\frac{\partial y}{\partial x}\right)^-\right]$$

$$+ \frac{1}{2}\frac{c}{g}\left[\lambda^+\left(\frac{\partial V}{\partial x}\right)^+ - \lambda^-\left(\frac{\partial V}{\partial x}\right)^-\right] = 0 \tag{14-29}$$

in which we use superscripts + and − to indicate evaluation of the partial derivatives along the positive and negative characteristics, respectively.

Subtracting Eq. 14-28 from Eq. 14-27 and multiplying throughout by g/c give

$$\frac{\partial V}{\partial t} + \frac{1}{2}\frac{g}{c}\left[\lambda^+\left(\frac{\partial y}{\partial x}\right)^+ - \lambda^-\left(\frac{\partial y}{\partial x}\right)^-\right]$$

$$+ \frac{1}{2}\left[\lambda^+\left(\frac{\partial V}{\partial x}\right)^+ + \lambda^-\left(\frac{\partial V}{\partial x}\right)^-\right] = g(S_o - S_f) \tag{14-30}$$

These equations are referred to as the equations in the λ-form.

We use the backward or forward finite differences for the spatial partial derivatives according to the sign of λ; i.e., we use the backward finite difference for the positive direction and the forward finite difference for the negative direction. In calculus, the partial derivatives are equal, i.e., $V_x^+ = V_x^- = V_x$ and $y_x^+ = y_x^- = y_x$. If we substitute these relationships and expressions for λ^+ and λ^- and simplify the resulting equations, we end up with Eqs. 13-1 and 13-2. However, while writing the difference form of the governing equations, we use different finite-difference approximations for each of them to take into account correct signal directions. The spatial finite differences are written in terms of values at three grid points and the scheme comprises the predictor and corrector parts.

Predictor

We use the following finite differences for the spatial derivatives:

$$f_x^+ = \frac{2f_i^k - 3f_{i-1}^k + f_{i-2}^k}{\Delta x}$$

$$f_x^- = \frac{f_{i+1}^k - f_i^k}{\Delta x}$$

(14-31)

in which, for brevity, f refers to both V and y. Let us designate the quantities computed during the predicted part by the superscript *. Then, by substituting $\partial y/\partial t = (y_i^* - y_i^k)/\Delta t$ and the preceding finite-difference approximations into Eqs. 14-29 and 14-30 and simplifying the resulting equations, we obtain

$$y_i^* = y_i^k - \frac{1}{2}\frac{\Delta t}{\Delta x}\left[\lambda^+\left(2y_i^k - 3y_{i-1}^k + y_{i-2}^k\right) + \lambda^-\left(y_{i+1}^k - y_i^k\right)\right]$$

$$- \frac{1}{2}\frac{c_i^k}{g}\frac{\Delta t}{\Delta x}\left[\lambda^+\left(2V_i^k - 3V_{i-1}^k + V_{i-2}^k\right) - \lambda^-\left(V_{i+1}^k - V_i^k\right)\right]$$

(14-32)

Proceeding similarly,

$$V_i^* = V_i^k - \frac{1}{2}\frac{g}{c_i^k}\frac{\Delta t}{\Delta x}\left[\lambda^+\left(2y_i^k - 3y_{i-1}^k + y_{i-2}^k\right) - \lambda^-\left(y_{i+1}^k - y_i^k\right)\right]$$

$$- \frac{1}{2}\frac{\Delta t}{\Delta x}\left[\lambda^+\left(2V_i^k - 3V_{i-1}^k + V_{i-2}^k\right) + \lambda^-\left(V_{i+1}^k - V_i^k\right)\right] + g\Delta t(S_o - S_f^k)$$

(14-33)

Corrector

In the corrector part, we use the following finite-difference approximations:

$$f_x^+ = \frac{f_i^* - f_{i-1}^*}{\Delta x}$$

$$f_x^- = \frac{-2f_i^* + 3f_{i+1}^* - f_{i+2}^*}{\Delta x}$$

(14-34)

By using these finite differences and $\partial f/\partial t = (f_i^{**} - f_i^k)/\Delta t$ and using the values of different variables computed during the predictor part, we obtain

$$y_i^{**} = y_i^k - \frac{1}{2}\frac{\Delta t}{\Delta x}\left[\lambda^+\left(y_i^* - y_{i-1}^*\right) + \lambda^-\left(-2y_i^* + 3y_{i+1}^* - y_{i+2}^*\right)\right]$$

$$- \frac{1}{2}\frac{\Delta t}{\Delta x}\frac{c_i^*}{g}\left[\lambda^+\left(V_i^* - V_{i-1}^*\right) - \lambda^-\left(-2V_i^* + 3V_{i+1}^* - V_{i+2}^*\right)\right]$$

(14-35)

Proceeding similarly,

$$V_i^{**} = V_i^k \ - \frac{1}{2} \frac{g}{c_i^*} \frac{\Delta t}{\Delta x} \left[\lambda^+ \left(y_i^* - y_{i-1}^* \right) - \lambda^- \left(-2y_i^* + 3y_{i+1}^* - y_{i+2}^* \right) \right]$$

$$- \frac{1}{2} \frac{\Delta t}{\Delta x} \left[\lambda^+ \left(V_i^* - V_{i-1}^* \right) + \lambda^- \left(-2V_i^* + 3V_{i+1}^* - V_{i+2}^* \right) \right] \qquad (14\text{-}36)$$

$$+ g \Delta t (S_o - S_f^*)$$

Now the values at $k+1$ time step may be determined from the following equations:

$$y_i^{k+1} = \frac{1}{2} \left(y_i^* + y_i^{**} \right)$$

$$V_i^{k+1} = \frac{1}{2} \left(V_i^* + V_i^{**} \right) \qquad (14\text{-}37)$$

Boundary conditions.   The preceding equations are for the interior grid points. The boundary grid points may be included in the analysis in the same manner as we discussed for the Lax scheme. At the nodes adjacent to the boundary, first-order, one-sided finite differences may be used whenever three points are not available.

Stability.   For the scheme to be stable, C_n must be less than or equal to 1 at all grid points during each time step.

Gabutti Scheme

The *Gabutti scheme* is an extension of the Lambda scheme (Gabutti, 1983). It allows analysis of sub- and super-critical flows and has been used for such analyses by Fennema and Chaudhry (1987).

General formulation.   This scheme comprises the predictor and corrector parts, and the predictor part is subdivided into two parts. We use the λ-form of the equations and replace the partial derivatives as follows, taking into consideration the correct signal direction.

Predictor

Part 1: If f refers to both y and V, the spatial derivatives are approximated as follows:

$$f_x^+ = \frac{f_i^k - f_{i-1}^k}{\Delta x}$$

$$f_x^- = \frac{f_{i+1}^k - f_i^k}{\Delta x} \qquad (14\text{-}38)$$

By substituting $\partial f / \partial t = (f_i^* - f_i^k)/\Delta t$ and the finite-difference approximations of Eq. 14-38 into Eqs. 14-29 and 14-30 and simplifying the resulting equations, we obtain

$$y_i^* = y_i^k \; - \frac{1}{2}\frac{\Delta t}{\Delta x}\left[\lambda^+(y_i^k - y_{i-1}^k) + \lambda^-(y_{i+1}^k - y_i^k)\right]$$

$$- \frac{1}{2}\frac{\Delta t}{\Delta x}\frac{c_i^k}{g}\left[\lambda^+(V_i^k - V_{i-1}^k) - \lambda^-(V_{i+1}^k - V_i^k)\right]$$

$$V_i^* = V_i^k \; - \frac{1}{2}\frac{\Delta t}{\Delta x}\frac{g}{c_i^k}\left[\lambda^+(y_i^k - y_{i-1}^k) - \lambda^-(y_{i+1}^k - y_i^k)\right] \qquad (14\text{-}39)$$

$$- \frac{1}{2}\frac{\Delta t}{\Delta x}\left[\lambda^+(V_i^k - V_{i-1}^k) + \lambda^-(V_{i+1}^k - V_i^k)\right]$$

$$+ g\Delta t(S_o - S_{fi}^k)$$

in which the superscript * refers to predicted values.

Part 2: In part 2 of the predictor part, we use the following finite-difference approximations:

$$f_x^+ = \frac{2f_i^k - 3f_{i-1}^k + f_{i-2}^k}{\Delta x}$$

$$f_x^- = \frac{-2f_i^k + 3f_{i+1}^k - f_{i+2}^k}{\Delta x} \qquad (14\text{-}40)$$

By substituting these finite-difference approximations into Eqs. 14-29 and 14-30, we determine the predicted values of y_t^* and V_t^*. Note that we are using the values at the known time level, k. This process yields

$$y_t^* = -\frac{1}{2}\frac{1}{\Delta x}\left[\lambda^+(2y_i^k - 3y_{i-1}^k + y_{i-2}^k) + \lambda^-(-2y_i^k + 3y_{i+1}^k - y_{i+2}^k)\right]$$

$$- \frac{1}{2}\frac{1}{\Delta x}\frac{c_i^k}{g}\left[\lambda^+(2V_i^k - 3V_{i-1}^k + V_{i-2}^k) - \lambda^-(-2V_i^k + 3V_{i+1}^k - V_{i+2}^k)\right]$$

$$V_t^* = -\frac{1}{2}\frac{1}{\Delta x}\frac{g}{c_i^k}\left[\lambda^+(2y_i^k - 3y_{i-1}^k + y_{i-2}^k) - \lambda^-(-2y_i^k + 3y_{i+1}^k - y_{i+2}^k)\right] \qquad (14\text{-}41)$$

$$- \frac{1}{2}\frac{1}{\Delta x}\left[\lambda^+(2V_i^k - 3V_{i-1}^k + V_{i-2}^k)\right.$$

$$\left. + \lambda^-(-2V_i^k + 3V_{i+1}^k - V_{i+2}^k)\right] + g(S_o - S_{fi}^k)$$

Corrector

In the corrector part, we use the predicted values y_i^*, V_i^* and the corresponding values of coefficients and approximate the spatial derivatives by the following finite differences:

$$f_x^+ = \frac{f_i^* - f_{i-1}^*}{\Delta x}$$

$$f_x^- = \frac{f_{i+1}^* - f_i^*}{\Delta x} \qquad (14\text{-}42)$$

By substituting these finite differences into Eqs. 14-29 and 14-30, we obtain

$$y_t^{**} = -\frac{1}{2}\frac{1}{\Delta x}\left[\lambda^+(y_i^* - y_{i-1}^*) + \lambda^-(y_{i+1}^* - y_i^*)\right]$$

$$-\frac{1}{2}\frac{1}{\Delta x}\frac{c_i^k}{g}\left[\lambda^+(V_i^* - V_{i-1}^*) - \lambda^-(V_{i+1}^* - V_i^*)\right]$$

$$V_t^{**} = -\frac{1}{2}\frac{1}{\Delta x}\frac{g}{c_i^k}\left[\lambda^+(y_i^* - y_{i-1}^*) - \lambda^-(y_{i+1}^* - y_i^*)\right] \qquad (14\text{-}43)$$

$$-\frac{1}{2}\frac{1}{\Delta x}\left[\lambda^+(V_i^* - V_{i-1}^*) + \lambda^-(V_{i+1}^* - V_i^*)\right]$$

$$+ g(S_o - S_{fi}^*)$$

The values of y_i^{k+1} and V_i^{k+1} may now be determined from the following equations:

$$y_i^{k+1} = y_i^k + \frac{1}{2}\Delta t(y_t^* + y_t^{**})$$

$$V_i^{k+1} = V_i^k + \frac{1}{2}\Delta t(V_t^* + V_t^{**}) \qquad (14\text{-}44)$$

Boundary conditions.   The preceding equations are for the interior grid points. The boundary grid points may be included in the analysis in the same manner as we discussed for the Lax scheme. At the nodes adjacent to the boundary, first-order, one-sided finite differences may be used whenever three points are not available.

Stability.   For the scheme to be stable, C_n must be less than or equal to 1 at all grid points during each time step. However, if the unsteady flow equations are solved until the solution converges to a steady state, then the stability criterion may be relaxed to $C_n \leq 2$.

14-4 Implicit Finite-Difference Schemes

In implicit finite-difference schemes, the spatial partial derivatives and/or the coefficients are replaced in terms of the values at the unknown time level. The unknown variables, therefore, appear implicitly in the algebraic equations, and the methods are called *implicit* methods. The algebraic equations for the entire system have to be solved simultaneously in these methods.

Several implicit finite-difference schemes have been used for the analysis of unsteady open-channel flows (Amein and Fang 1970; Strelkoff 1970; Terzidis and Strelkoff 1970; Liggett and Cunge 1975; Abbott 1979; Cunge, et al. 1980; and Fennema and Chaudhry 1987). We present details of three of these schemes.

Preissmann Scheme

The *Preissmann scheme* has been extensively used since the early 1960s (Preissmann and Cunge 1961; Liggett and Cunge 1975; Cunge, et al. 1980). It has the advantages that a variable spatial grid may be used; steep wave fronts may be properly simulated by varying the weighting coefficient; and the scheme yields an exact solution of the linearized form of the governing equations for a particular value of Δx and Δt.

General formulation.   The partial derivatives and other coefficients are approximated as follows:

$$\frac{\partial f}{\partial t} = \frac{(f_i^{k+1} + f_{i+1}^{k+1}) - (f_i^k + f_{i+1}^k)}{2\Delta t}$$

$$\frac{\partial f}{\partial x} = \frac{\alpha(f_{i+1}^{k+1} - f_i^{k+1})}{\Delta x} + \frac{(1 - \alpha)(f_{i+1}^k - f_i^k)}{\Delta x} \qquad (14\text{-}45)$$

$$f = \frac{1}{2}\alpha(f_{i+1}^{k+1} + f_i^{k+1}) + \frac{1}{2}(1 - \alpha)(f_{i+1}^k + f_i^k)$$

in which α is a weighting coefficient; in the partial derivatives f refers to both V and y, and f as a coefficient stands for S_f and V. By selecting a suitable value for α, the scheme may be made totally explicit ($\alpha = 0$) or implicit ($\alpha = 1$). The scheme is stable if $0.55 < \alpha \le 1$. Steep wave fronts are properly simulated for low values of α, but there are oscillations behind the wave front. These oscillations are eliminated for α close to unity; however, steep wave fronts are somewhat smeared (Fig. 14-4). For typical applications, $\alpha = 0.6\text{–}0.7$ may be used.

By substituting the preceding finite-difference approximations and the coefficients into Eqs. 14-15 and rearranging the terms of the resulting equation, we obtain

$$\mathbf{U}_i^{k+1} + \mathbf{U}_{i+1}^{k+1} + 2\frac{\Delta t}{\Delta x}\left[\alpha(\mathbf{F}_{i+1}^{k+1} - \mathbf{F}_i^{k+1}) + (1 - \alpha)(\mathbf{F}_{i+1}^k - \mathbf{F}_i^k)\right]$$

$$+ \Delta t\left[\alpha(\mathbf{S}_i^{k+1} + \mathbf{S}_{i+1}^{k+1}) + (1 - \alpha)(\mathbf{S}_{i+1}^k + \mathbf{S}_i^k)\right] = \mathbf{U}_i^k + \mathbf{U}_{i+1}^k \qquad (14\text{-}46)$$

In Eq. 14-46, we have four unknowns, namely, V_i^{k+1}, y_i^{k+1}, V_{i+1}^{k+1}, and y_{i+1}^{k+1}. If we write these two equations for each grid point, we have $2N$ equations (N = number of reaches on the channel). We cannot write these equations for the grid points at the downstream end. However, we have $2(N + 1)$ unknowns, i.e., two unknowns for each grid point. Thus, for a unique solution we need two more equations. These are provided by the *end conditions*, as discussed in the following paragraphs.

Boundary conditions.   Unlike the explicit schemes, we include directly in the system of equations the equations describing the end conditions (equations describing a number of typical boundaries are listed in Eq. 14-18). In other words, we do not have to use the characteristic equations or the reflection procedures.

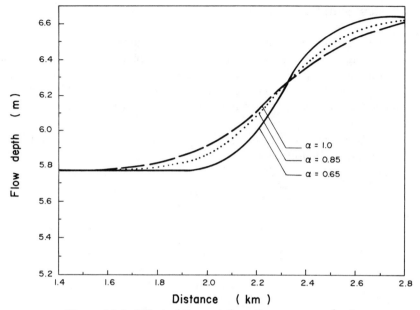

Figure 14-4   Effect of the variation of α on a wave front.

For a constant-level reservoir at the downstream end, the equation representing the downstream boundary may be written as

$$y_{i,n+1}^{k+1} = y_{resd}$$

in which y_{resd} is the depth in the downstream reservoir.

The following equation may be used to represent the boundary for a rating curve:

$$Q_i^{k+1} = a_0 + a_1 y_i^{k+1} + a_2\left(y_i^{k+1}\right)^2$$

in which the coefficients a_0, a_1, and a_2 may be determined by a regression analysis of the recorded data.

To simulate a series junction boundary, one may use the following equations:

$$V_{i,n+1}^{k+1} A_{i,n+1}^{k+1} = V_{i+1,1}^{k+1} A_{i+1,1}^{k+1}$$

and

$$z_i \left| y_{i,n+1}^{k+1} \right| \frac{\left(V_{i,n+1}^{k+1}\right)^2}{2g} = z_{i+1} + y_{i+1,1}^{k+1} + \frac{(1+K)\left(V_{i+1,1}^{k+1}\right)^2}{2g}$$

in which K = coefficient of junction losses.

Similarly, other boundary conditions may be incorporated in the system of equations.

Stability.   The scheme is unconditionally stable provided $\alpha > 0.5$, i.e., the flow variables are weighted toward the $k + 1$ time level. An unconditional stability means that there is no restriction on the size of Δx and Δt for stability. However, for accuracy C_n, should be close to 1. The preceding unconditional stability criterion is derived assuming a frictionless channel. If a linearized form of the friction-loss term is included in the analysis (Samuels and Skeels 1990) then Vedernikov number, $\mathbf{V}_n \leq 1$, for the scheme to be stable, where

$$\mathbf{V}_n = \frac{p A \mathbf{F}_r}{m R} \frac{dR}{dA}$$

The exponents m and p are as defined in Eq. 12-10e and $\mathbf{F}_r$ = Froude number. Through numerical experimentation, Evans (1977) and Yen and Lin (1986) investigated the stability of this scheme for the full St. Venant equations for specific flow parameters. These studies have the advantage that they are not based on the linearized analyses; however, they are not general, since the conclusions are based on a specific range of parameters.

Solution Procedure.   Expansion of Eq. 14-46 yields

$$A_i^{k+1} + A_{i+1}^{k+1} + 2\frac{\Delta t}{\Delta x}\left\{\alpha\left[(VA)_{i+1}^{k+1} - (VA)_i^{k+1}\right]\right.$$

$$\left. + (1 - \alpha)\left[(VA)_{i+1}^k - (VA)_i^k\right]\right\} = A_i^k + A_{i+1}^k \tag{14-47}$$

$$(VA)_i^{k+1} + (VA)_{i+1}^{k+1} + 2\frac{\Delta t}{\Delta x}\left\{\alpha\left[(V^2A + gA\bar{y})_{i+1}^{k+1} - (V^2A + gA\bar{y})_i^{k+1}\right]\right\}$$

$$- gA\Delta t\left\{\alpha\left[(S_o - S_f)_{i+1}^{k+1} + (S_o - S_f)_i^{k+1}\right]\right\}$$

$$= gA\Delta t\left\{(1 - \alpha)\left[(S_o - S_f)_{i+1}^k + (S_o - S_f)_i^k\right]\right\} \tag{14-48}$$

$$+ (VA)_i^k + (VA)_{i+1}^k - (1 - \alpha)\frac{2\Delta t}{\Delta x}\left\{(V^2A + gA\bar{y})_{i+1}^k\right.$$

$$\left. - (V^2A + gA\bar{y})_i^k\right\}$$

The system of equations (Eqs. 14-47 and 14-48 for each node and the boundary conditions) is a set of nonlinear algebraic equations. These may be solved by an iterative technique. The solution by the Newton-Raphson method is discussed in this section; readers interested in the solution by the double-sweep method should see Cunge, et al. (1980).

In the Newton-Raphson method, we estimate values for the unknown variables V and y at each node and then iterate to refine the solution as we discussed in Chapter 6. To determine the corrections for each iteration, we need the partial derivatives of Eqs. 14-47 and 14-48 and of the boundary conditions with respect to y_i^{k+1}, y_{i+1}^{k+1}, V_i^{k+1}, and V_{i+1}^{k+1}. If we derive these partial derivatives and arrange the governing equations, we obtain

$$\mathbf{A}\mathbf{x} = \mathbf{b} \qquad (14\text{-}49)$$

in which $\mathbf{A}$ is a matrix comprising the partial derivatives and $\mathbf{x}$ is a column vector comprising the corrections Δy_i and ΔV_i ($i = 1, 2, \ldots, N + 1$). Thus, we have $2(n + 1)$ equations in $2(n + 1)$ unknowns. These unknowns are the corrections. Now, we check that $\sum_i^{N+1} |\Delta y_i| + |\Delta V_i| \leq \epsilon$, where ϵ is the specified tolerance. If the sum of the corrections is less than the specified tolerance, then we proceed to the next step after applying the correction. Otherwise, we apply the correction and iterate the given procedure.

Note that matrix $\mathbf{A}$ is banded with a bandwidth of 4. We may utilize this fact while solving Eq. 14-49, since a banded-matrix solution routine requires less storage and gives more accurate results.

For branching or parallel systems, the nodes may be numbered (Chaudhry 1987) such that the resulting matrix is banded.

Beam and Warming Scheme

Several different formulations of the Beam and Warming scheme have been presented (Anderson, et al., 1984). However, we shall present formulations, which are based on splitting the coefficients. Thus, a correct signal transmission can be forced thereby allowing the analysis of supercritical flows. Fennema and Chaudhry (1987) used these schemes for the simulation of dam-break flows. A general description of these schemes for application to two-dimensional flows is presented in the next chapter; details for one-dimensional analysis are given in this section.

By using trapezoidal rule, the time difference may be approximated as

$$\mathbf{U}^{k+1} = \mathbf{U}^k + \frac{\Delta t}{2}\left[\left(\frac{\partial \mathbf{U}}{\partial t}\right)^{k+1} + \left(\frac{\partial \mathbf{U}}{\partial t}\right)^k\right] \qquad (14\text{-}50)$$

By substituting the value of $\mathbf{U}_t$ from Eq. 14-15 into this equation

$$\mathbf{U}^{k+1} = \mathbf{U}^k - \frac{\Delta t}{2}\left[\left(\frac{\partial \mathbf{F}}{\partial x} + \mathbf{S}\right)^{k+1} + \left(\frac{\partial \mathbf{F}}{\partial x} + \mathbf{S}\right)^k\right] \qquad (14\text{-}51)$$

The terms $\mathbf{F}^{k+1}$ and $\mathbf{S}^{k+1}$ in Eq. 14-51 are nonlinear and may be linearized as follows. The Taylor series expansion of $\mathbf{F}^{k+1}$ may be written as

$$\mathbf{F}^{k+1} = \mathbf{F}^k + \Delta t \left(\frac{\partial \mathbf{F}}{\partial t}\right)^k + O(\Delta t)^2$$

$$= \mathbf{F}^k + \Delta t \left(\frac{\partial \mathbf{F}}{\partial \mathbf{U}} \frac{\partial \mathbf{U}}{\partial t}\right)^k + O(\Delta t)^2 \tag{14-52}$$

$$= \mathbf{F}^k + \mathbf{A}^k \frac{\partial \mathbf{U}}{\partial t} \Delta t$$

$$= \mathbf{F}^k + \mathbf{A}^k (\mathbf{U}^{k+1} - \mathbf{U}^k)$$

Similarly

$$\mathbf{S}^{k+1} = \mathbf{S}^k + \mathbf{B}^k (\mathbf{U}^{k+1} - \mathbf{U}^k) \tag{14-53}$$

where $\mathbf{A}$ and $\mathbf{B}$ are the Jacobians of $\mathbf{F}$ and $\mathbf{S}$ respectively and are given as

$$\mathbf{A} = \begin{pmatrix} 0 & 1 \\ gD - V^2 & 2V \end{pmatrix}$$

$$\mathbf{B} = \begin{pmatrix} 0 & 0 \\ \dfrac{-gS_o - 1.33gn^2V|V|}{R^{1.33}} & \dfrac{gn^2|V|}{R^{1.33}} \end{pmatrix} \tag{14-54}$$

Substituting Eqs. 14-52 and 14-53 into Eq. 14-51 and simplifying

$$\left[\mathbf{I} + \frac{\Delta t}{2} \left(\frac{\partial \mathbf{A}}{\partial x} + \mathbf{B}\right)^k\right] \Delta_t \mathbf{U}^{k+1} = -\Delta t \left(\frac{\partial \mathbf{F}}{\partial x} + \mathbf{S}\right)^k \tag{14-55}$$

For correct signal transmission, the matrices $\mathbf{A}$ and $\mathbf{F}$ may by split as

$$\mathbf{A} = \mathbf{A}^+ + \mathbf{A}^-$$
$$\mathbf{F} = \mathbf{F}^+ + \mathbf{F}^- \tag{14-56}$$

in which $\mathbf{A}^+ = \mathbf{MD}^+\mathbf{M}^{-1}$; $\mathbf{A}^- = \mathbf{MD}^-\mathbf{M}^{-1}$; $\mathbf{F}_x^+ = \mathbf{A}^+\mathbf{U}_x$; $\mathbf{F}_x^- = \mathbf{A}^-\mathbf{U}_x$; $\mathbf{D}$ is the diagonal matrix of eigenvalues of $\mathbf{A}$ and

$$\mathbf{M} = \begin{pmatrix} \dfrac{1}{2c} & \dfrac{-1}{2c} \\ \dfrac{V+c}{2c} & \dfrac{V-c}{2c} \end{pmatrix}$$

By substituting Eq 14-56 into 14-55, we obtain

$$\left[\mathbf{I} + \frac{\Delta t}{2}\left\{\frac{\partial}{\partial x}(\mathbf{A}^+ + \mathbf{A}^-) + \mathbf{B}\right\}^k\right] \Delta_t \mathbf{U}^{k+1} = -\Delta t \left[\frac{\partial}{\partial x}(\mathbf{F}^+ + \mathbf{F}^-) + \mathbf{S}\right]^k$$

This equation in finite-difference form may be written as

$$\left[\mathbf{I} + \frac{1}{2}\frac{\Delta t}{\Delta x}\left\{(\nabla_x\mathbf{A}^+ + \Delta_x\mathbf{A}^-) + \frac{\Delta t}{2}\mathbf{B}_i\right\}^k\right]\Delta_t\mathbf{U}^{k+1}$$

$$= -\frac{\Delta t}{\Delta x}\left[\mathbf{A}_i^+\nabla_x\mathbf{U} + \mathbf{A}_i^-\Delta_x\mathbf{U}\right]^k - \Delta t\mathbf{S}_i^k \tag{14-57}$$

in which $\nabla_x \mathbf{A}^+ = \mathbf{A}_i^+ - \mathbf{A}_{i-1}^+$ and $\Delta_x \mathbf{A}^- = \mathbf{A}_{i+1}^- - \mathbf{A}_i^-$. The left-hand side of Eq. 14-57 constitute the block tridiagonal system for each time step and may be solved by using special algorithm.

Vasiliev Scheme

The *Vasiliev* scheme was developed by a team of researchers at the Institute of Hydro-dynamics, Novosibirsk, Russia (Vasiliev et al. 1965). It uses the following finite-difference approximations:

$$\frac{\partial f}{\partial t} = \frac{f_i^{k+1} - f_i^k}{\Delta t}$$

$$\frac{\partial f}{\partial x} = \frac{f_{i+1}^{k+1} - f_{i-1}^{k+1}}{2\Delta x} \qquad (14\text{-}58)$$

Vasiliev et al. applied this discretization to the continuity equation (Eq. 12-4) and to the momentum equation converted to the following form:

$$\frac{\partial Q}{\partial t} + 2V\frac{\partial Q}{\partial x} + (c^2 - V^2)B\frac{\partial y}{\partial x} = gA(S_o - S_f) \qquad (14\text{-}59)$$

If we substitute these finite differences in the governing equations, we obtain $2N - 2$ equations for a channel divided into N reaches, since we can not write these equations for the boundary nodes. Two equations are provided by the end conditions. Thus, we need two more equations for a unique solution of the resulting system of equations. The characteristic equations for the upstream and the downstream end provide these equations.

14-5 Consistency

A finite-difference scheme is said to be *consistent* if the finite-difference form of the equation tends to the original differential equation as Δx and Δt tend to zero.

To check the consistency of the Lax scheme, let us expand V_i^{k+1}, V_{i+1}^k, V_{i-1}^k in a Taylor series about values at grid point (i, k):

$$V_i^{k+1} = V_i^k + \frac{\partial V}{\partial t}\Delta t + \frac{1}{2!}\frac{\partial^2 V}{\partial t^2}(\Delta t)^2 + \frac{1}{3!}\frac{\partial^3 V}{\partial t^3}(\Delta t)^3 + \cdots$$

$$V_{i+1}^k = V_i^k + \frac{\partial V}{\partial x}\Delta x + \frac{1}{2!}\frac{\partial^2 V}{\partial x^2}(\Delta x)^2 + \frac{1}{3!}\frac{\partial^3 V}{\partial x^3}(\Delta x)^3 + \cdots \qquad (14\text{-}60)$$

$$V_{i-1}^k = V_i^k - \frac{\partial V}{\partial x}\Delta x + \frac{1}{2!}\frac{\partial^2 V}{\partial x^2}(\Delta x)^2 - \frac{1}{3!}\frac{\partial^3 V}{\partial x^3}(\Delta x)^3 + \cdots$$

Similarly, we can expand y_i^{k+1}, y_{i+1}^k, and y_{i-1}^k in terms of y_i^k. Substituting these expansions into Eq. 14-14, we obtain

$$V_i^k + \frac{\partial V}{\partial t}\Delta t + \frac{1}{2}\frac{\partial^2 V}{\partial t^2}(\Delta t)^2 + \frac{1}{3!}\frac{\partial^3 V}{\partial t^3}(\Delta t)^3 + \cdots$$

$$-\frac{1}{2}\left[V_i^k - \frac{\partial V}{\partial x}\Delta x + \frac{1}{2}\frac{\partial^2 V}{\partial x^2}(\Delta x)^2 - \frac{1}{3!}\frac{\partial^3 V}{\partial x^3}(\Delta x)^3 + \cdots \right.$$

$$\left. + V_i^k + \frac{\partial V}{\partial x}\Delta x + \frac{1}{2}\frac{\partial^2 V}{\partial x^2}(\Delta x)^2 + \frac{1}{3!}\frac{\partial^3 V}{\partial x^3}(\Delta x)^3 + \cdots \right]$$

$$+\frac{1}{2}g\frac{\Delta t}{\Delta x}\left[y_i^k + \frac{\partial y}{\partial x}\Delta x + \frac{1}{2}\frac{\partial^2 y}{\partial x^2}(\Delta x)^2 + \frac{1}{3!}\frac{\partial^3 y}{\partial x^3}(\Delta x)^3 + \cdots \right. \tag{14-61}$$

$$\left. - y_i^k + \frac{\partial y}{\partial x}\Delta x - \frac{1}{2}\frac{\partial^2 y}{\partial x^2}(\Delta x)^2 + \frac{1}{3!}\frac{\partial^3 y}{\partial x^3}(\Delta x)^3 + \cdots \right]$$

$$+\frac{1}{2}V_i^k\frac{\Delta t}{\Delta x}\left[V_i^k + \frac{\partial V}{\partial x}\Delta x + \frac{1}{2}\frac{\partial^2 V}{\partial x^2}(\Delta x)^2 + \frac{1}{3!}\frac{\partial^3 V}{\partial x^3}(\Delta x)^3 + \cdots \right.$$

$$\left. - V_i^k + \frac{\partial V}{\partial x}\Delta x - \frac{1}{2}\frac{\partial^2 V}{\partial x^2}(\Delta x)^2 + \frac{1}{3!}\frac{\partial^3 V}{\partial x^3}(\Delta x)^3 + \cdots \right] = 0$$

Simplifying and dividing throughout by Δt, this equation becomes

$$\frac{\partial V}{\partial t} + g\frac{\partial y}{\partial x} + V\frac{\partial V}{\partial x} + \frac{1}{2}\frac{\partial^2 V}{\partial t^2}\Delta t + \frac{1}{3!}\frac{\partial^3 V}{\partial t^3}(\Delta t)^2$$

$$-\frac{1}{2}\frac{\partial^2 V}{\partial x^2}\frac{(\Delta x)^2}{\Delta t} + \frac{1}{3!}V_i^k\frac{\partial^3 V}{\partial x^3}(\Delta x)^2 \tag{14-62}$$

$$+\frac{1}{3!}g\frac{\partial^3 y}{\partial x^3}(\Delta x)^2 = 0$$

In order to have the space and time derivatives converge uniformly as Δx and Δt become smaller so that both of these derivatives have similar errors, the first error term of time derivatives, $\frac{1}{2}(\partial^2 V/\partial t^2)\Delta t$, should be equivalent to the first error term of the space derivative, $\frac{1}{2}(\partial^2 V/\partial x^2)(\Delta x)^2/\Delta t$. Thus, the computational grid has to be constructed so that $(\Delta x)^2 = k\Delta t$, where k is a constant (Liggett and Cunge 1975). Holding this ratio between Δx and Δt constant, letting Δt and Δx approach zero, and neglecting terms of third and higher order, this equation takes the form

$$\frac{\partial V}{\partial t} + g\frac{\partial y}{\partial x} + V\frac{\partial V}{\partial x} - \frac{1}{2}k\frac{\partial^2 V}{\partial x^2} = 0 \tag{14-63}$$

In this equation there is an additional diffusion like term, $\frac{1}{2}k(\partial^2 V/\partial x^2)$, which is not present in the original equation. Similarly, we can show that an additional term, $\frac{1}{2}k(\partial^2 y/\partial x^2)$, is introduced in the continuity equation. Therefore, the finite-difference equations in the Lax scheme are not consistent with the original governing equations.

14-6 Stability

We investigate the stability of a numerical scheme by studying whether an error grows or decays as the solution progresses in a marching procedure. Rigorous procedures are not presently available to determine the stability of nonlinear equations. However, by neglecting or linearizing the nonlinear terms, stability may be studied. If the nonlinearities are not strong, then the criteria developed for the linear equations may be assumed to be valid for the nonlinear equations as well. The effects of boundaries on the stability of a scheme are not included in such analyses.

For illustration purposes, we analyze the stability of the Lax scheme in the following paragraphs. The analysis procedure we use is referred to as *von Neumann or Fourier analysis.*

To do this analysis, let us linearize the governing equations, Eqs. 13-1 and 13-2, by neglecting the friction term and by using the coefficients V_o and D_o, in which the subscript o indicates steady-state value. In addition, let us drop S_o for simplicity and replace y by Y. Then, the linearized set of governing equations becomes

$$\frac{\partial V}{\partial t} + V_o \frac{\partial V}{\partial x} + g \frac{\partial Y}{\partial x} = 0 \tag{14-64}$$

$$\frac{\partial Y}{\partial t} + D_o \frac{\partial V}{\partial x} + V_o \frac{\partial Y}{\partial x} = 0 \tag{14-65}$$

Substitution of the finite-difference approximations of Eq. 14-13 into these equations and rearrangement of the resulting equations yield

$$
\begin{aligned}
Y_i^{k+1} &= \frac{1}{2}\left(Y_{i-1}^k + Y_{i+1}^k \right) - \frac{1}{2}\frac{\Delta t}{\Delta x} D_o \left(V_{i+1}^k - V_{i-1}^k \right) \\
&\quad - \frac{1}{2}\frac{\Delta t}{\Delta x} V_o \left(Y_{i+1}^k - Y_{i-1}^k \right) \\
V_i^{k+1} &= \frac{1}{2}\left(V_{i-1}^k + V_{i+1}^k \right) - \frac{1}{2}\frac{\Delta t}{\Delta x} g \left(Y_{i+1}^k - Y_{i-1}^k \right) \\
&\quad - \frac{1}{2}\frac{\Delta t}{\Delta x} V_o \left(V_{i+1}^k - V_{i-1}^k \right)
\end{aligned}
\tag{14-66}
$$

Let the exact solution of these equations be V and Y. Such a solution will be obtained by a computer having an infinite accuracy. However, real machines have only finite accuracy. Let us call the solutions obtained on a real machine as V_{comp} and Y_{comp} in which roundoff errors v and y have been introduced. By substituting $V_{exact} = V_{comp} + v$ and $Y_{exac} = Y_{comp} + y$ into Eqs. 14-66 and noting that V_{exac} and Y_{exac} must also satisfy these equations, we obtain

$$
\begin{aligned}
v_i^{k+1} &- \frac{1}{2}(v_{i-1}^k + v_{i+1}^k) + \frac{1}{2}V_o \frac{\Delta t}{\Delta x}(v_{i+1}^k - v_{i-1}^k) \\
&+ \frac{1}{2}g\frac{\Delta t}{\Delta x}(y_{i+1}^k - y_{i-1}^k) = 0
\end{aligned}
\tag{14-67}
$$

$$y_i^{k+1} - \frac{1}{2}(y_{i+1}^k + y_{i-1}^k) + \frac{1}{2}D_o\frac{\Delta t}{\Delta x}(v_{i+1}^k - v_{i-1}^k)$$

$$+ \frac{1}{2}\frac{\Delta t}{\Delta x}V_o(y_{i+1}^k - y_{i-1}^k) = 0 \tag{14-68}$$

For the scheme to be stable, these errors must decay as the solution progresses from one time step to the next. To investigate this, let us express v_i^k and y_i^k in Fourier series as

$$v_i^k = \sum_{n=0}^{N} A'_n e^{j\theta_n x_i}$$

$$y_i^k = \sum_{n=0}^{N} B'_n e^{j\theta_n x_i} \tag{14-69}$$

in which $j = \sqrt{-1}$, the wave number, $\theta_n = n\pi/L$ $(n = 0, 1, 2, \ldots, N)$, and the interval of interest is of length L. Now, $x_{i+1} = x_i + \Delta x$ and $x_{i-1} = x_i - \Delta x$. Hence, we may write

$$v_{i-1}^k = \sum_{n=0}^{N} A'_n e^{j\theta_n x_i} e^{-j\theta_n \Delta x}$$

$$v_{i+1}^k = \sum_{n=0}^{N} A'_n e^{j\theta_n x_i} e^{j\theta_n \Delta x}$$

$$\tag{14-70}$$

$$y_{i-1}^k = \sum_{n=0}^{N} B'_n e^{j\theta_n x_i} e^{-j\theta_n \Delta x}$$

$$y_{i+1}^k = \sum_{n=0}^{N} B'_n e^{j\theta_n x_i} e^{j\theta_n \Delta x}$$

Since the system is linear, we may consider only one term of the series instead of the sum of N terms. In addition, we may write

$$v_i^{k+1} = e^{\alpha t} v_i^k = \xi v_i^k = \xi A' e^{j\theta x_i} \tag{14-71}$$

in which ξ is called the *amplification factor*. Depending upon the value of ξ, an error introduced at any time grows or decays as the computations progress in time. If $|\xi| < 1$, the error decays as the computations progress and the scheme is called *stable*; however, if $|\xi| > 1$, then the error grows with time and the scheme is called *unstable*. The scheme is said to be *neutrally stable* if $|\xi| = 1$.

Similarly, we may express the error in the flow depth as

$$y_i^{k+1} = \xi B' e^{j\theta x_i} \tag{14-72}$$

Substituting these expressions into Eqs. 14-67 and letting $r = \Delta t/\Delta x$, we obtain

$$\xi A'e^{j\theta x_i} - \frac{1}{2}\left(A'e^{j\theta x_i}e^{-j\theta \Delta x} + A'e^{j\theta x_i}e^{j\theta \Delta x}\right)$$

$$+ \frac{1}{2}rV_o\left(A'e^{j\theta x_i}e^{j\theta \Delta x} - A'e^{j\theta x_i}e^{-j\theta \Delta x}\right) \tag{14-73}$$

$$+ \frac{1}{2}rg\left(B'e^{j\theta x_i}e^{j\theta \Delta x} - B'e^{j\theta x_i}e^{-j\theta \Delta x}\right) = 0$$

Canceling out $e^{j\theta x_i}$ and rearranging the terms, we obtain

$$\left[\xi - \frac{1}{2}\left(e^{j\theta \Delta x} + e^{-j\theta \Delta x}\right) + \frac{1}{2}rV_o\left(e^{j\theta \Delta x} - e^{-j\theta \Delta x}\right)\right]A'$$

$$+ \frac{1}{2}rg\left(e^{j\theta \Delta x} - e^{-j\theta \Delta x}\right)B' = 0 \tag{14-74}$$

Let $\delta = \theta \Delta x$. Then the preceding equation may be written as

$$(\xi - \cos \delta + jrV_o \sin \delta)A' + jrg \sin \delta B' = 0 \tag{14-75}$$

Proceeding similarly, Eq. 14-68 becomes

$$(\xi - \cos \delta + jrV_o \sin \delta)B' + jD_o r \sin \delta A' = 0 \tag{14-76}$$

For a nontrivial solution of A' and B', it follows from Eqs. 14-75 and 14-76 that

$$\begin{vmatrix} \xi - \cos \delta + jrV_o \sin \delta & jrg \sin \delta \\ jD_o r \sin \delta & \xi - \cos \delta + jV_o \sin \delta \end{vmatrix} = 0 \tag{14-77}$$

It follows from this equation that

$$(\xi - \cos \theta + jrV_o \sin \delta)^2 = -D_o gr^2 \sin^2 \delta \tag{14-78}$$

Taking the square root of both sides,

$$\xi - \cos \delta + jrV_o \sin \delta = \pm j\sqrt{D_o g}r \sin \delta \tag{14-79}$$

or

$$\xi = \cos \delta - j(V_o \pm \sqrt{D_o g})r \sin \delta \tag{14-80}$$

For the error to decay, $|\xi| < 1$, i.e.,

$$\left(\cos^2 \delta + (V_o \pm \sqrt{D_o g})^2 r^2 \sin^2 \delta\right)^{\frac{1}{2}} < 1$$

or

$$\cos^2 \delta + (V_o \pm \sqrt{D_o g})^2 r^2 \sin^2 \delta < 1 \tag{14-81}$$

or

$$(1 - \sin^2 \delta) + (V_o \pm \sqrt{g D_o})^2 r^2 \sin^2 \delta < 1 \tag{14-82}$$

or

$$(V_o \pm \sqrt{g D_o})^2 r^2 < 1 \tag{14-83}$$

Noting that $c = \sqrt{g D_o}$, it follows from this equation that

$$\frac{\Delta t}{\Delta x} < \frac{1}{V_o \pm c} \tag{14-84}$$

It is clear from Eq. 14-80 that the amplification factor depends upon the mesh size and the wave number or frequency. Based on this equation, we may write

$$\xi = |\xi| e^{j\phi} \tag{14-85}$$

in which $|\xi|$ = amplitude of the amplification factor and ϕ = phase angle. The expressions for the amplitude and the phase angle are

$$|\xi| = \sqrt{\cos^2\delta + C_n^2 \sin^2\delta} \tag{14-86}$$

and

$$\phi = \tan^{-1}(-C_n \tan\delta) \tag{14-87}$$

in which C_n = Courant number = $(V_o + \sqrt{g D_o})\Delta t / \Delta x$.

Figure 14-5, taken from Anderson, et al. (1984), shows the amplitude and the phase angle for the amplification factor for the Lax scheme for several values of C_n. It is clear from this figure that all frequency components are propagated without attenuation if $C_n = 1$. However, if $C_n < 1$, the attenuation is small for the low- and high-frequency components and it is severe for the midrange frequency components.

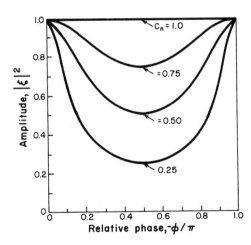

Figure 14-5 Amplitude and phase angle for amplification factor (After Anderson, et al. 1984).

Note that the Courant condition is satisfied even for larger values of Δt if the value of Δx is large. However, for the Lax diffusive scheme, Koren (1967) showed that the following additional stability criterion must be satisfied for a small-amplitude disturbance applied to an initial steady-state flow to damp out:

$$\Delta t \leq \frac{\sqrt{1 + 2\left|\dfrac{V_o}{c_o}\right| - 1}}{\left|\dfrac{V_o}{c_o}\right|\dfrac{g S_o}{V_o}} \qquad (14\text{-}88)$$

in which V_o and c_o are the initial velocity and the celerity of a shallow water wave, respectively. Numerical experimentation of Huang and Song (1985) confirm the validity of this criterion in addition to showing that it is applicable to the method of characteristics as well.

Example 14-1

A trapezoidal channel having a bottom width of 6.1 m and side slopes of 1.5H to 1 V is carrying a flow of 126 m^3/s at a flow depth of 5.79 m. The bottom slope is 0.00008, Manning n = 0.013 and the channel length is 5 km. There is constant level reservoir at the upstream end of the channel. A sluice gate at the downstream end is suddenly closed at time t = 0,

a. *Compute the transient conditions until t = 2000 s by using:*
   i. *Method of specified intervals*
  ii. *Lax diffusive scheme*
 iii. *MacCormack scheme*
 iv. *Gabutti scheme*
  v. *Preissmann scheme.*
b. *Plot the computed flow depth in the channel at t = 0, 500, 1000, 1500, and 2000 s.*
c. *Plot the variation of the flow depth with time at distances of 1.5, 2.5, 3, and 5 km from the reservoir.*

Solution

A computer program was written based on these finite-difference schemes for the interior grid points. In the first four schemes, the positive characteristic equation and the condition that flow velocity at the downstream end is always zero for $t > 0$ were utilized to simulate the downstream end condition. At the upstream end, the negative characteristic equation was used along with the condition that the flow depth is equal to the reservoir depth. In the Preissmann scheme, the upstream- and downstream-end conditions were directly incorporated into the solution. The downstream end condition specified the flow velocity at the downstream end to be always zero following the gate closure, and the upstream end condition specified the flow depth to be constant and equal to the reservoir depth at the channel entrance.

The channel length was divided into 50 equal-length reaches and the computational time interval, Δt, was selected so that the stability condition was always satisfied in the explicit schemes at every grid point. If this was not the case, then the computational time interval was reduced by 20 percent and the flow conditions were recalculated. However, if the time step was considerably smaller than that required by the Courant condition, then the time step for the next interval was increased by 15 percent.

The computed flow depths in the channel at different times by using these schemes are shown in Fig. 14-6. The variation of flow depth with respect to time at different locations is plotted in Fig. 14-7.

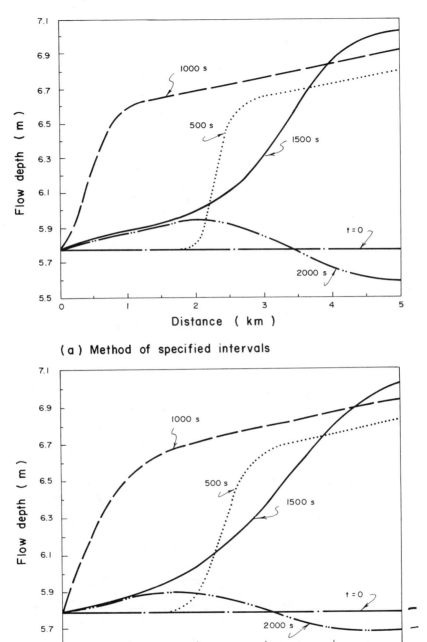

(a) Method of specified intervals

(b) Lax diffusive scheme

Figure 14-6 Computed flow depths at different times.

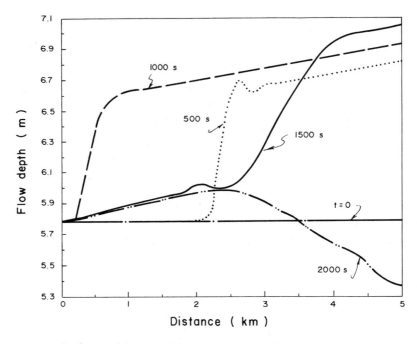

(c) MacCormack scheme

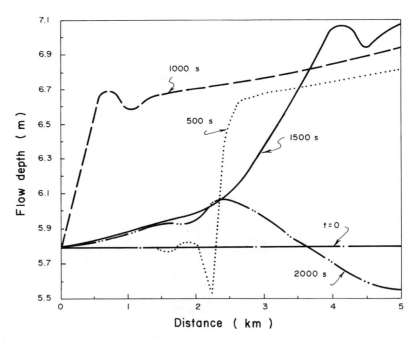

(d) Gabutti scheme

Figure 14-6 (continued)

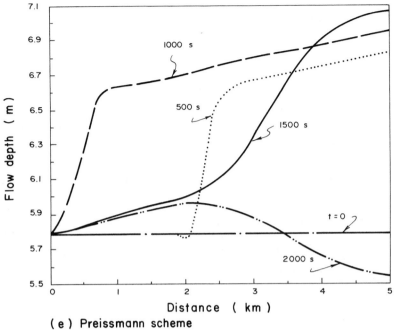

(e) Preissmann scheme

Figure 14-6 (concluded)

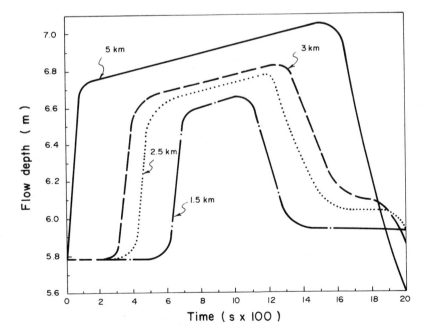

(a) Method of specified intervals

Figure 14-7 Variation of flow depth with time.

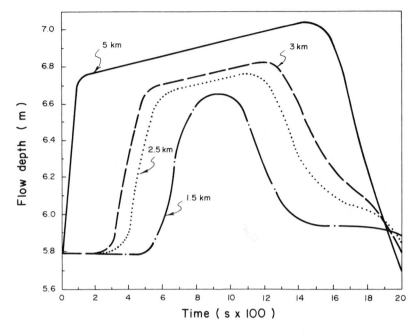

(b) Lax diffusive scheme

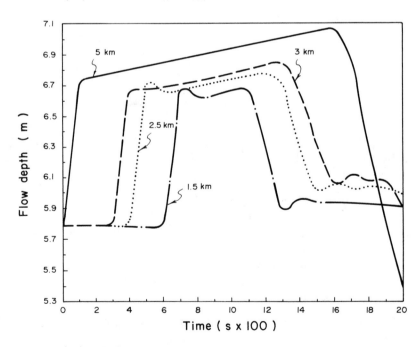

(c) MacCormack scheme

Figure 14-7 (continued)

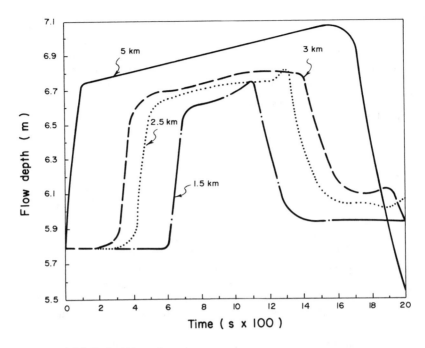

(d) Gabutti scheme

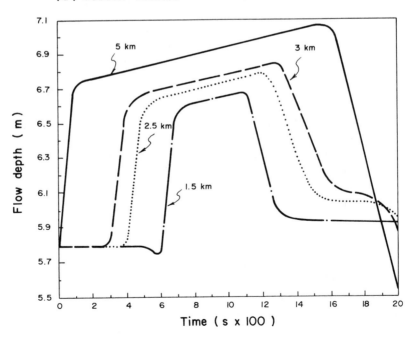

(e) Preissmann scheme

Figure 14-7 (concluded)

14-7 Summary

In this chapter, we presented several explicit and implicit finite- difference methods for the numerical integration of shallow-water equations. The specification of the initial and boundary conditions was outlined. The consistency and stability of numerical schemes were discussed. The von Neumann stability analysis was demonstrated by applying it to the Lax scheme.

Problems

14-1. Based on the finite-difference schemes of this chapter, write a computer program to analyze unsteady flows in a trapezoidal channel having a constant-level reservoir at the upstream end, bottom width of 10 m, side slopes of 2H : 1V and bottom slope of 0.0001. Initially, the flow is uniform at a flow depth of 3 m. At t = 0, a control gate is instantaneously closed at the downstream end. Assume Manning n for the channel surface is 0.010.

14-2. Write a computer program by using the conservation and nonconservation forms of the governing equations and the Lax and MacCormack schemes. For the data of Prob. 14-1, compare the velocity of propagation of the surge wave and the conservation of mass at different times for the different forms of the equations and the numerical schemes.

14-3. By using von Neumann analysis, show that the finite-difference scheme given by Eq. 14-12 is unstable irrespective of the size of grid spacing.

14-4. A 3-m-wide rectangular channel is carrying a discharge of 30 m^3/s at a flow depth of 1 m. There is a constant-level reservoir at the upper end and the flow is uniform during the initial conditions. The downstream reservoir level is suddenly raised to 6.5 m at $t = 0$. Apply different numerical schemes of this chapter to compute the final steady-state water-surface profile in the channel. (*Hint:* First specify the initial conditions for the supercritical flow. Then change the dowstream-end condition and continue the computations till they converge to a steady state.)

14-5. Set up a computational procedure to solve the governing equations at a series junction.

14-6. Derive the stability conditions for the MacCormack and Gabutti schemes.

14-7. Show that the Preissmann scheme is unconditionally stable if the friction-loss term is neglected. Derive the stability condition if a linearized form of the friction term is included.

REFERENCES

Abbott, M. B. 1979. *Computational hydraulics; elements of the theory of free surface flows*, Pitman, London.

Amein, M., and Fang, C. S. 1970. Implicit flood routing in natural channels, *Jour., Hydraulics Div.*, Amer. Soc. of Civil Engrs., 96, no. 12:2481–2500.

Anderson, D. A.; Tannehill, J. C.; and Pletcher, R. H. 1984. *Computational fluid mechanics and heat transfer*, McGraw-Hill, New York, NY.

Beam, R. M., and Warming, R. F. 1976. An implicit finite-difference algorithm for hyperbolic systems in conservation-law form, *Jour. Comp. Phys.*, 22:87–110.

Chaudhry, M. H. 1987. *Applied hydraulic transients,* 2nd ed., Van Nostrand Reinhold, New York, NY.

Cunge, J.; Holly, F. M.; and Verwey, A. 1980. *Practical aspects of computational river hydraulics,* Pitman, London.

Dammuller, D.; Bhallamudi, S. M.; and Chaudhry, M. H. 1989. Modeling of unsteady flow in curved channels, *Jour. Hydraulic Engineering,* Amer. Soc. Civil Engrs., 115, no. 11:1479–95.

Evans, E. P. 1977. The behaviour of a mathematical model of open-channel flow, Paper A97, *17th Congress,* Inter. Assoc. for Hydraulic Research, Baden Baden, Germany.

Fennema, R. J., and Chaudhry, M. H. 1986. Explicit numerical schemes for unsteady free-surface flows with shocks, *Water Resources Research,* 22, no. 13:1923–30.

———. 1987. Simulation of one-dimensional dam-break flows, *Jour., Hydraulic Research,* Inter. Assoc. for Hydraulic Research 25, no. 1:41–51.

Fread, D. L., and Harbaugh, T. E. 1973. Transient simulation of breached earth dams, *Jour. Hydraulics Division,* Amer. Soc. Civil Engrs., no. 1:139–154.

French, R. H. 1985. Open-channel hydraulics, McGraw-Hill, New York, NY.

Gabutti, B. 1983. On two upwind finite-difference schemes for hyperbolic equations in nonconservation form, *Computers and Fluids,* 11, no. 3:207–30.

Galperin, B.; Blumberg, A. F.; and Weisberg, R. H. 1992. The importance of density driven circulation in well-mixed estuaries: The Tampa Bay Experience, *Proc. Water Forum,* Amer. Soc. Civil Engrs., 11pp.

Huang, J., and Song, C. C. S. 1985. Stability of dynamic flood routing schemes, *Jour. Hydraulic Engineering,* Amer. Soc. Civil Engrs., 111, no. 12:1497–1505.

Isaacson, E.; Stoker, J. J.; and Troesch, B. A. 1954. Numerical solution of flood prediction and river regulation problems (Ohio-Mississippi Floods), *Report II,* Inst. Math. Sci. Rept. IMM-NYU-205, New York University, New York, NY.

Jameson, A.; Schmidt, W.; and Turkel, E. 1981. Numerical solutions of the euler equations by finite volume methods using Runge-Kutta time-stepping schemes, *AIAA 14th Fluid And Plasma Dynamics Conference,* Palo Alto, Calif., AIAA-81-1259.

Katopodes, N., and Wu, C-T. 1986. Explicit computation of discontinous channel flow, *Jour. Hydraulic Engineering,* Amer Soc. Civil Engrs., 112, no. 6:456–75.

Koren, V. I. 1967. The analysis of stability of some explicit finite difference schemes for the integration of St. Venant equations, *Meterologiya i Gidrologiya,* no. 1 (in Russian).

Lai, C. 1986. "Numerical modeling of unsteady open-channel flow." In *Advances in hydroscience,* vol. 14, Academic Press, New York, NY:161–333.

Lax, P. D. 1954. Weak solutions of nonlinear hyperbolic partial differential equations and their numerical computation, *Communications on Pure and Applied Mathematics,* 7:159–63.

Leendertse, J. J. 1967. Aspects of a computational model for long-period water-wave propagation, *Memo RM-5294-PR,* Rand Corporation, May.

Liggett, J. A., and Cunge, J. A. 1975. "Numerical methods of solution of unsteady flow equations." In *Unsteady open channel flow,* edited by K. Mahmood and V. Yevjevich, Water Resources Publications, Fort Collins, Colo.

Liggett, J. A., and Woolhiser, D. A. 1967. Difference solutions of shallow-water equations, *Jour. Engineering Mech. Division,* Amer. Soc. Civil Engrs. 93, no. EM2:39–71.

Lyn, D. A. and Goodwin, P. 1987. Stability of a general Preissmann scheme, *Jour. Hydraulic Engineering,* Amer. Soc. Civil Engrs., 113, no. 1:16–27.

MacCormack, R. W. 1969. The effect of viscosity in hypervelocity impact cratering. *Amer. Inst. Aero. Astro.,* Paper 69–354, Cincinnati, OH.

Moretti, G. 1979, The lambda scheme, *Computers and Fluids,* 7:191–205.

Pandolfi, M. 1975. Numerical experiments on free surface water motion with bores, *Proc. 4th Int. Conf. on Numerical Methods in Fluid Dynamics,* Lecture Notes in Physics No. 35, Springer-Verlag, New York, NY. 304–12.

Preissmann, A., and Cunge, J. 1961. Calcul du mascaret sur machines electroniques, *La Houille Blanche,* no. 5:588–96.

Price, R. K. 1974. Comparison of four numerical flood routing methods, *Jour. Hydraulics Division,* Amer. Soc. Civil Engrs., 100, no. 7:879–99.

Richtmyer, R. D., and Morton, K. W. 1967. *Difference methods for initial value problems,* 2nd. ed., Wiley, New York, NY.

Roache, P. J. 1972. *Computational fluid dynamics,* Hermosa Publishers, Albuquerque, NM.

Samuels, P. G., and Skeels, C. P. 1990. Stability limits for Preissmann's scheme, *Jour. Hydraulic Engineering,* Amer. Soc. Civil Engrs., 116, no. 8:997–1012.

Stoker, J. J. 1957. *Water waves,* Interscience, Wiley, New York, NY.

Strelkoff, T. 1970. Numerical solution of St. Venant equations. *Jour. Hydraulics Div.,* Amer. Soc. of Civil Engrs. 96, no. 1: pp. 223–52.

Terzidis, G., and Strelkoff, T. 1970. Computation of open channel surges and shocks, *Jour. Hydraulics Division,* Amer. Soc. Civil Engrs., 96, no. 12:2581–2610.

Vasiliev, O. F.; Gladyshev, M. T.; Pritvits, N. A.; and Sudobiocher, V. G. 1965. Numerical method for the calculation of shock wave propagation in open channels, *Proc., 11th Congress,* Inter. Assoc. for Hydraulic Research, vol. 3, paper 3.44. 14 pp.

Verboom, G. K.; Stelling, G. S.; and Officier, M. J. 1982. Boundary conditions for the shallow water equations. In *Engineering applications of computational hydraulics,* vol. 1. Edited by M. B. Abbott and J. A. Cunge., Pitman, Boston.

Yen, C-L., and Lin, C.-H. 1986. Numerical stability in unsteady supercritical flow simulation, *5th Congress Asian and Pacific Regional Division,* International Assoc. for Hydraulic Research August.

15 Two-Dimensional Flow

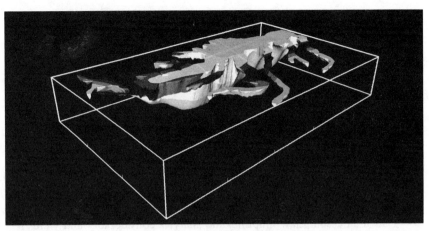

Salinity field in the Chesapeake Bay computed by a three-dimensional, unsteady, finite-difference mathematical model (Courtesy of Billy Johnson, U.S. Army Waterways Experiment Station, Vicksburg, Miss.).

15-1 Introduction

In the previous chapters, we considered one-dimensional flows. However, the assumption of one-dimensional flow may not be valid in many situations—e.g., flow in a nonprismatic channel (i.e., a channel with varying cross section and alignment), flow downstream of a partially breached dam, or lateral flow from a failed dyke. Although flow in these situations is three-dimensional, we may simplify their analysis by considering them as two-dimensional flows by using vertically averaged quantities. Such an assumption simplifies the analysis considerably and yields results of reasonable accuracy.

In this chapter, we discuss the analysis of two-dimensional flows. First, we derive the equations describing unsteady two-dimensional flows. Then, we present explicit and implicit finite-difference methods for their solution.

15-2 Governing Equations

We will derive the governing equations (Jimenez 1987) by integrating the Navier-Stokes equations for an incompressible fluid over the flow depth (Lai 1986).

Except for one-dimensional flow, the same assumptions as those given in Sec. 12-2 are used. The effects of large bottom slope are included, and it is assumed that the bottom of the channel is an inclined plane. A Cartesian orthogonal coordinate system with the x-y plane parallel to the plane of the channel bottom is considered. For a right-hand system, the positive z direction points upward and is perpendicular to the x-y plane.

The Navier-Stokes equations for an incompressible fluid are as follows.

Continuity equation

$$\frac{\partial u}{\partial x} + \frac{\partial v}{\partial y} + \frac{\partial w}{\partial z} = 0 \tag{15-1}$$

Momentum equation

$$\frac{\partial u}{\partial t} + u\frac{\partial u}{\partial x} + v\frac{\partial u}{\partial y} + w\frac{\partial u}{\partial z} = g_x - \frac{1}{\rho}\frac{\partial p}{\partial x} + \frac{\mu}{\rho}\nabla^2 u \tag{15-2}$$

$$\frac{\partial v}{\partial t} + u\frac{\partial v}{\partial x} + v\frac{\partial v}{\partial y} + w\frac{\partial v}{\partial z} = g_y - \frac{1}{\rho}\frac{\partial p}{\partial y} + \frac{\mu}{\rho}\nabla^2 v \tag{15-3}$$

$$\frac{\partial w}{\partial t} + u\frac{\partial w}{\partial x} + v\frac{\partial w}{\partial y} + w\frac{\partial w}{\partial z} = g_z - \frac{1}{\rho}\frac{\partial p}{\partial z} + \frac{\mu}{\rho}\nabla^2 w \tag{15-4}$$

in which u, v, and w are the components of the velocity along the x, y, and z directions; $\mathbf{g} = (g_x, g_y, g_z)^T$ is the gravitational force per unit mass; μ is the dynamic viscosity; p is pressure; and the symbol ∇^2 stands for the Laplace operator,

$$\nabla^2 = \frac{\partial^2}{\partial x^2} + \frac{\partial^2}{\partial y^2} + \frac{\partial^2}{\partial z^2}$$

We will integrate these equations over the flow depth to obtain the depth-averaged equations.

Continuity equation

The depth-averaged continuity equation for two-dimensional flow may be obtained by integrating Eq. 15-1 over the flow depth, i.e.,

$$\int_{Z_b}^{Z} \frac{\partial u}{\partial x}\, dz + \int_{Z_b}^{Z} \frac{\partial v}{\partial y}\, dz + w(Z) - w(Z_b) = 0 \tag{15-5}$$

in which Z and Z_b are the z-coordinates of the water surface and the channel bottom respectively (these are measured perpendicular to the plane of the channel bottom). The integrals of Eq. 15-5 may be evaluated by using the Leibnitz rule:

$$\int_{Z_b}^{Z} \frac{\partial u}{\partial x}\, dz = \frac{\partial}{\partial x}\int_{Z_b}^{Z} u\, dz - u(Z)\frac{\partial Z}{\partial x} + u(Z_b)\frac{\partial Z_b}{\partial x}$$
$$\int_{Z_b}^{Z} \frac{\partial v}{\partial y}\, dz = \frac{\partial}{\partial y}\int_{Z_b}^{Z} v\, dz - v(Z)\frac{\partial Z}{\partial y} + v(Z_b)\frac{\partial Z_b}{\partial y} \tag{15-6}$$

If the function $Z(x, y, t)$ specifies the z-coordinate of the free-water surface and if it is assumed that any particle on the surface does not leave it, then the vertical velocity of a particle on the water surface, $w(Z)$, is given by

$$w(Z) = \frac{DZ}{Dt} = \frac{\partial Z}{\partial t} + u(Z)\frac{\partial Z}{\partial x} + v(Z)\frac{\partial Z}{\partial y} \tag{15-7}$$

Similarly, if the bottom of the channel is rigid, then $F_b = Z_b(x, y) - z = 0$, in which $Z_b(x, y)$ gives the z-coordina'. of the channel bottom. Hence,

$$w(Z_b) = \frac{DF_b}{Dt} = u(Z_b)\frac{\partial Z_b}{\partial x} + v(Z_b)\frac{\partial Z_b}{\partial y} \tag{15-8}$$

Substitution of Eqs. 15-6 through 15-8 into Eq. 15-5 leads to

$$\frac{\partial Z}{\partial t} + \frac{\partial(\bar{u}d)}{\partial x} + \frac{\partial(\bar{v}d)}{\partial y} = 0 \tag{15-9}$$

in which $\bar{u}$ and $\bar{v}$ are the mean values of u and v over the depth of the channel:

$$\bar{u} = \frac{1}{d} \int_{Z_b}^{Z} u \, dz; \qquad \bar{v} = \frac{1}{d} \int_{Z_b}^{Z} v \, dz \tag{15-10}$$

in which $d = Z - Z_b$ is the water depth measured perpendicular to the bottom of the channel.

Momentum equations

Since we are assuming the vertical acceleration to be negligible,

$$\frac{Dw}{Dt} \approx 0 \qquad \mu \nabla^2 w \approx 0 \tag{15-11}$$

Therefore, Eq. 15-4 reduces to

$$g_z - \frac{1}{\rho} \frac{\partial p}{\partial z} = 0 \tag{15-12}$$

Integrating Eq. 15-12 in the z-direction and considering the atmospheric pressure to be zero, we obtain

$$p = \rho g_z (z - Z) \tag{15-13}$$

Hence, it follows that

$$-\frac{1}{\rho} \frac{\partial p}{\partial x} = g_z \frac{\partial Z}{\partial x} \tag{15-14}$$

$$-\frac{1}{\rho} \frac{\partial p}{\partial y} = g_z \frac{\partial Z}{\partial y} \tag{15-15}$$

By multiplying Eq. 15-1 by u, adding to Eq. 15-2, substituting the expression for $-(1/\rho)(\partial p/\partial x)$ from Eq. 15-14, and rearranging the terms of the resulting equation, we obtain

$$\frac{\partial u}{\partial t} + \frac{\partial u^2}{\partial x} + \frac{\partial (uv)}{\partial y} + \frac{\partial (uw)}{\partial z} = g_x + g_z \frac{\partial Z}{\partial x} + \frac{\mu}{\rho} \nabla^2 u \tag{15-16}$$

Similarly, multiplying Eq. 15-1 by v, adding it to Eq. 15-3, substituting the expression for $-(1/\rho)(\partial p/\partial y)$ from Eq. 15-15, we obtain

$$\frac{\partial v}{\partial t} + \frac{\partial (uv)}{\partial x} + \frac{\partial v^2}{\partial y} + \frac{\partial (vw)}{\partial z} = g_y + g_z \frac{\partial Z}{\partial y} + \frac{\mu}{\rho} \nabla^2 v \tag{15-17}$$

Let us integrate Eqs. 15-16 and 15-17 in the z-direction. To simplify presentation, we consider the left- and right-hand sides of these equations separately. Integration of the left-hand side of Eq. 15-16 and application of the Leibnitz rule yield

$$\frac{\partial}{\partial t}\int_{Z_b}^{Z} u\, dz - u(Z)\frac{\partial Z}{\partial t} + \frac{\partial}{\partial x}\int_{Z_b}^{Z} u^2\, dz - u^2(Z)\frac{\partial Z}{\partial x}$$

$$+ u^2(Z_b)\frac{\partial Z_b}{\partial x} + \frac{\partial}{\partial y}\int_{Z_b}^{Z} uv\, dz - u(Z)v(Z)\frac{\partial Z}{\partial y} \tag{15-18}$$

$$+ u(Z_b)v(Z_b)\frac{\partial Z_b}{\partial y} + u(Z)w(Z) - u(Z_b)w(Z_b)$$

Based on the assumption of uniform velocity distribution (i.e., u and v are constants in the z-direction) and substituting Eqs. 15-7 and 15-8, Expression 15-18 simplifies as

$$\frac{\partial}{\partial t}(\bar{u}d) + \frac{\partial}{\partial x}(\bar{u}^2 d) + \frac{\partial}{\partial y}(\bar{u}\bar{v}d) \tag{15-19}$$

Similarly, the left-hand side of Eq. 15-17 becomes

$$\frac{\partial}{\partial t}(\bar{v}d) + \frac{\partial}{\partial x}(\bar{u}\bar{v}d) + \frac{\partial}{\partial y}(\bar{v}^2 d) \tag{15-20}$$

Integration of the right-hand side of Eqs. 15-16 and 15-17 yields

$$\left(g_x + g_z\frac{\partial Z}{\partial x}\right)d + \int_{Z_b}^{Z} \frac{\mu}{\rho}\nabla^2 u\, dz \tag{15-21}$$

$$\left(g_y + g_z\frac{\partial Z}{\partial y}\right)d + \int_{Z_b}^{Z} \frac{\mu}{\rho}\nabla^2 v\, dz \tag{15-22}$$

Since the x-y plane is parallel to the channel bottom, Z_b is constant. Therefore,

$$\frac{\partial Z}{\partial x} = \frac{\partial(Z_b + d)}{\partial x} = \frac{\partial d}{\partial x} \tag{15-23}$$

Similarly,

$$\frac{\partial Z}{\partial y} = \frac{\partial d}{\partial y} \tag{15-24}$$

Now, let us consider the shear stress terms. In turbulent flow, the dynamic viscosity is replaced by an eddy viscosity coefficient. Moreover, a distinction is made between the stresses acting in the x-y plane and the stresses acting in the x-z and y-z planes. For example, the shear-stress term of the momentum equation in the x-direction may be written as

$$\epsilon_{xy}\left(\frac{\partial^2 u}{\partial x^2} + \frac{\partial^2 u}{\partial y^2}\right) + \epsilon_{zx}\frac{\partial^2 u}{\partial z^2} \tag{15-25}$$

in which ϵ_{xy} and ϵ_{zx} are the eddy-viscosity coefficients. In addition, it is assumed that the effective stresses are dominated by the bottom shear stresses. This means that the first term in Eq. 15-25 is negligible as compared to the second term. Therefore, the shear stress term of Eq. 15-25 reduces to $\epsilon_{zx}\partial^2 u/\partial z^2$. Integration of this expression with respect to z yields

$$\int_{Z_b}^{Z} \epsilon_{zx} \frac{\partial^2 u}{\partial z^2}\, dz = \epsilon_{zx}\left(\frac{\partial u}{\partial z}\right)_{z=Z} - \epsilon_{zx}\left(\frac{\partial u}{\partial z}\right)_{z=Z_b} \tag{15-26}$$

$$= \tau_{s_x} - \tau_{b_x}$$

in which τ_{s_x} and τ_{b_x} are the shear stresses at the water surface and at the channel bottom acting in the x-direction. Similarly, the shear stress term of Eq. 15-22 reduces to

$$\tau_{s_y} - \tau_{b_y} \tag{15-27}$$

The shear stresses acting at the water surface due to wind velocity, τ_{s_x} and τ_{s_y}, are neglected and the shear stresses at the channel bottom, τ_{b_x} and τ_{b_y} are evaluated by using empirical formulas. For example, the Chezy formula gives

$$\tau_b = \frac{\rho g}{C^2} V^2 \tag{15-28}$$

where V is the amplitude of flow velocity (i.e., $V = \sqrt{\bar{u}^2 + \bar{v}^2}$) and C is the Chezy coefficient. It follows from Eq. 15-28 that

$$\tau_{b_x} = \tau_b \cos\theta = \frac{\rho g}{C^2}\bar{u} V$$

$$\tau_{b_y} = \tau_b \sin\theta = \frac{\rho g}{C^2}\bar{v} V \tag{15-29}$$

in which θ is the angle between the velocity vector and the x-axis.

Different terms of the depth-integrated momentum equations may now be assembled together. Substitution of Eqs. 15-19 through 15-24, 15-26, 15-27, and 15-29 into Eqs. 15-16 and 15-17 gives

$$\frac{\partial}{\partial t}(\bar{u}d) + \frac{\partial}{\partial x}(\bar{u}^2 d) + \frac{\partial}{\partial y}(\bar{u}\bar{v}d) = \left(g_x + g_z \frac{\partial d}{\partial x}\right)d - \frac{g}{C^2}\bar{u}\sqrt{\bar{u}^2 + \bar{v}^2} \tag{15-30}$$

$$\frac{\partial}{\partial t}(\bar{v}d) + \frac{\partial}{\partial x}(\bar{u}\bar{v}d) + \frac{\partial}{\partial y}(\bar{v}^2 d) = \left(g_y + g_z \frac{\partial d}{\partial y}\right)d - \frac{g}{C^2}\bar{v}\sqrt{\bar{u}^2 + \bar{v}^2} \tag{15-31}$$

Eqs. 15-30 and 15-31 are the momentum equations with respect to a coordinate system x-y parallel to the bottom of the channel. For the case of one-dimensional flow, the preceding equations reduce to those given by Yen (1986, p. 46) for rectangular channels. For example, if $\bar{v} = 0$ and $\partial d/\partial y = 0$, Eq. 15-30 yields

$$\frac{\partial}{\partial t}(\bar{u}d) + \frac{\partial}{\partial x}(\bar{u}^2 d) + gd\cos\alpha_x \frac{\partial d}{\partial x} = gd(\sin\alpha_x - S_{f_x}) \tag{15-32}$$

where α_x is the angle of inclination of the channel bottom.

Equations 15-9, 15-30, and 15-31 may be expressed in a horizontal system of coordinates, $\tilde{x}$, $\tilde{y}$, and $\tilde{z}$ (Fig. 15-1). In this coordinate system, a channel may have piecewise constant-bottom slope. In order to transform from the inclined system x-y-z to the horizontal system $\tilde{x}$-$\tilde{y}$-$\tilde{z}$, the former coordinate system must be rotated. A rotation

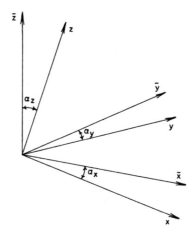

Figure 15-1 Definition of α_x, α_y, and α_z.

of this type is generally defined by using the direction cosines that give the angles between the axes in both systems. However, in this case it is better to express the rotation as a function of the angles between the bottom of the channel and the x- and y-axes (α_x and α_y in Fig. 15-1), since these angles are generally known. According to Fig. 15-1 and after some vectorial manipulations, the transformation between both systems of coordinates is given by

$$
\begin{pmatrix} \tilde{x} \\ \tilde{y} \\ \tilde{z} \end{pmatrix} = \begin{pmatrix} \cos\alpha_x & -\cos\varphi\cos\alpha_x/\sin\varphi & \tan\alpha_x\cos\alpha_z \\ 0 & \cos\alpha_y/\sin\varphi & \tan\alpha_y\cos\alpha_z \\ -\sin\alpha_x & -\sin\alpha_y\cos^2\alpha_x/\sin\varphi & \cos\alpha_z \end{pmatrix} \begin{pmatrix} x \\ y \\ z \end{pmatrix} \tag{15-33}
$$

where $\cos\alpha_z = 1/\sqrt{1+\tan^2\alpha_x+\tan^2\alpha_y}$; $\cos\varphi = \sin\alpha_x\cos\alpha_y$; and $\sin\varphi = \sqrt{1-\sin^2\alpha_x\sin^2\alpha_y}$. The terms g_x, g_y, and g_z may be computed from Eq. 15-33. For example,

$$
g_x = \mathbf{g}\cdot\hat{\mathbf{e}}_1 = -g\tilde{\mathbf{e}}_3\cdot\hat{\mathbf{e}}_1 = g\sin\alpha_x \tag{15-34}
$$

in which $\hat{\mathbf{e}}_i$ are the unit vectors of the x-y-z system, and $\tilde{\mathbf{e}}_i$ are the unit vectors of the $\tilde{x}$-$\tilde{y}$-$\tilde{z}$ system. Note that the term $-\sin\alpha_x$ in Eq. 15-34 is the $(3,1)$ element of the transformation matrix. Hence,

$$
\begin{aligned}
g_x &= g\sin\alpha_x \\
g_y &= g\frac{\sin\alpha_y\cos^2\alpha_x}{\sin\varphi} \\
g_z &= -g\cos\alpha_z
\end{aligned} \tag{15-35}
$$

where α_x, α_y, and α_z are as defined in Fig. 15-1. Some simplification is needed before carrying out the transformation; otherwise the algebra becomes unwieldy. Let us assume that although $\sin\alpha_x$ and $\sin\alpha_y$ may not be small, their product is small, i.e.,

$$
\sin\alpha_x\sin\alpha_y \approx \sin^2\alpha_x \approx \sin^2\alpha_y \approx 0 \tag{15-36}
$$

This approximation introduces a small error (< 3 percent) if $|\alpha_x|$, $|\alpha_y| < 10°$. It follows from Eq. 15-36 that

$$\sin \varphi = 1; \qquad \cos \varphi = 0 \tag{15-37}$$

Based on these approximations and Eq. 15-33, we may write

$$\begin{aligned}
\tilde{x} &= x \cos \alpha_x + z \tan \alpha_x \cos \alpha_z \\
\tilde{y} &= y \cos \alpha_y + z \tan \alpha_y \cos \alpha_z
\end{aligned} \tag{15-38}$$

The transformed dependant variables become

$$\begin{aligned}
h &= \frac{d}{\cos \alpha_z} \\
\tilde{u} &= \bar{u} \cos \alpha_x \\
\tilde{v} &= \bar{v} \cos \alpha_y
\end{aligned} \tag{15-39}$$

in which h is the flow depth measured *vertically* and $\tilde{u}$ and $\tilde{v}$ are the velocity components along the $\tilde{x}$- and $\tilde{y}$-directions respectively. Also, note that according to Eq. 15-38,

$$\begin{aligned}
\frac{\partial}{\partial x} &= \cos \alpha_x \frac{\partial}{\partial \tilde{x}} + \tan \alpha_x \cos \alpha_z \frac{\partial}{\partial \tilde{z}} \\
\frac{\partial}{\partial y} &= \cos \alpha_y \frac{\partial}{\partial \tilde{y}} + \tan \alpha_y \cos \alpha_z \frac{\partial}{\partial \tilde{z}}
\end{aligned} \tag{15-40}$$

The presence of derivatives along the $\tilde{z}$-direction is undesirable, since the basic idea is to eliminate a spatial dimension out of the problem. However, it can be shown that terms like $\tan \alpha_x \cos \alpha_z (\partial/\partial \tilde{z})$ are of the order of $\sin^2 \alpha_x$ and, therefore, are negligible. Introduction of Eqs. 15-35, 15-38, 15-39, and 15-40 into Eqs. 15-9, 15-30, and 15-31, simplifying, and dropping the $\sim$ symbols yield

$$\frac{\partial h}{\partial t} + \frac{\partial}{\partial x}(uh) + \frac{\partial}{\partial y}(vh) = 0$$

$$\frac{\partial}{\partial t}(uh) + \frac{\partial}{\partial x}(u^2 h) + \frac{\partial}{\partial y}(uvh) = gh \left[\cos \alpha_x S_{o_x} - (\cos \alpha_x \cos \alpha_z)^2 \frac{\partial h}{\partial x} - S_{f_x} \right]$$

$$\frac{\partial}{\partial t}(vh) + \frac{\partial}{\partial x}(uvh) + \frac{\partial}{\partial y}(v^2 h) = gh \left[\cos \alpha_y S_{o_y} - (\cos \alpha_y \cos \alpha_z)^2 \frac{\partial h}{\partial y} - S_{f_y} \right]$$

$$\tag{15-41}$$

in which

$$\begin{aligned}
S_{o_x} &= \sin \alpha_x; & S_{o_y} &= \sin \alpha_y; \\
S_{f_x} &= \frac{u\sqrt{u^2 + v^2}}{C^2 \cos \alpha_z h}; & S_{f_y} &= \frac{v\sqrt{u^2 + v^2}}{C^2 \cos \alpha_z h}
\end{aligned} \tag{15-42}$$

For small channel-bottom slope, Eqs. 15-41 may be written as

$$\mathbf{U}_t + \mathbf{E}_x + \mathbf{F}_y + \mathbf{S} = 0 \tag{15-43}$$

in which

$$\mathbf{U} = \begin{pmatrix} h \\ uh \\ vh \end{pmatrix}$$

$$\mathbf{E} = \begin{pmatrix} uh \\ u^2h + \frac{1}{2}gh^2 \\ uvh \end{pmatrix}$$

$$\mathbf{F} = \begin{pmatrix} vh \\ uvh \\ v^2h + \frac{1}{2}gh^2 \end{pmatrix} \tag{15-44}$$

$$\mathbf{S} = \begin{pmatrix} 0 \\ -gh(S_{o_x} - S_{f_x}) \\ -gh(S_{o_y} - S_{f_y}) \end{pmatrix}$$

and (uh) and (vh) are momenta convected in the x- and y-directions. If the Manning equation is used to compute the friction terms instead of the Chezy equation, then

$$S_{f_x} = \frac{n^2 u \sqrt{u^2 + v^2}}{C_o^2 h^{1.33}} \qquad S_{f_y} = \frac{n^2 v \sqrt{u^2 + v^2}}{C_o^2 h^{1.33}} \tag{15-45}$$

in which $n =$ Manning coefficient and $C_o =$ a dimensional constant ($C_o = 1$ for SI units and $C_o = 1.49$ for customary units).

In terms of the primitive flow variables $h, u,$ and v, the governing equations may also be written as

$$\mathbf{V}_t + \mathbf{P}_x + \mathbf{R}_y + \mathbf{T} = 0 \tag{15-46}$$

in which

$$\mathbf{V} = \begin{pmatrix} h \\ u \\ v \end{pmatrix}$$

$$\mathbf{P} = \begin{pmatrix} uh \\ \frac{1}{2}u^2 + gh \\ uv \end{pmatrix}$$

$$\mathbf{R} = \begin{pmatrix} vh \\ uv \\ \frac{1}{2}v^2 + gh \end{pmatrix} \tag{15-47}$$

$$\mathbf{T} = \begin{pmatrix} 0 \\ -g(S_{o_x} - S_{f_x}) \\ -g(S_{o_y} - S_{f_y}) \end{pmatrix}$$

It is necessary for certain numerical schemes that the equations be in the nonconservation form. The nonconservation form of Eq.15-43 is

$$\mathbf{U}_t + \mathbf{A}\mathbf{U}_x + \mathbf{B}\mathbf{U}_y + \mathbf{S} = 0 \tag{15-48}$$

in which **A** and **B** are the Jacobians of **E** and **S**:

$$\mathbf{A} = \begin{pmatrix} 0 & 1 & 0 \\ -u^2 + gh & 2u & 0 \\ -uv & v & u \end{pmatrix} \qquad \mathbf{B} = \begin{pmatrix} 0 & 0 & 1 \\ -uv & v & u \\ -v^2 + gh & 0 & 2v \end{pmatrix} \tag{15-49}$$

Similarly, the nonconservation form of Eqs. 15-47 is

$$\mathbf{V}_t + \mathbf{G}\mathbf{V}_x + \mathbf{H}\mathbf{V}_y + \mathbf{T} = 0 \tag{15-50}$$

in which

$$\mathbf{G} = \begin{pmatrix} u & h & 0 \\ g & u & 0 \\ 0 & 0 & u \end{pmatrix} \qquad \mathbf{H} = \begin{pmatrix} v & 0 & h \\ 0 & v & 0 \\ g & 0 & v \end{pmatrix} \tag{15-51}$$

Matrices **A**, **B**, **G**, and **H** of Eqs. 15-49 and 15-51 have the important property that their eigenvalues or characteristic directions are identical and are given by

$$\mathbf{A} \text{ and } \mathbf{G} \begin{cases} \lambda_1 & = u \\ \lambda_2 & = u + c \\ \lambda_3 & = u - c \end{cases} \qquad \mathbf{B} \text{ and } \mathbf{H} \begin{cases} \omega_1 & = v \\ \omega_2 & = v + c \\ \omega_3 & = v - c \end{cases} \tag{15-52}$$

where c is the wave celerity ($c = \sqrt{gh}$).

The conservation form of the equations, Eqs. 15-43 and 15-46, has the advantage of being superior in conserving the flow variables as compared to the case when the equations are not in the conservation form. However, note that Eqs. 15-43 and 15-46, are not in full conservation form because of the vectors **S** and **T**. When the source terms are not equal to zero, they act as sources or sinks. Since the contribution of these terms is usually small, the conservative properties are not significantly impaired.

15-3 Numerical Solution

The governing equations, Eqs. 15-43 and 15-46, are nonlinear, first-order, hyperbolic partial differential equations for which analytical solutions are not available, except for very simplified one-dimensional cases. Therefore, they are solved numerically.

For gradually varied two-dimensional, unsteady flows, characteristics, explicit, and implicit finite-difference methods have been used by a number of investigators (Leendertse 1967; Katopodes and Strelkoff, 1978; Benque et al. 1982; Abbott 1979; Lai, 1986). The method of characteristics was used by Katopodes and Strelkoff (1978) to simulate the two-dimensional propagation of dam-break flood waves. The bore was isolated and tracked explicitly along the downstream channel. Matsutomi (1983) used the leapfrog scheme to compute dam-break flow profiles over a dry bed. He employed shock-fitting to track the bore, as done by Sakkas and Strelkoff (1973) and used Ritter's solution for the first few time steps. Katopodes (1984b) used a finite-element technique based on the Petrov-Galerkin formulation to solve several discontinuous flow situations in two dimensions. Most of these computational procedures fail if subcritical and supercritical flows are present either simultaneously in different parts of the channel or if they occur in sequence at different times.

The notation used here for the finite-difference mesh in x, y, and t space is shown in Fig. 15-2. The x direction is designated by the subscript i, the y direction, by the subscript j, and the t direction, by the superscript k. The time level where the flow variables are known is denoted by superscript k and the unknown time level is shown by superscript $k + 1$.

Finite-difference methods involving the splitting of flux matrices **A**, **B**, **G** and **H** (Eqs. 15-49 and 15-51) have recently been introduced. Algorithms using this concept write finite-difference approximations along the characteristic directions and transmit information only in the correct directions, i.e., from the upwind direction in supercritical flow and from opposite sides in subcritical flow. Then the compatability equations are written in finite-difference form and integrated in the characteristic directions. These methods have an advantage over other methods when both subcritical and supercritical flows are present.

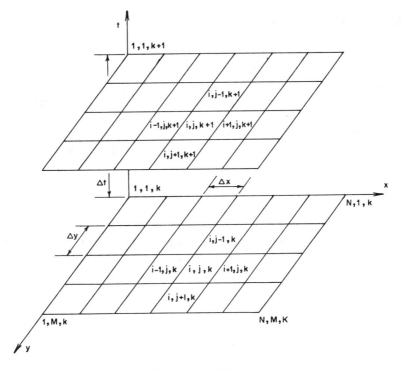

Figure 15-2 Finite-difference grid.

It is natural to assume that a good algorithm will be obtained if this information is incorporated into the scheme. Basically this requires that the spatial difference operators should be one-sided and obtain information only from the upwind side in supercritical flow and from both sides in subcritical flow. In situations where both types of flows are present, the spatial operators should be switched appropriately to account for the correct signal direction. The method of characteristics has the capability of preserving directional information, but it is quite cumbersome to develop and program in two or more dimensions. In comparison, fixed-grid capabilities of finite-difference schemes are attractive. By incorporating information on signal propagation in finite-difference schemes, algorithms may be developed that are simpler to program but emulate the method of characteristics.

Explicit and implicit finite-difference schemes are introduced in this chapter for the solution of the equations describing two-dimensional unsteady flow. These schemes are second-order accurate in both space and time, predict the location and height of the bores without the use of shock-fitting, and allow initial conditions with discontinuities. The steep wave fronts are usually spread only over a few mesh points. Fennema and Chaudhry used these schemes for the solution of one-dimensional (1986, 1987) and two-dimensional (1989, 1990) open-channel flows. Some of these schemes are capable of simulating both subcritical and supercritical flows.

15-4 MacCormack Scheme

Difference methods investigated by Lax and Wendroff (1960) have become very popular for solving hyperbolic systems. These methods, known as two-step schemes, are based on second-order Taylor series expansions in time. An interesting and simpler variation of the Lax-Wendroff scheme was introduced by MacCormack (1969) and has been widely used in computational fluid dynamics. For linear problems, the scheme is identical to the original Lax-Wendroff method. For the analysis of two-dimensional open-channel flows, the scheme has been used by Fennema and Chaudhry (1989, 1990) and Gracia and Kahawita (1986).

General Formulation

This scheme consists of a two-step predictor-corrector sequence. The flow variables are known at time level k and their values are to be determined at time level $k + 1$. Then for grid points i, j, the following finite-difference equations may be written to approximate Eq. 15-43.

Predictor

$$\mathbf{U}_{i,j}^* = \mathbf{U}_{i,j}^k - \frac{\Delta t}{\Delta x}\nabla_x\mathbf{E}_{i,j}^k - \frac{\Delta t}{\Delta y}\nabla_y\mathbf{F}_{i,j}^k - \Delta t\,\mathbf{S}_{i,j}^k \quad \left\{ \begin{array}{l} 2 \leq i \leq N \\ 2 \leq j \leq M \end{array} \right. \qquad (15\text{-}53)$$

Corrector

$$\mathbf{U}_{i,j}^{**} = \mathbf{U}_{i,j}^{k} - \frac{\Delta t}{\Delta x}\Delta_x \mathbf{E}_{i,j}^{*} - \frac{\Delta t}{\Delta y}\Delta_y \mathbf{F}_{i,j}^{*} - \Delta t\, \mathbf{S}_{i,j}^{*} \left\{ \begin{array}{l} 1 \leq i \leq N - 1 \\ 1 \leq j \leq M - 1 \end{array} \right. \qquad (15\text{-}54)$$

in which $\mathbf{U}^*$ and $\mathbf{U}^{**}$ are intermediate values for $\mathbf{U}$. The new values for the vector $\mathbf{U}$ at time $k + 1$ are obtained from

$$\mathbf{U}_{i,j}^{k+1} = \frac{1}{2}(\mathbf{U}_{i,j}^{*} + U_{i,j}^{**}) \qquad (15\text{-}55)$$

The grid points i, j, and k are as defined in Fig. 15-2. The scheme first uses the backward space differences (∇_x and ∇_y) to predict an intermediate solution from the known information at time level k. The forward space differences (Δ_x and Δ_y) are used in the second step to correct the predicted values. The forward (Δ) and backward (∇) difference operators are defined as

$$\begin{aligned} \Delta_x \mathbf{U}_{i,j} &= \mathbf{U}_{i+1,j} - \mathbf{U}_{i,j} \\ \nabla_x \mathbf{U}_{i,j} &= \mathbf{U}_{i,j} - \mathbf{U}_{i-1,j} \end{aligned} \qquad (15\text{-}56)$$

where the subscript indicates the direction of differencing. The corrector step always uses one-sided differences opposite to the ones used in the predictor part; in this case forward finite-differences are used.

The differencing may be reversed or it may be alternated each time step. An example of a sequence that repeats itself every fourth time step is given in Fig. 15-3.

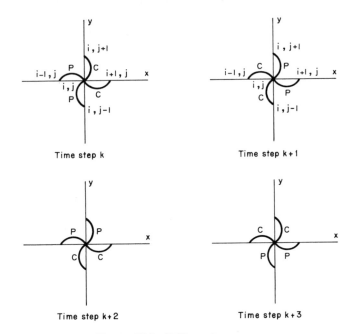

Figure 15-3 Differencing sequence.

This sequence removes most of the directional bias of this scheme that would otherwise be present if steps were not alternated. The differencing shown in the first sketch, the k time step, is equivalent to the differencing used in Eqs. 15-53 and 15-54.

The values of primitive variables are determined from the computed value of U at each step as follows:

$$h^{k+1} = h^{k+1}$$
$$u^{k+1} = \frac{(uh)^{k+1}}{h^{k+1}} \tag{15-57}$$
$$v^{k+1} = \frac{(vh)^{k+1}}{h^{k+1}}$$

in which $k+1$ refers to an intermediate value obtained during a current predictor or corrector sequence.

Boundary Conditions

Reflection boundaries may be easily incorporated in the MacCormack scheme. In this procedure, fictitious points in the solid wall are replaced by immediate interior points. Antisymmetric reflection is incorporated by changing the sign of the normal component of velocity. To illustrate, consider a solid boundary along the y-direction with the computational domain in the positive x-direction (Fig. 15-4). For the predictor step, the difference equations using the conservation form become

$$h_{i,j}^* = h_{i,j}^k \quad - \frac{\Delta t}{\Delta x}\left[(uh)_{i,j}^k + (uh)_{i+1,j}^k\right] - \frac{\Delta t}{\Delta y}\left[(vh)_{i,j}^k - (vh)_{i,j-1}^k\right]$$

$$(uh)_{i,j}^* = (uh)_{i,j}^k - \frac{\Delta t}{\Delta x}\left[(u^2 h + \frac{1}{2}gh^2)_{i,j}^k - (u^2 h + \frac{1}{2}gh^2)_{i+1,j}^k\right]$$
$$\quad - \frac{\Delta t}{\Delta y}\left[(uvh)_{i,j}^k - (uvh)_{i,j-1}^k\right]$$
$$\quad + gh_{i,j}^k \Delta t (S_{o_x} - S_{f_x})_{i,j}^k \tag{15-58}$$

$$(vh)_{i,j}^* = (vh)_{i,j}^k - \frac{\Delta t}{\Delta x}\left[(uvh)_{i,j}^k + (uvh)_{i+1,j}^k\right]$$
$$\quad - \frac{\Delta t}{\Delta y}\left[(v^2 h + \frac{1}{2}gh^2)_{i,j}^k - (v^2 h + \frac{1}{2}gh^2)_{i,j-1}^k\right]$$
$$\quad + gh_{i,j}^k \Delta t (S_{o_y} - S_{f_y})_{i,j}^k$$

In these equations the values at all the points $(i-1, j)$ have been replaced by values at points $(i+1, j)$ and the sign of the normal velocity component, u, has been switched. No difficulty arises during the corrector step because none of the differences are from the interior of the solid boundary.

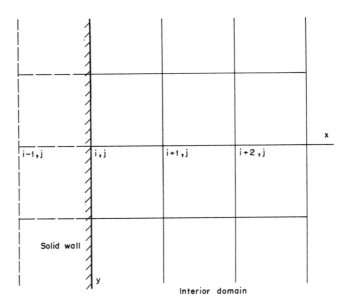

Figure 15-4 Reflection boundary.

15-5 Gabutti Scheme

An explicit scheme based on the characteristic relations was introduced by Moretti (1979); he called it the λ-scheme. The algorithm is based on the nonconservative equations (Eq. 15-50) and relies on the theory of characteristics. Gabutti (1983) analyzed its numerical properties and presented an improved scheme. The resulting algorithm is an explicit scheme that consists of three sequential steps. This class of finite-difference schemes is also referred to as the class of split-coefficient matrix (SCM) methods. The objective is to write difference equations that transmit information from the relevant characteristic directions. This is accomplished by diagonalizing and splitting the coefficient matrices so that each part incorporates only positive or negative contributions of the eigenvalues.

General Formulation

The algorithm is developed in the following manner. The Jacobians of Eqs. 15-51 may be diagonalized by transformation matrices (see, for example, Anton, 1981), that are made up of the right eigenvectors of $\mathbf{G}$ and $\mathbf{H}$. Eqs. 15-50 may then be written as

$$\mathbf{V}_t + \mathbf{M}\mathbf{D}_G\mathbf{M}^{-1}\mathbf{V}_x + \mathbf{N}\mathbf{D}_H\mathbf{N}^{-1}\mathbf{V}_y + \mathbf{T} = 0 \tag{15-59}$$

in which $\mathbf{M}$ and $\mathbf{N}$ are defined by

$$
\mathbf{M} = \begin{pmatrix} 0 & \dfrac{h}{2c} & \dfrac{h}{2c} \\ 0 & \frac{1}{2} & -\frac{1}{2} \\ 1 & 0 & 0 \end{pmatrix} \qquad \mathbf{N} = \begin{pmatrix} 0 & \dfrac{h}{2c} & \dfrac{h}{2c} \\ 1 & 0 & 0 \\ 0 & \frac{1}{2} & -\frac{1}{2} \end{pmatrix} \tag{15-60}
$$

where $\mathbf{D}_G$ and $\mathbf{D}_H$ are diagonal matrices of the eigenvalues presented in Eqs. 15-52. Diagonalization of the flux matrices leads to

$$
\mathbf{G} = \mathbf{M}\mathbf{D}_G\mathbf{M}^{-1} = \begin{pmatrix} \frac{1}{2}(\lambda_2 + \lambda_3) & \frac{h}{2c}(\lambda_2 - \lambda_3) & 0 \\ \frac{c}{2h}(\lambda_2 - \lambda_3) & \frac{1}{2}(\lambda_2 + \lambda_3) & 0 \\ 0 & 0 & \lambda_1 \end{pmatrix} \tag{15-61}
$$

$$
\mathbf{H} = \mathbf{N}\mathbf{D}_H\mathbf{N}^{-1} = \begin{pmatrix} \frac{1}{2}(\omega_2 + \omega_3) & 0 & \frac{h}{2c}(\omega_2 - \omega_3) \\ 0 & \omega_1 & 0 \\ \frac{c}{2h}(\omega_2 - \omega_3) & 0 & \frac{1}{2}(\omega_2 + \omega_3) \end{pmatrix} \tag{15-62}
$$

The diagonal matrices $\mathbf{D}_G$ and $\mathbf{D}_H$ may be split into positive and negative parts, $\mathbf{D}_G = \mathbf{D}_G^+ + \mathbf{D}_G^-$ and $\mathbf{D}_H = \mathbf{D}_H^+ + \mathbf{D}_H^-$. This can be accomplished in a number of ways. For example, one alternative is to test each eigenvalue by

$$
\begin{aligned}
\lambda_l^+ &= \max(\lambda_l, 0) & \omega_l^+ &= \max(\omega_l, 0) \\
\lambda_l^- &= \min(\lambda_l, 0) & \omega_l^- &= \min(\omega_l, 0)
\end{aligned} \tag{15-63}
$$

In this manner matrices are obtained which contain only positive or negative parts. The flux terms are thus split into

$$
\begin{aligned}
\mathbf{G} &= \mathbf{G}^+ + \mathbf{G}^- = \mathbf{M}\mathbf{D}_G^+\mathbf{M}^{-1} + \mathbf{M}\mathbf{D}_G^-\mathbf{M}^{-1} \\
\mathbf{H} &= \mathbf{H}^+ + \mathbf{H}^- = \mathbf{N}\mathbf{D}_H^+\mathbf{N}^{-1} + \mathbf{N}\mathbf{D}_H^-\mathbf{N}^{-1}
\end{aligned} \tag{15-64}
$$

Incorporating these split flux terms in the system leads to the equation

$$
\mathbf{V}_t + \mathbf{G}^+\mathbf{V}_x + \mathbf{G}^-\mathbf{V}_x + \mathbf{H}^+\mathbf{V}_y + \mathbf{H}^-\mathbf{V}_y + \mathbf{T} = 0 \tag{15-65}
$$

Each split Jacobian matrix multiplies a spatial derivative, $\mathbf{V}_x$ or $\mathbf{V}_y$; thus with the established direction of the fluxes, appropriate differencing can be taken. With a positive coefficient matrix, a backward difference is used, e.g., $\mathbf{G}^+\mathbf{V}_x \approx \mathbf{G}_i^+ (\mathbf{V}_i - \mathbf{V}_{i-1})/\Delta x$; with a negative coefficient matrix, a forward difference is used, e.g., $\mathbf{G}^-\mathbf{V}_x \approx \mathbf{G}_i^- (\mathbf{V}_{i+1} - \mathbf{V}_i)/\Delta x$. The propagation of the signal is now correctly applied along the characteristics. In subcritical flow, both positive and negative characteristics are used, whereas in supercritical flow the information is carried only along the characteristics from the direction of flow. The spatial and temporal differencing can be implemented in many ways. In the Gabutti scheme the algorithm takes the following form.

Predictor

Part A

$$\widetilde{\mathbf{V}}_{i,j} = \mathbf{V}_{i,j}^k - \frac{\Delta t}{\Delta x}\left(\mathbf{G}^+\nabla_x\mathbf{V}_{i,j}^k + \mathbf{G}^-\Delta_x\mathbf{V}_{i,j}^k\right)$$
$$- \frac{\Delta t}{\Delta y}\left(\mathbf{H}^+\nabla_y\mathbf{V}_{i,j}^k + \mathbf{H}^-\Delta_y\mathbf{V}_{i,j}^k\right) - \Delta t\,\mathbf{T}_{i,j}^k \tag{15-66}$$

Part B

$$\widehat{\mathbf{V}}_{i,j} = \mathbf{V}_{i,j}^k - \frac{\Delta t}{\Delta x}\left(\mathbf{G}^+(1+\nabla_x)\nabla_x\mathbf{V}_{i,j}^k + \mathbf{G}^-(1-\Delta_x)\Delta_x\mathbf{V}_{i,j}^k\right)$$
$$- \frac{\Delta t}{\Delta y}\left(\mathbf{H}^+(1+\nabla_y)\nabla_y\mathbf{V}_{i,j}^k + \mathbf{H}^-(1-\Delta_y)\Delta_y\mathbf{V}_{i,j}^k\right) - \Delta t\,\mathbf{T}_{i,j}^k \tag{15-67}$$

Corrector

$$\overline{\mathbf{V}}_{i,j} = \widetilde{\mathbf{V}}_{i,j} - \frac{\Delta t}{\Delta x}\left(\widetilde{\mathbf{G}}^+\nabla_x\widetilde{\mathbf{V}}_{i,j} + \widetilde{\mathbf{G}}^-\Delta_x\widetilde{\mathbf{V}}_{i,j}\right)$$
$$- \frac{\Delta t}{\Delta y}\left(\widetilde{\mathbf{H}}^+\nabla_y\widetilde{\mathbf{V}}_{i,j} + \widetilde{\mathbf{H}}^-\Delta_y\widetilde{\mathbf{V}}_{i,j}\right) - \Delta t\,\widetilde{\mathbf{T}}_{i,j} \tag{15-68}$$

where $\widetilde{\mathbf{V}}$, $\widehat{\mathbf{V}}$ and $\overline{\mathbf{V}}$ are intermediate values for $\mathbf{V}$. The combination of the operators in the predictor step, part B, are the difference approximations given by

$$(1 + \nabla_x)\nabla_x\mathbf{V}_{i,j} = 2\mathbf{V}_{i,j} - 3\mathbf{V}_{i-1,j} + \mathbf{V}_{i-2,j}$$
$$(1 - \Delta_x)\Delta_x\mathbf{V}_{i,j} = -2\mathbf{V}_{i,j} + 3\mathbf{V}_{i+1,j} - \mathbf{V}_{i+2,j} \tag{15-69}$$

At the end of each sequence the new values for $\mathbf{V}$ are obtained from

$$\mathbf{V}_{i,j}^{k+1} = \frac{1}{2}(\mathbf{V}_{i,j}^k + \widehat{\mathbf{V}}_{i,j} + \overline{\mathbf{V}}_{i,j} - \widetilde{\mathbf{V}}_{i,j}) \tag{15-70}$$

Boundary Conditions

Boundaries based on the characteristic principles may be included in the Gabutti scheme. The compatability equations valid along each characteristic aligned with the axes of the computational domain are obtained first, and the appropriate one is replaced by a specified boundary condition. The system is multiplied by the eigenvectors associated with the respective characteristic directions. In the case of a solid boundary, shown in Fig. 15-4, the relevant eigenvectors corresponding to the eigenvalues given by Eqs. 15-52 are contained in $\mathbf{M}^{-1}$. Premultiplying each term of Eq. 15-59 by $\mathbf{M}^{-1}$ gives

$$\mathbf{M}^{-1}\mathbf{V}_t + \mathbf{D}_G\mathbf{M}^{-1}\mathbf{V}_x + \mathbf{M}^{-1}\mathbf{N}\mathbf{D}_H\mathbf{N}^{-1}\mathbf{V}_y + \mathbf{M}^{-1}\mathbf{T} = 0 \qquad (15\text{-}71)$$

The incoming characteristic, $\lambda_2 = u + c$, is associated with the second eigenvector, $\mathbf{M}^{-1}$. The information along λ_2 comes from the wall and is replaced by the boundary condition $u = 0$ or $u_t = 0$. Thus the equations at the wall, after simplification, may be written as follows:

$$h_t + \lambda_3 h_x^- - \lambda_3 \frac{h}{c} u_x^- + \frac{1}{2}(\omega_2 + \omega_3)h_y - \frac{h}{c}\omega_1 u_y$$
$$+ \frac{h}{2c}(\omega_2 - \omega_3)v_y + \frac{gh}{c}(S_{o_x} - S_{f_x}) = 0 \qquad (15\text{-}72)$$
$$u_t = 0$$
$$v_t + \frac{c}{2h}(\omega_2 - \omega_3)h_y + \frac{1}{2}(\omega_2 + \omega_3)v_y - g(S_{o_y} - S_{f_y}) = 0$$

where the superscript indicates that a forward difference should be used. The first equation is solved for the head h^{k+1} using information along the backward characteristic $\lambda_1 = u - c$. The third equation is solved for v^{k+1} by using the information along the path or world line, $\lambda_3 = u$. These are the equations for solving the flow variables h and v at the solid boundaries oriented parallel to the y-axis with the computational domain facing the positive x-direction. Similar equations may be written for the boundaries with different orientations.

15-6 Artificial Viscosity

A characteristic of many second-order finite-difference schemes is that they produce numerical oscillations near discontinuities for Courant numbers less than 1. These oscillations are due to truncation errors and are part of the diffusive properties of the scheme. Generally, if the leading term of the truncation error has odd derivatives, dispersive errors occur in the form of wiggles near the bore. It may be necessary to add an explicit damping term to smooth these oscillations. The procedure used here was developed by Jameson, et al. (1981) and has the advantage of smoothing regions of sharp gradients while leaving relatively smooth areas undisturbed. These artificial dissipative terms are added to Eq. 15-43 in the form

$$\mathbf{U}_t + \mathbf{E}_x + \mathbf{F}_y + \mathbf{S} - \mathbf{D}\mathbf{U} = 0 \qquad (15\text{-}73)$$

where $\mathbf{D}$ is a dissipative operator along the principal axes defined by $\mathbf{D}\mathbf{U} = \mathbf{D}_x\mathbf{U} + \mathbf{D}_y\mathbf{U}$. Using second-order differences, the operator in the x-direction becomes

$$\mathbf{D}_x\mathbf{U} = \left[\epsilon_{x_{i+\frac{1}{2},j}}(\mathbf{U}_{i+1,j} - \mathbf{U}_{i,j}) - \epsilon_{x_{i-\frac{1}{2},j}}(\mathbf{U}_{i,j} - \mathbf{U}_{i-1,j}) \right] \qquad (15\text{-}74)$$

where ϵ_x is a parameter defined from a normalized form of the gradients of one variable (e.g., h) as

$$v_{x_{i,j}} = \frac{|h_{i+1,j} - 2h_{i,j} + h_{i-1,j}|}{|h_{i+1,j}| + |2h_{i,j}| + |h_{i-1,j}|}$$

$$\epsilon_{x_{i-\frac{1}{2},j}} = \kappa \max \left(v_{x_{i-1,j}}, v_{x_{i,j}} \right)$$

(15-75)

where κ is used to regulate the amount of dissipation. The computed variables are then modified by

$$\mathbf{U}_{i,j}^{k+1} = \mathbf{U}_{i,j}^{k+1} + \mathbf{D}_x \mathbf{U}_{i,j}^{k+1} + \mathbf{D}_y \mathbf{U}_{i,j}^{k+1}$$

(15-76)

This statement should be viewed as a FORTRAN replacement statement. The terms are added after a predetermined number of time steps using the latest values of h, u, and v.

15-7 Beam and Warming Schemes

In this section, implicit finite-difference schemes are presented for the solution of the governing equations. Beam and Warming (1976) developed them for the solution of hyperbolic systems in the conservation form, and they have been used in computational fluid dynamics. The schemes are noniterative, which results in a considerable saving in computer time, especially in multidimensional problems. Most formulations of this scheme, discussed in the following paragraphs, are second-order accurate in time and can be made second- or fourth-order accurate in space.

The general formulation of the time differencing is presented first, followed by an efficient solution algorithm using an alternating-direction implicit (ADI) procedure. Switching techniques incorporated in the schemes allow the analysis of flows if both subcritical and supercritical flows are present simultaneously. The inclusion of the boundary conditions in the numerical schemes is discussed.

General Formulation

The system of Eqs. 15-43 may be solved by using time-difference approximations of the general form

$$\mathbf{U}^{k+1} = \mathbf{U}^k + \Delta t \left[\frac{\theta}{1+\xi} \left(\frac{\partial \mathbf{U}}{\partial t} \right)^{k+1} + \frac{1-\theta}{1+\xi} \left(\frac{\partial \mathbf{U}}{\partial t} \right)^k + \frac{\xi}{1+\xi} \left(\frac{\partial \mathbf{U}}{\partial t} \right)^{k-1} \right]$$

(15-77)

in which θ and ξ are parameters leading to a variety of schemes. Some of the common formulations (Richtmyer and Morton 1967) are listed in Table 15-1.

**Table 15-1 Different formulations of beam
and warming schemes**

Scheme	θ	ξ
Euler Implicit (Backward Euler)	1	0
Three-Point Backward	1	$\frac{1}{2}$
Trapezoidal Formula (Crank Nicolson)	$\frac{1}{2}$	0

Substitution for $\partial U/\partial t$ from Eq. 15-43 in terms of the flux and source terms $\mathbf{E}$, $\mathbf{F}$, and $\mathbf{S}$ yields

$$\mathbf{U}^{k+1} = \mathbf{U}^k - \Delta t \left[\frac{\theta}{1+\xi} \left(\frac{\partial \mathbf{E}}{\partial x} + \frac{\partial \mathbf{F}}{\partial y} + \mathbf{S} \right)^{k+1} \right.$$
$$\left. + \frac{1-\theta}{1+\xi} \left(\frac{\partial \mathbf{E}}{\partial x} + \frac{\partial \mathbf{F}}{\partial y} + \mathbf{S} \right)^{k} \right] + \frac{\xi \Delta t}{1+\xi} \left(\frac{\partial \mathbf{U}}{\partial t} \right)^{k-1} \tag{15-78}$$

Spatial finite-difference approximations will be substituted for the space derivatives later in the development, after a convenient form of the time discretization is obtained. The nonlinearity of the flux vectors, $\mathbf{E}^{k+1}$, $\mathbf{F}^{k+1}$, and $\mathbf{S}^{k+1}$, presents some difficulty, since they exist at the advanced time level. However, these are all functions of the flow variables, $\mathbf{U}$, for which solutions are to be obtained. These flux vectors may be linearized by using a local Taylor series expansion. For instance, the expansion of $\mathbf{E}^{k+1}$ yields

$$\mathbf{E}^{k+1} = \mathbf{E}^k + \Delta t \frac{\partial \mathbf{E}^k}{\partial t} + \frac{(\Delta t)^2}{2} \frac{\partial^2 \mathbf{E}^k}{\partial t^2} + \cdots \tag{15-79}$$

Using the chain rule, $\partial \mathbf{E}^k/\partial t = (\partial \mathbf{E}^k/\partial \mathbf{U}) (\partial \mathbf{U}^k/\partial t)$. Since $\partial \mathbf{E}^k/\partial \mathbf{U}$ is simply the Jacobian $\mathbf{A}^k$, we may write $\partial \mathbf{E}^k/\partial t = \mathbf{A}^k \partial \mathbf{U}^k/\partial t$. Second-order accuracy is obtained if only the first two terms on the right-hand side of Eq. 15-79 are retained. Thus, retaining these two terms, substituting $\partial \mathbf{E}^k/\partial t = \mathbf{A}^k \partial \mathbf{U}^k/\partial t$ and writing $\partial \mathbf{U}^k/\partial t$ in difference form, we obtain

$$\mathbf{E}^{k+1} = \mathbf{E}^k + \mathbf{A}^k(\mathbf{U}^{k+1} - \mathbf{U}^k) \tag{15-80}$$

Similarly, expansions for $\mathbf{F}^{k+1}$ and $\mathbf{S}^{k+1}$ may be written as

$$\mathbf{F}^{k+1} = \mathbf{F}^k + \mathbf{B}^k(\mathbf{U}^{k+1} - \mathbf{U}^k)$$
$$\mathbf{S}^{k+1} = \mathbf{S}^k + \mathbf{Q}^k(\mathbf{U}^{k+1} - \mathbf{U}^k) \tag{15-81}$$

where $\mathbf{B}$ and $\mathbf{Q}$ are the Jacobian of $\mathbf{F}$ and $\mathbf{S}$ respectively. Substitution of Eqs. 15-80 and 15-81 into Eqs. 15-78 and combining terms of the same time level transforms the system into the following set of equations which contain $\mathbf{U}$ terms at the $(k+1)$ and k time levels

$$\mathbf{U}^{k+1} = \mathbf{U}^k - \Delta t \left[\frac{\theta}{1+\xi} \left(\frac{\partial}{\partial x} \mathbf{A}^k \mathbf{U}^{k+1} + \frac{\partial}{\partial y} \mathbf{B}^k \mathbf{U}^{k+1} + \mathbf{Q}^k \mathbf{U}^{k+1} \right) \right.$$

$$- \frac{\theta}{1+\xi} \left(\frac{\partial}{\partial x} \mathbf{A}^k \mathbf{U}^k + \frac{\partial}{\partial y} \mathbf{B}^k \mathbf{U}^k + \mathbf{Q}^k \mathbf{U}^k \right) \qquad (15\text{-}82)$$

$$\left. + \frac{1}{1+\xi} \left(\frac{\partial \mathbf{E}}{\partial x} + \frac{\partial \mathbf{F}}{\partial y} + \mathbf{S} \right)^k \right] + \Delta t \frac{\xi}{1+\xi} \left(\frac{\partial \mathbf{U}}{\partial t} \right)^{k-1}$$

Transposing the dependent variables at the advanced time level to the left-hand side of this equation yields a linear system for $\mathbf{U}^{k+1}$:

$$\left[\mathbf{I} + \Delta t \, \frac{\theta}{1+\xi} \left(\frac{\partial}{\partial x} \mathbf{A}^k + \frac{\partial}{\partial y} \mathbf{B}^k + \mathbf{Q}^k \right) \right] \mathbf{U}^{k+1}$$

$$= \left[\mathbf{I} + \Delta t \frac{\theta}{1+\xi} \left(\frac{\partial}{\partial x} \mathbf{A}^k + \frac{\partial}{\partial y} \mathbf{B}^k + \mathbf{Q}^k \right) \right] \mathbf{U}^k \qquad (15\text{-}83)$$

$$- \Delta t \frac{1}{1+\xi} \left(\frac{\partial \mathbf{E}}{\partial x} + \frac{\partial \mathbf{F}}{\partial y} + \mathbf{S} \right)^k + \Delta t \frac{\xi}{1+\xi} \left(\frac{\partial \mathbf{U}}{\partial t} \right)^{k-1}$$

in which $\mathbf{I}$ is the unit matrix. The notation $(\partial/\partial x \, \mathbf{A}^k + \partial/\partial y \, \mathbf{B}^k)\mathbf{U}^{k+1}$ is to be interpreted as $\partial/\partial x \, (\mathbf{A}^k \mathbf{U}^{k+1}) + \partial/\partial x \, (\mathbf{B}^k \mathbf{U}^{k+1})$, i.e., the vector $\mathbf{U}$ is to be evaluated inside the derivatives. Note that the terms inside the brackets on both the left-hand and right-hand side of Eqs. 15-83 are identical. By using a forward-difference operator, $\Delta_t \mathbf{U}^{k+1} = \mathbf{U}^{k+1} - \mathbf{U}^k$, and replacing the last terms of Eqs. 15-83 by a forward-difference operator, Eqs. 15-83 may be written as

$$\left[\mathbf{I} + \Delta t \frac{\theta}{1+\xi} \left(\frac{\partial}{\partial x} \mathbf{A}^k + \frac{\partial}{\partial y} \mathbf{B}^k + \mathbf{Q}^k \right) \right] \Delta_t \mathbf{U}^{k+1}$$

$$= -\Delta t \frac{1}{1+\xi} \left(\frac{\partial \mathbf{E}}{\partial x} + \frac{\partial \mathbf{F}}{\partial y} + \mathbf{S} \right)^k + \frac{\xi}{1+\xi} \Delta_t \mathbf{U}^k \qquad (15\text{-}84)$$

The above algorithm is said to be in the delta form; the flow variables, $\mathbf{U}$, exist only in increments of $\mathbf{U}$ between two time levels. The principal advantage of this formulation is the computational efficiency due to a reduction in number of terms.

The order of the spatial differencing may be different between the right-hand and left-hand side of the equations. The preceding schemes lead to an unwieldy inversion (solution of a linear system) problem. The coefficients in the brackets have a bandwidth of $2N$ in the matrix due to the addition of the components of $\mathbf{B}$. For example, an inversion of an 80-band matrix is needed for 40 mesh points in the x-direction. A more efficient solution algorithm is obtained by factoring the left-hand side of Eqs. 15-84; as discussed in the following paragraphs.

Factored Schemes

A matrix with a small bandwidth can be obtained by reducing the two-dimensional problem to two one-dimensional problems. By the method of approximate factorization (AF), the left-hand side of the implicit schemes ($\theta \neq 0$) described by Eqs. 15–84 can be rewritten as a product of two components, each containing the terms of a specific direction:

$$
\left[\mathbf{I} + \Delta t \frac{\theta}{1+\xi} \frac{\partial}{\partial x} \mathbf{A}^k\right]\left[\mathbf{I} + \Delta t \frac{\theta}{1+\xi}\left(\frac{\partial}{\partial y}\mathbf{B} + \mathbf{Q}\right)\right]^k \Delta_t \mathbf{U}^{k+1}
$$
$$
= -\Delta t \frac{1}{1+\xi}\left(\frac{\partial \mathbf{E}}{\partial x} + \frac{\partial \mathbf{F}}{\partial y} + \mathbf{S}\right)^k + \frac{\xi}{1+\xi}\Delta_t \mathbf{U}^k
$$

(15-85)

This formulation contains two additional terms obtained by the multiplication on the left-hand side. Since the lowest order of accuracy of the scheme is $O(\Delta t)^2$ and since the additional terms are of the same order, the formal accuracy of the schemes is not affected by the new terms. Thus, without compromising the accuracy of the solution, the equations are factored in a series of steps that are directionally dependent. These steps are

$$
\left[\mathbf{I} + \Delta t \frac{\theta}{1+\xi}\frac{\partial}{\partial x}\mathbf{A}^k\right]\Delta_t \widehat{\mathbf{U}} = -\Delta t \frac{1}{1+\xi}\left(\frac{\partial \mathbf{E}}{\partial x} + \frac{\partial \mathbf{F}}{\partial y} + \mathbf{S}\right)^k + \frac{\xi}{1+\xi}\Delta_t \mathbf{U}^k
$$
$$
\left[\mathbf{I} + \Delta t \frac{\theta}{1+\xi}\left(\frac{\partial}{\partial y}\mathbf{B} + \mathbf{Q}\right)\right]^k \Delta_t \mathbf{U}^{k+1} = \Delta_t \widehat{\mathbf{U}}
$$

(15-86)

$$
\mathbf{U}^{k+1} = \mathbf{U}^k + \Delta_t \mathbf{U}^{k+1}
$$

where $\Delta_t \widehat{\mathbf{U}}$ is an intermediate value obtained by solving the system first along the rows (x-direction). This is an alternating-direction implicit ADI procedure. With these sequential steps, the inversion is reduced to solving a small-band width matrix along each row and column, which is more efficient. To allow for the presence of both subcritical and supercritical flows, a switching technique is used to obtain an appropriate spatial differencing. This switching technique is incorporated into the scheme by splitting the flux matrices into components containing either the positive or negative parts.

Implicit Split-Flux Factoring

Flux splitting may be incorporated in the approximate-factored algorithm, Eqs. 15-86, as follows. The submatrices are used to operate on the flow variables with appropriate space differences. Substitution of the diagonalized matrices into the algorithm gives an implicit split-flux-factored scheme, which may be implemented by the sequence

$$\left[\mathbf{I} + \Delta t \frac{\theta}{1 + \xi} \frac{\partial}{\partial x} \left(\mathbf{A}^+ + \mathbf{A}^- \right)^k \right] \Delta_t \widehat{\mathbf{U}}$$

$$= -\Delta t \frac{1}{1 + \xi} \left(\frac{\partial \mathbf{E}^+}{\partial x} + \frac{\partial \mathbf{E}^-}{\partial x} + \frac{\partial \mathbf{F}^+}{\partial y} + \frac{\partial \mathbf{F}^-}{\partial y} + \mathbf{S} \right)^k + \frac{\xi}{1 + \xi} \Delta_t \mathbf{U}^k \quad (15\text{-}87)$$

$$\left[\mathbf{I} + \Delta t \frac{\theta}{1 + \xi} \left(\frac{\partial}{\partial y} \mathbf{B}^+ + \frac{\partial}{\partial y} \mathbf{B}^- + \mathbf{Q} \right)^k \right] \Delta_t \mathbf{U}^{k+1} = \Delta_t \widehat{\mathbf{U}}$$

$$\mathbf{U}^{k+1} = \mathbf{U}^k + \Delta_t \mathbf{U}^{k+1}$$

where the sign of the flux components indicates the use of the split form—e.g., $\mathbf{A}^+ = \mathbf{M} \mathbf{D}^+ \mathbf{M}^{-1}$;

$$\mathbf{M} = \begin{pmatrix} 0 & \dfrac{h}{2c} & \dfrac{h}{2c} \\ 0 & \dfrac{1}{2} & -\dfrac{1}{2} \\ 1 & 0 & 0 \end{pmatrix} \quad (15\text{-}88)$$

and $\mathbf{D}$ is the diagonal matrix of the eigenvalues of $\mathbf{A}$. The right-hand side of Eqs. 15-87 are evaluated as $\mathbf{E}_x = \mathbf{A} \mathbf{U}_x$ and $\mathbf{F}_y = \mathbf{B} \mathbf{U}_y$.

Further factorization of Eqs. 15-87 is possible, since the left-hand side of the first step can be split into terms containing $\partial \mathbf{A}^+ / \partial x$ and $\partial \mathbf{A}^- / \partial x$. This leads to a further reduction in the bandwidth of the coefficient matrix. When using second- or third-order-accurate space differencing, some advantage may be gained by this additional factoring. However, when using first-order-accurate spatial differencing, the nonfactored coefficient matrix is block-tridiagonal, for which efficient solution algorithms are available. The complete finite-difference equations with first-order spatial differencing (Fig. 15-2) become

$$\left[\mathbf{I} + \frac{\Delta t}{\Delta x} \frac{\theta}{1 + \xi} \left(\nabla_x \mathbf{A}_{i,j}^+ + \Delta_x \mathbf{A}_{i,j}^- \right)^k \right] \Delta_t \widehat{\mathbf{U}}_{i,j}$$

$$= -\frac{\Delta t}{\Delta x} \frac{1}{1 + \xi} \left(\mathbf{A}_{i,j}^+ \nabla_x \mathbf{U}_{i,j} + \mathbf{A}_{i,j}^- \Delta_x \mathbf{U}_{i,j} \right)^k$$

$$- \frac{\Delta t}{\Delta y} \frac{1}{1 + \xi} \left(\mathbf{B}_{i,j}^+ \nabla_y \mathbf{U}_{i,j} + \mathbf{B}_{i,j}^- \Delta_y \mathbf{U}_{i,j} \right)^k - \Delta t \frac{1}{1 + \xi} \mathbf{S}_{i,j}^k + \frac{\xi}{1 + \xi} \Delta_t \mathbf{U}_{i,j}^k$$

$$\left[\mathbf{I} + \Delta t \frac{\theta}{1 + \xi} \left(\frac{1}{\Delta y} \left(\nabla_y \mathbf{B}_{i,j}^+ + \Delta_y \mathbf{B}_{i,j}^- \right) + \mathbf{Q}_{i,j} \right)^k \right] \Delta_t \mathbf{U}_{i,j}^{k+1} = \Delta_t \widehat{\mathbf{U}}_{i,j}$$

$$\mathbf{U}_{i,j}^{k+1} = \mathbf{U}_{i,j}^k + \Delta_t \mathbf{U}_{i,j}^{k+1} \quad (15\text{-}89)$$

Each step in the algorithm leads to a system of equations that have tridiagonal coefficient matrices. The first set of equations gives intermediate values of the flow variables and is a set of simultaneous equations along each row (x-direction). The coefficient matrix for each row has the structure

$$
\begin{pmatrix}
\mathbf{b}_1 & \mathbf{c}_1 & 0 & 0 & \cdots & 0 & 0 & 0 \\
\mathbf{a}_2 & \mathbf{b}_2 & \mathbf{c}_2 & 0 & \cdots & 0 & 0 & 0 \\
0 & \mathbf{a}_3 & \mathbf{b}_3 & \mathbf{c}_3 & \cdots & 0 & 0 & 0 \\
\vdots & \vdots & \vdots & \vdots & \ddots & \vdots & \vdots & \vdots \\
0 & 0 & 0 & 0 & \cdots & \mathbf{a}_{N-1} & \mathbf{b}_{N-1} & \mathbf{c}_{N-1} \\
0 & 0 & 0 & 0 & \cdots & 0 & \mathbf{a}_N & \mathbf{b}_N
\end{pmatrix}
\tag{15-90}
$$

$$
\times
\begin{pmatrix}
\Delta_t \widehat{\mathbf{U}}_i \\
\vdots \\
\Delta_t \widehat{\mathbf{U}}_N
\end{pmatrix}
=
\begin{pmatrix}
\mathrm{RHS}_1 \\
\vdots \\
\mathrm{RHS}_N
\end{pmatrix}
$$

where $\mathbf{a}_i$, $\mathbf{b}_i$, and $\mathbf{c}_i$ are 3×3 matrices of coefficients, $\Delta_t \widehat{\mathbf{U}}$ are three-element vectors of flow variables, h, uh and vh, in delta form and RHS are three-element vectors containing the terms of the right-hand side of Eqs. 15-89. Solution of this system may be obtained efficiently by special block-tridiagonal solvers.

After a solution is obtained along the rows, the next set of equations solves a similar system of block-tridiagonal equations along the columns (y-direction). Finally, in the last calculation of Eqs. 15-89 the flow variables, $\mathbf{U}_{i,j}^{k+1}$, are obtained by adding $\Delta_t \mathbf{U}_{i,j}^{k+1}$ to $\mathbf{U}_{i,j}^k$ values for the last time-step values. The flow variables obtained from this sequence are in the form h, uh and, vh. The primitive flow variables $h, u,$ and v are then obtained by solving the following equations:

$$
h_{i,j}^{k+1} = h_{i,j}^k + \Delta_t h_{i,j}^{k+1}
$$

$$
u_{i,j}^{k+1} = \frac{(uh)_{i,j}^k + \Delta_t (uh)_{i,j}^{k+1}}{h_{i,j}^{k+1}}
\tag{15-91}
$$

$$
v_{i,j}^{k+1} = \frac{(vh)_{i,j}^k + \Delta_t (vh)_{i,j}^{k+1}}{h_{i,j}^{k+1}}
$$

The solution at the next time step may now be obtained by repeating the whole process. Second-order spatial differencing may be used in Eqs. 15-89, but the coefficient matrix structure will be less efficient. However, this type of differencing may be used on the right-hand side without affecting the formal accuracy of the system.

Boundary Conditions

Boundary conditions in an implicit formulation may be used by directly including the physical boundary specifications (Anderson, et al., 1984). Extrapolation techniques can be used to calculate values at the boundary nodes for which explicit boundary conditions are not given. For solid walls the normal velocity is zero and, depending on the orientation, either $\Delta_t (uh)$ or $\Delta_t (vh)$ is zero. The remaining two equations use either the positive part of $\mathbf{A}$ and $\mathbf{B}$ for boundaries facing the positive x- or y-direction or the negative part of $\mathbf{A}$ or $\mathbf{B}$ for boundaries facing the negative x- or y-direction. For example, the solid boundary as shown in Fig. 15-4 is evaluated using the equations

$$\left[\mathbf{I} - \frac{\Delta t}{\Delta x}\frac{\theta}{1+\xi}\mathbf{A}_{i,j}^{-}\right]\Delta_t\widehat{\mathbf{U}}_{i,j} + \frac{\Delta t}{\Delta x}\frac{\theta}{1+\xi}\mathbf{A}_{i+1,j}^{-}\Delta_t\widehat{\mathbf{U}}_{i+1,j}$$

$$= -\Delta t\frac{1}{1+\xi}\left[\frac{1}{\Delta x}\,\mathbf{A}_{i,j}^{-}(\mathbf{U}_{i+1,j} - \mathbf{U}_{i,j}) + \frac{1}{\Delta y}\mathbf{B}_{i,j}^{+}(\mathbf{U}_{i,j} - \mathbf{U}_{i,j-1})\right. \tag{15-92}$$

$$\left. + \frac{1}{\Delta y}\mathbf{B}_{i,j}^{-}(\mathbf{U}_{i,j+1} - \mathbf{U}_{i,j}) + \mathbf{S}_{i,j}\right]^k + \frac{\xi}{1+\xi}\Delta_t\mathbf{U}_{i,j}^k$$

The second equation is replaced by the boundary condition $\Delta_t\mathbf{U} = 0$. Matrix $\mathbf{A}_{i,j}^{-}$ becomes

$$\mathbf{A}_{i,j}^{-} = \frac{1}{2c}\begin{pmatrix} (u+c)\lambda_3 & 0 & 0 \\ 0 & 1 & 0 \\ (u+c)v\lambda_3 & 0 & 0 \end{pmatrix} \tag{15-93}$$

The spatial differencing for this example is in the positive x-direction, and normal switching techniques are used along the boundary in the y-direction. Similar equations may be written for boundaries facing other directions. Inflow and outflow boundaries may also be handled in this manner if an appropriate boundary condition is specified.

15-8 Applications

To demonstrate the application of the MacCormack and Gabutti schemes in hydraulic engineering, two typical problems are solved. In one of these problems, a strong bore is formed; the second problem deals with gradually varied flows. This is done intentionally to demonstrate robustness of the schemes for solving diversified types of problems. The governing equations may not be valid in the vicinity of the bore, where there are sharp curvatures. However, the computed results, such as maximum water levels and arrival time of a wave, may be used with confidence for typical engineering applications even though the details of the bore itself are not simulated in a rigorous manner. The inclusion of boundary conditions and the addition of artificial viscosity to smooth oscillations caused by dispersive errors of the finite-difference schemes are investigated.

The first example analyzes a partial dam breach, structural failure or instantaneous opening of sluice gates. The breach is nonsymmetrical to demonstrate the analysis of a general case. The second problem is the passage of a flood wave through a channel contraction. Such flow conditions occur for flow through a bridge opening, cofferdams or structures occupying partial channel widths, etc. The boundaries in these examples are taken parallel to the coordinate axes, which would usually be the case in actual problems of this kind — e.g., flow through a bridge opening or sudden opening of sluices. The 90° corner imposes a rather severe test on the schemes and on the inclusion of boundaries. Other boundary shapes may be approximated by a staircase-type configuration. Since the grid size is usually small, this approximation should not introduce significant errors in the computed results.

Application of these schemes yields flow velocities and flow depths throughout the computational domain at specified grid points. Such information may be utilized to assess the scour or erosion potential, to assess the effectiveness of preventive measures to reduce scour and erosion, to determine the height of walls and dykes, and to determine the size of rocks needed for cofferdam closure during river diversion. In addition, these applications show that these schemes may be employed to determine flow conditions in the immediate vicinity of hydraulic structures, or they may be used to generate data for input to one-dimensional flow models away from the regions where flow may be assumed as one-dimensional. With some ingenuity, any typical problem involving two-dimensional flows may be analyzed using the experience gained from these applications.

Although only unsteady flows are investigated, steady flows may be analyzed with the MacCormack and Gabutti schemes by letting the computations converge to a steady state subject to the specified end conditions. It becomes necessary to use this procedure if both sub- and supercritical flows are present in the flow field simultaneously. This is because the steady equations are hyperbolic if the flow is supercritical; but they are not hyperbolic if the flow is subcritical. For this purpose, application of a scheme like the Gabutti scheme becomes imperative, since most other finite-difference schemes presently used in hydraulic engineering either fail or give incorrect results. Because of the shock-capturing capabilities of the schemes presented herein, they become attractive for solving flow situations where hydraulic jump is formed. This procedure not only directly yields the location of the jump but gives an approximate length of the jump as well.

Computations for the problems included here were done on an IBM 3090 mainframe computer, and three-dimensional graphs were prepared by using a standard package available for this purpose at the Computing Center of Washington State University.

Partial Breach or Opening of Sluice Gates

In this problem, the dam is assumed to fail instantaneously or the sluice gates are assumed to be opened instantly. The discontinous initial conditions impose severe difficulties in starting the computations, and most of the presently used numerical schemes fail under such conditions. In the simulations included herein, the channel downstream of the dam or the gates is assumed to have some finite flow depth. This is quite normal for usual applications, where a downstream control keeps the downstream channel in a "wet" condition. To simulate a dry channel, however, a very small flow depth may be assumed in the analysis. This procedure is much easier than to track the bore propagation explicitly and should give results that are of the same, if not better, accuracy than that of the other input variables.

The computational domain comprises a 200-m-long and 200-m-wide channel. The nonsymmetrical breach or sluice gates are 75 m wide, and the structure or dam is 10 m thick in the direction of flow. The grid is 41 by 41 points, which results in an individual mesh size of 5 m by 5 m. Additional details are shown in Fig. 15-5. To prevent any damping by the source terms, a frictionless, horizontal channel was used, and initial conditions (Fig. 15-6) had a tailwater/reservoir ratio $h_t/h_r = 0.5$ in the initial few runs. Flow conditions were analyzed for a wide variation of flow parameters, such as including the friction losses (Manning n from 0 to 0.15), assuming a sloping channel (bottom slope from 0 to 0.07), different ratios of the tailwater to reservoir depths (as low as 0.2), and symmetrical and unsymmetrical breach. However, only typical results are included here to conserve space. The flow conditions were computed for 7.1 s after the dam failure or opening of the sluices. The presented results are for $7.1 \le t < 7.1 + \Delta t$ seconds. At this time, the bore is well developed in the central portion of the downstream channel and the wave front has reached one bank of the channel.

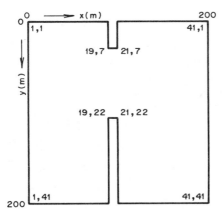

Figure 15-5 Definition sketch for partial dam breach.

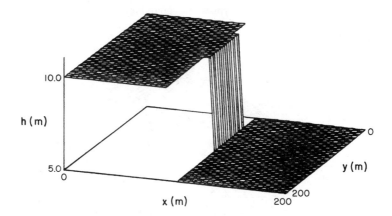

Figure 15-6 Initial conditions.

Two types of figures are used to present the computed results. The first figure is a perspective plot of the water surface. Remnants of the dam are represented by gaps near the middle of the plot. The vertical scale is exaggerated with respect to the horizontal scales. The second is a velocity vector plot. At each node the velocity is indicated by an arrow, with the magnitude represented by the length of the arrow. For esthetic reasons, velocities below a specified tolerance are not drawn (magnitude with a length less than 20 percent of the mesh size). The velocity vectors on the boundaries, parallel to the solid boundaries, and at right angles to inflow and outflow boundaries are also not drawn.

To illustrate the difference between symmetric and antisymmetric boundary conditions, both were incorporated in the MacCormack scheme. Figure 15-7 shows the perspective views of the water surface. The profile near boundaries, particularly in the reservoir area, illustrate the difference between the boundary conditions. Dispersion errors of this scheme, in the form of oscillations, are noticeable on the backside of the bore. Without affecting the quality of the profile, the solution may be smoothed by the addition of artificial viscosity. This is done in the results shown in Fig. 15-8(a).

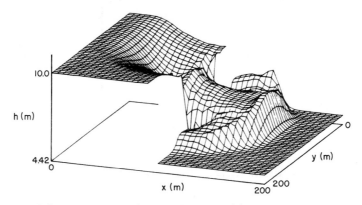

(a) Anti-symmetric boundary conditions

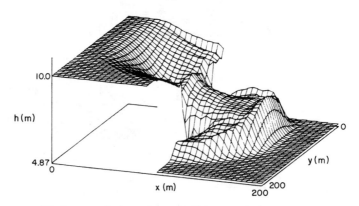

(b) Symmetric boundary conditions

Figure 15-7 Water-surface profile computed by the MacCormack scheme with no artificial viscosity.

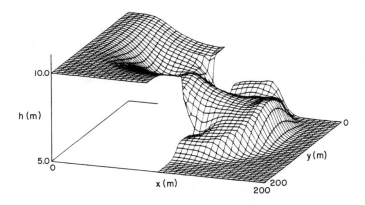

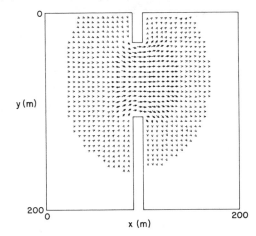

(a) Water surface profile

(b) Flow velocities

Figure 15-8   Results computed by MacCormack scheme; artificial viscosity added.

Antisymmetric boundaries are used in this run, and the computed velocity vectors are plotted in Fig. 15-8(b). In addition to eliminating the wiggles near the bore, the artificial viscosity term also reduces the undershoot near sharp corners. These steep depressions in the water surface are especially noticeable downstream of the breach. In these runs the sharp corners are modeled by assuming they are boundaries parallel with the x-direction.

The solution of this problem by using the Gabutti scheme and with no artificial viscosity is shown in Fig. 15-9(a). Since the diffusive properties of this scheme are different than those of the MacCormack scheme, the oscillations occur at different locations. The addition of artificial viscosity leads to the profile shown in Fig. 15-9(b).

Figures 15-6 through 15-9 are pictorial views of the computed water-surface profiles. These results show a qualitive comparison of these schemes, comparison of the procedures used for the inclusion of boundary conditions, and the effectiveness of the
artificial viscosity to smooth the oscillations. For a quantitative comparison of the
MacCormack and Gabutti schemes, computed results are shown in Figs. 15-10 and
15-11. Fig. 15-10 shows the comparison of a computed transverse water-surface profile
at $i = 16$ (upstream of the breach), at $i = 20$ (inside the breach), and at $i = 24$
(downstream of the breach). Figure 15-11 shows the time variation of flow depth at grid
point (20, 15), i.e., inside the breach, and at grid point (24, 15), i.e., downstream of the
breach. It is clear from these comparisons that both schemes give comparable results.

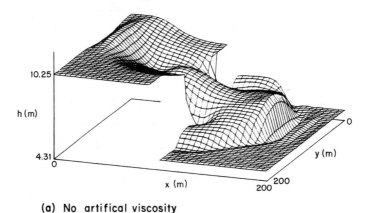

(a) No artifical viscosity

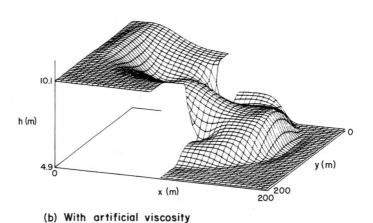

(b) With artificial viscosity

Figure 15-9 Water-surface profiles computed by Gabutti scheme.

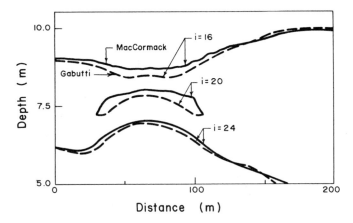

Figure 15-10   Comparison of transverse profiles.

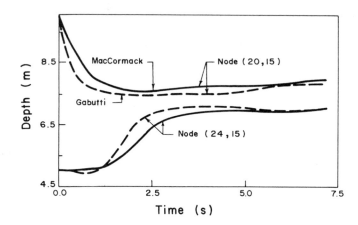

Figure 15-11   Comparison of variation of flow depth with time.

Propagation of a Flood Wave Through a Channel Contraction

The second example involves the propagation of a flood wave through a channel contraction (Fig. 15-12). The flood wave is shown in Fig. 15-13. The channel is 2000 m wide, and the flow conditions are computed in a 2000 m length of the channel. The flow domain is divided into 50 x 50 grid. The unsymmetrical channel contraction is located near the midlength of the channel. The flow depth at all grid points at $t = 0$ is 5 m, and the flow velocity is zero. The flood wave of Fig. 15-13 is introduced at the upstream end over the full channel width. The flow depth at the downstream end is kept constant and equal to the initial flow depth.

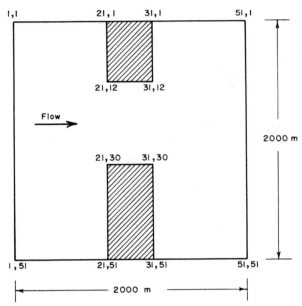

Figure 15-12 Definition sketch for flow through contracted opening.

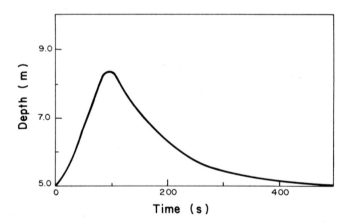

Figure 15-13 Flood wave.

The flow depths computed by the MacCormack scheme at node (20, 21) and at node (25, 21) are shown in Fig. 15-14. Node (20, 21) is located upstream of the contraction and node (25, 21) is located downstream of the contraction. It is clear from this figure how the wave height and wave shape are modified as it travels through the computational flow domain. Figure 15-15 shows the transverse water-surface profiles computed by the MacCormack scheme at $t = 220$ s at three locations. These locations are at $i = 20$ (upstream of the contraction), at $i = 25$ (inside contraction), and at $i = 35$ (downstream of contraction). It is clear from this figure how the wave is modified as it

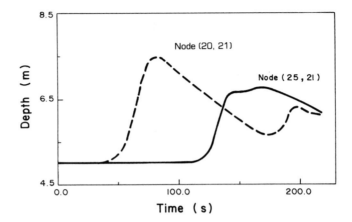

Figure 15-14 Computed variation of flow depth.

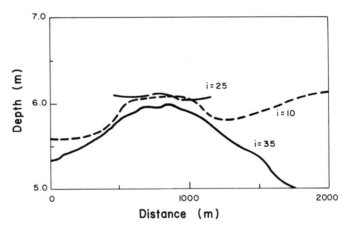

Figure 15-15 Computed transverse flow profile.

passes through the contraction. Since the contraction is asymmetrical, the water surface is also asymmetrical with respect to the centerline of the channel.

Comparison with Other Methods

Figure 15-16 compares the longitudinal water-surface profile at $t = 7.1$ s computed by the two explicit schemes presented here and the Beam and Warming implicit schemes (Fennema and Chaudhry 1989). These water-surface profiles are for the case of partial failure of a dam or sudden opening of sluice gates (Fig. 15-5). A 41-by-41 grid was used to simulate the flow conditions in the computational domain. In the MacCormack and Gabutti schemes, artificial viscosity was added ($\kappa = 0.25$) to smooth the high-frequency oscillations in the computed flow depth. Antisymmetric boundary conditions were used

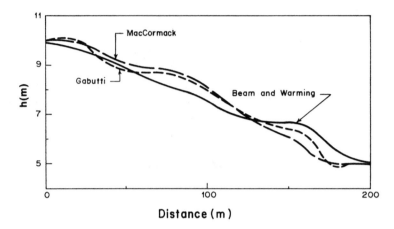

Figure 15-16 Comparison of computed water-surface profiles for dam-breach of Fig. 15-5.

in the analysis. Flow was supercritical at several grid points downstream of the breach after a short time following the opening of the breach. The MacCormack scheme failed for h_t/h_r ratios less than 0.25, the Gabutti scheme failed for this ratio less than 0.2, and the Beam and Warming schemes failed for this ratio less than 0.001.

It is clear from this figure that the agreement between the results computed by these schemes is satisfactory.

15-9 Summary

Depth-averaged equations describing two-dimensional, unsteady, free-surface flow were derived starting with the Navier-Stokes equations. The friction losses were included using empirical relationships.

Explicit and implicit finite-difference schemes were presented for the analysis of two-dimensional, unsteady, free-surface flows. Numerical oscillations occuring near the sharp-fronted waves may be controlled by adding artificial viscosity. In these schemes, boundary conditions may be easily incorporated, initial conditions may have discontinuities, and explicit tracking of the bore is not necessary. In addition, these schemes are relatively easy to program and give good results.

The split-flux schemes allow the analysis of both subcritical and supercritical flows, with the Beam and Warming schemes especially able to handle flows with large shocks and bores.

The schemes described in this chapter may be used to provide reliable solutions to problems associated with events such as dam breaks, dike breaches, and surge waves.

REFERENCES

Abbott, M. B. 1979. *Computational hydraulics: Elements of the theory of free surface flows,* Pitman, London.

Anderson, D. A.; Tannehill J. C.; and Pletcher, R. H. 1984. *Computational fluid mechanics and heat transfer,* McGraw-Hill, New York, NY.

Anton, H. 1981. *Elementary linear algebra,* John Wiley, New York, NY.

Beam, R. M., and Warming, R. F. 1976. An implicit finite-difference algorithm for hyperbolic systems in conservation-saw form, *Jour. Comp. Phys.,* 22:87–110.

Benque, J. P.; Hauguel, A.; and Viollet, P. L. 1982. *Engineering applications of computational hydraulics,* Pitman, London, England.

Chaudhry, M. H. 1987. *Applied hydraulic transients,* 2nd ed., Chapter 12, Van Nostrand Reinhold, New York, NY.

Courant, R. 1936. *Differential and integral calculus,* vol. II, Interscience, Wiley, New York, NY.

Cunge, J. A., Holly F. M., Jr., and Verwey, A. 1980. *Practical aspects of computational river hydraulics,* Pitman Publishing, London.

Fennema, R. J. 1985. Numerical solution of two-dimensional transient free-surface flows. Ph.D. diss., Washington State University, Pullman, WA.

Fennema, R. J., and Chaudhry, M. H. 1986. Second-order numerical schemes for unsteady free-surface flows with shocks, *Water Resources Research,* 22, no. 13:1923–30.

———. 1987. Simulation of one-dimensional dam-break flows, *Jour. of Hydraulic Research,* International Assoc. for Hydraulic Research, 25, no 1:41–51.

———. 1989. Implicit methods for two-dimensional unsteady free-surface flows, *Jour. of Hydraulic Research,* International Assoc. for Hydraulic Research, 27, no. 3:321–32.

———. 1990. Explicit methods for two-dimensional unsteady free-surface flows, *Jour. Hydraulic Engineering,* Amer. Soc. of Civil Engrs., 116, no. 8:1013–34.

Gabutti, B. 1983. On two upwind finite-difference schemes for hyperbolic equations in nonconservative form, *Computers and Fluids,* 11, no. 3:207–30.

Garcia, R. and Kahawita, R.A. 1986. Numerical solution of the St. Venant equations with the MacCormack finite-difference scheme, *International Jour. for Numerical Methods in Fluids,* 6:259–274.

Jameson, A.; Schmidt, W., and Turkel, E. 1981. Numerical solutions of the Euler equations by finite volume methods using Runge-Kutta time-stepping schemes, *Proc., AIAA 14th Fluid and Plasma Dynamics Conference,* Palo Alto, Calif. AIAA–81–1259.

Jimenez, O. 1987. Personal communications with M. H. Chaudhry.

Katopodes, N. D. 1984a. Two-dimensional surges and shocks in open channels, *Jour. Hydraulic Engineering,* Amer. Soc. Civil Engrs., 110, no. 6:794–812.

———. 1984b. A dissipative Galerkin scheme for open-channel flow, *Jour. Hydralic Division,* Amer. Soc. Civil Engrs., 110, no. HY6:450–66.

Katopodes, N. D., and Strelkoff, T. 1978. Computing two-dimensional dam-break flood waves, *Jour. Hydralic Division,* Amer. Soc. Civ. Engrs., 104, no. HY9:1269–88.

Lai, C. 1986. "Numerical modeling of unsteady open-channel flow." *In Advances in Hydroscience,* 14:161–333, Academic Press, New York, NY.

Lax, P. D. and Wendroff, B. 1960. Systems of conservation laws, *Com. Pure Appl. Math.*, 13:217–37.

Leendertse, J. J. 1967, Aspects of a computational model for long period water-wave propagation, *Memo* RM-5294-PR, Rand Corporation, Santa Monica, CA.

MacCormack, R. W. 1969. The effect of viscosity in hypervelocity impact cratering, *Amer. Institute of Aeronautics and Astronautics*, Paper 69–354. Cincinnati, Ohio.

Matsutomi, H. 1983. Numerical computations of two-dimensional inundation of rapidly varied flows due to breaking of dams, *Proc.*, XX Congress of the IAHR, Moscow, Subject A, II, September:479–88.

Moretti, G. 1979. The λ-scheme, *Computer and Fluids*, 7:191–205.

Richtmyer, R. D., and Morton, K. W. 1967. *Difference methods for initial-value problems,* 2nd Edition, John Wiley and Sons, New York, NY.

Sakkas, J. G., and Strelkoff, T. 1973. Dam-break flood in a prismatic dry channel, *Jour. Hydralic Division*, Amer. Soc. Civil Engrs., 99, no. HY12.

Warming, R. F., and Beam, R. M. 1978. On the construction and application of implicit factored schemes for conservation laws, *Proc., Symposium on Computational Fluid Dynamics,* SIAM-AMS 11:85–129.

16 Finite-Element Methods

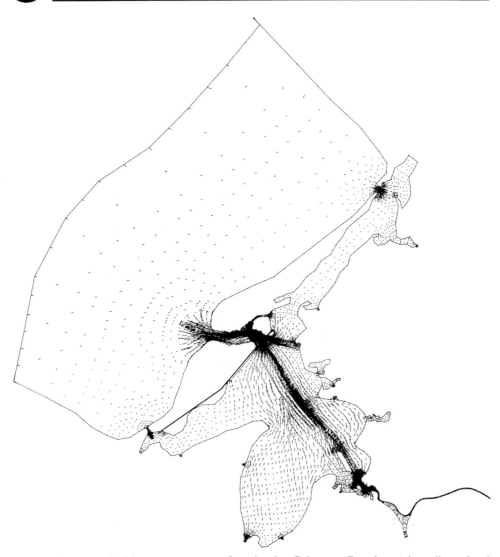

Computed flow velocities near water surface in the Galveston Bay by a three-dimensional finite-element mathematical model (Courtesy of C. Burger, U.S. Army Waterways Experiment Station, Vicksburg, Miss.).

16-1 Introduction*

In Chapters 14 and 15 we introduced the finite-difference methods and applied them to both one- and two-dimensional flow problems. The finite-element method is an alternative technique for solving these same differential equations. It is a newer technique, has not been as extensively used as the finite-difference method, and does not offer any significant advantage for one-dimensional problems.

The main advantage of the finite-element method is its ability to handle irregular boundaries and grid refinement. The finite-difference solutions presented in this book have grids with square corners, straight edges, and mostly uniform grid spacings. The finite-element method adapts easily to problems where these boundary characteristics are unsuitable. The computational grids for the finite-difference method are usually defined by parallel lines and cannot easily be used to simulate natural boundaries. Smaller grid spacings are frequently required with either method for areas of rapid variable change (i.e., near edges and corners). These grid refinements are easily handled by the finite-element method but require additional effort when using the finite-difference method.

The methods discussed in this chapter are very general and broad in application and for that reason are also among the most complicated of the finite-element methods. This is because the equations being solved are nonlinear. For linear problems (such as stress in a beam with uniform homogeneous properties or saturated groundwater flow in a porous media), we may use the simplest finite elements, do not need numerical integration, and do not require iteration to arrive at a solution. The method discussed in this chapter is effective for both linear and nonlinear problems, although it is unnecessarily complicated for linear problems.

16-2 Domain Discretization

The solution domain in the finite-difference method is divided into a grid of individual points, whereas in the finite-element method it is divided into a grid of elements.These elements, in turn, are composed of patterns of grid points (called *nodes*). The variables and parameters of each differential equation are interpolated within the element by a polynomial. Finite elements can be one-, two-, or three-dimensional. In the following paragraphs, we discuss different finite elements.

*This chapter was written by Dr. John I. Finnie of the University of Idaho.

One-Dimensional Linear Shape Functions

For one-dimensional problems, each element can be composed of two or more grid points (or nodes). For example, one-dimensional channel flow may be divided into four distinct two-node elements, as shown in Fig. 16-1.

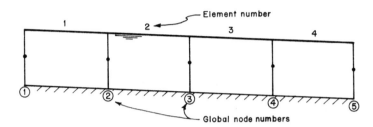

Figure 16-1   Finite-element grid.

In this figure, the element numbers are listed above the channel, whereas the individual (called *global*) node numbers are marked below the channel. We can also identify the local node numbers for each element. For example, the two global node numbers for element 4 are 4 and 5. These are also referred to as *local nodes* 1 and 2 for element number 4.

For each element, all variables within the element (in this case, channel slope and water depth) are interpolated as a linear combination of their values at each end of the element. If the variable of interest is ψ (see Fig. 16-2), then ψ at any point x_a within an element may be calculated by proportion as

$$
\begin{aligned}
\psi_a &= \psi_i + \frac{x_a - x_i}{x_j - x_i}(\psi_j - \psi_i) \\
&= \frac{x_j - x_a}{x_j - x_i}\psi_i + \frac{x_a - x_i}{x_j - x_i}\psi_j \\
&= N_i\psi_i + N_j\psi_j
\end{aligned}
\tag{16-1}
$$

in which N_i and N_j are called *shape functions*. Shape functions are also called *interpolation, trial, or basis functions.*

One-Dimensional Quadratic Shape Functions

We may want to use three-node, one-dimensional elements. In that case, we interpolate values with a quadratic shape function. Figure 16-3 shows a three-node element.

The value of ψ within this element is given by

$$
\psi = N_i\psi_i + N_j\psi_j + N_k\psi_k
$$

in which the shape functions for this element are (Segerlind 1976, 262)

$$N_i = \left[1 - \frac{2(x_a - x_i)}{x_k - x_i}\right]\left(1 - \frac{x_a - x_i}{x_k - x_i}\right)$$

$$N_j = \frac{4(x_a - x_i)}{x_k - x_i}\left(1 - \frac{x_a - x_i}{x_k - x_i}\right)$$

$$N_k = -\frac{x_a - x_i}{x_k - x_i}\left[1 - \frac{2(x_a - x_i)}{x_k - x_i}\right]$$

(16-2)

in which $x_k - x_j = x_j - x_i$.

Higher-order elements and shape functions (with more nodes) are also available for one-dimensional problems. See Segerlind (1976) for further examples.

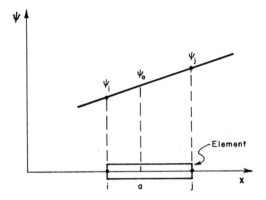

Figure 16-2 One-dimensional linear element.

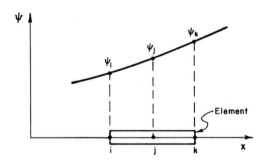

Figure 16-3 One-dimensional quadratic element.

Example 16-1

Calculate the value of the shape function for the one-dimensional quadratic element at nodes i, j, and k and the value of ψ at $x_a = 5$. The following values are given: $\psi_i = 2.5$ at $x_i = 2$; $\psi_j = 6.5$ at $x_j = 4$; and $\psi_k = 18.5$ at $x_k = 6$.

Solution

$$N_i = \left[1 - \frac{2(5-2)}{6-2}\right]\left(1 - \frac{5-2}{6-2}\right) = -0.125$$

$$N_j = \frac{4(5-2)}{6-2}\left(1 - \frac{5-2}{6-2}\right) = 0.75$$

$$N_k = -\frac{5-2}{6-2}\left[1 - \frac{2(5-2)}{6-2}\right] = 0.375$$

$$\psi_a = -0.125(2.5) + 0.75(6.5) + 0.375(18.5) = 11.5$$

Note that for any location x_a, the sum of the shape functions is 1.0. This also means that if $x_a = x_i$, then $N_i = 1$ and $N_j = N_k = 0$.

Two-dimensional Linear Shape Functions

Let us expand our set of elements to include two-dimensional elements. Like the one-dimensional problem, we may choose between linear, quadratic, or higher-order shape functions. Figure 16-4 shows the triangular and quadrilateral linear two-dimensional elements and their local node numbers.

For convenience, we will use local coordinates (ξ and η) for the quadrilateral element. A coordinate transformation allows us to convert from local coordinates to x_1, x_2 (global) coordinates. Both the local coordinates range from -1 to $+1$. Figure 16-5 illustrates local and global coordinate systems.

The shape functions are

$$N_1 = \frac{1}{4}(1 - \xi)(1 - \eta)$$

$$N_2 = \frac{1}{4}(1 + \xi)(1 - \eta)$$

$$N_3 = \frac{1}{4}(1 + \xi)(1 + \eta)$$

$$N_4 = \frac{1}{4}(1 - \xi)(1 + \eta)$$

$$(16\text{-}3)$$

The coordinate transformation is

$$x_1 = x_{1_o} + \frac{\partial x_1}{\partial \xi}\, d\xi + \frac{\partial x_1}{\partial \eta}\, d\eta \qquad (16\text{-}4)$$

$$x_2 = x_{2_o} + \frac{\partial x_2}{\partial \xi}\, d\xi + \frac{\partial x_2}{\partial \eta}\, d\eta \qquad (16\text{-}5)$$

in which x_{1_o} and x_{2_o} are the x_1 and x_2 coordinates of the origin of the ξ-η plane.

The coordinate transformation is discussed further in the section on Galerkin's method.

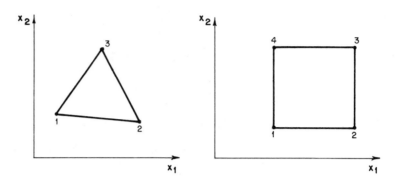

Figure 16-4 Two-dimensional linear elements.

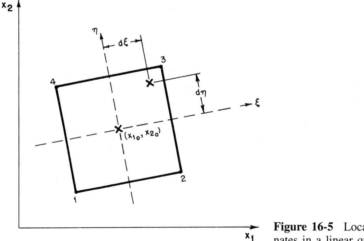

Figure 16-5 Local and global coordinates in a linear quadrilateral element.

Two-Dimensional Quadratic Shape Functions

Figure 16-6 shows some triangular and quadrilateral quadratic elements and their local node numbers. Both 8-node and 9-node quadrilaterals are available. Figure 16-7 shows how 8-node quadrilaterals may be combined into a grid.

The equations for these two-dimensional shape functions may be obtained from general finite-element texts, such as Segerlind (1976), or Zienkiewicz and Taylor (1989).

Order of interpolation. Higher-order shape functions are preferred if the variables are changing rapidly (Lee and Froehlich 1986, 6). It is possible to use linear elements if the element size is sufficiently small. However, it is usually more efficient to use higher-order elements.

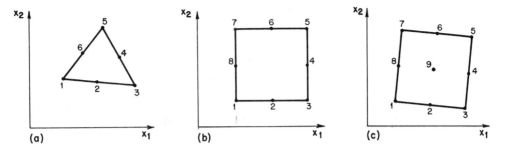

Figure 16-6 Two-dimensional quadratic elements.

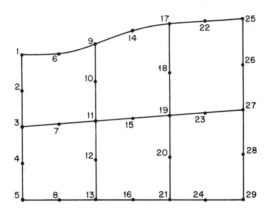

Figure 16-7 Finite-element grid.

It is sometimes necessary to use mixed interpolation. For example, to solve the Navier-Stokes equations, the velocity is interpolated with quadratic shape functions (8 or 9 nodes), whereas the pressure is interpolated within the same area using linear shape functions (at the 4 corner nodes). This is called the 8-4 or 9-4 element. The use of equal-order interpolation for velocity and pressure results in a singular (and unsolvable) set of equations. While there is disagreement about why it is needed (Lee and Froehlich 1986, 10; Chung 1978, 208) a mixed-order interpolation solves the problem.

Although both the 8-4 element and 9-4 element are used in two-dimensional flow problems, the 9-4 element is more popular. Some authors report that the pressure results of the 9-4 element are superior than that of the 8-4 element (Huyakorn et al. 1978). Others (Finnie and Jeppson 1991; Saez and Carbonell 1985) contend that either element may give erroneous pressure results and that this may be prevented by having a sufficient number of grid points.

Note that the shape functions may also be used to calculate derivatives; i.e., if $\psi = N_1\psi_1 + N_2\psi_2 + N_3\psi_3$, then

$$\frac{\partial \psi}{\partial x_i} = \frac{\partial N_1}{\partial x_i}\psi_1 + \frac{\partial N_2}{\partial x_i}\psi_2 + \frac{\partial N_3}{\partial x_i}\psi_3 \tag{16-6}$$

This makes sense, since the N are functions of x_i and the ψ_i are constants but are unknown. (Only one value of ψ satisfies the governing equation.)

Convergence

For the finite-element method to converge to the correct solution, the following requirements must be satisfied (Lee and Froehlich 1986, 6):

1. As the size of the element decreases to zero, the nodal values of ψ become identical, and the shape functions give constant values throughout the element. This is called the *completeness requirement*.
2. If the governing equation is of order n, then the variable and its derivative must be continuous across the boundary to the order of at least $n - 1$. This is the *compatibility requirement*.

The second requirement addresses the value of variables and their derivatives at the finite-element boundary (not within each element). All the shape functions introduced so far guarantee only that variables themselves are equal on adjacent boundaries (order of $n = 0$ on the boundary, which is also called C_o continuity). They do not guarantee that derivatives of variables are equal on the boundaries of adjacent elements ($n = 1$, or C_1 continuity). This does not present an insurmountable problem, since we will see later that our governing equation (of order $n = 2$) can be reduced to include only first derivatives ($n = 1$) by integration by parts. After integration by parts, the governing equations are of order $n = 1$, so our shape functions (with $n = 0$ on the boundary) will be adequate. See also Segerlind (1976, 79) for more information about element continuity.

16-3 Transformation of Differential Equation into Integral Equation

The first step in solving a differential equation is to convert it into an integral equation. For this purpose, the following three finite-element approaches are available to convert the governing differential equation into integral equations: direct, variational, and weighted residual methods. Each results in an identical integral equation for linear partial differential equations.

The weighted residual methods are general methods that can be applied in cases where direct and variational methods do not work. We will discuss this method in the following paragraphs.

Method of Weighted Residuals

The method of weighted residuals may be classified into the following three approaches: Galerkin, collocation, and least squares method. Galerkin method is the most widely used and is described here. This method is used on differential equations of the general form

$$Lu - f(x_j) = 0 \qquad (16\text{-}7)$$

in which L is a differential operator and u is the dependent variable. The method works by forcing the error of the approximation to zero. If the approximate solution is u_{approx} then the error, E', of approximation is defined as

$$E' = Lu_{approx} - f(x_j) \qquad (16\text{-}8)$$

The error is forced to zero by making it orthogonal to the set of r linearly independent weighting functions, N_r. These weighting functions should have the property that they span the solution space. This property can be interpreted to mean that every location in the solution space (domain) is reachable as a linear combination of the weighting functions and that the weighting functions are mutually orthogonal (perpendicular and independent). Since the weighting functions span the space, the only function that is orthogonal to all of the weighting functions is zero. In order to make the error of approximation zero, an inner product is formed between the error and all the weighting functions, which is then set equal to zero. That is

$$(N_r, E') = 0 \qquad (16\text{-}9)$$

in which the parentheses indicate inner product, or

$$\int_R N_r [Lu_{approx} - f(x_j)] \, dR = 0 \qquad (16\text{-}10)$$

It might be useful at this point to recall that by definition, an inner product of two functions is equal to the product of their magnitudes and the cosine of the angle θ between them, i.e.,

$$(N_r, E') = |N_r||E'| \cos \theta \qquad (16\text{-}11)$$

If N_r and E' are nonzero, then for the inner product to be zero, $\cos \theta$ must be zero. In other words N_r is orthogonal to E'.

Although the Galerkin finite-element method uses the same functions for both the weighting and shape function, it is not a necessity.

The next example shows how Galerkin method can be used to solve differential equations.

Example 16-2

Solve the following trivial ordinary differential equation from $x = 0$ to $x = 3$ with the finite elements shown in Fig. 16-8:

$$\frac{dy}{dx} = 3$$

for the initial condition that at $x = 0$, $y = -5$.

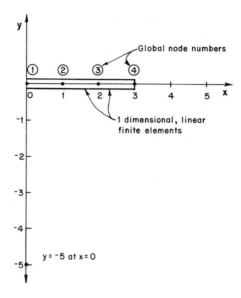

Figure 16-8 Linear finite elements.

Solution: In this case

$$Lu - f(x_i) = 0 = y' - 3$$

Galerkin method applied to the governing equation gives

$$\int_R N_r (y' - 3) \, dR = 0$$

Based on the shape functions (Eq. 16-1), $N_1 = (x_j - x)/(x_j - x_i)$ and $N_2 = (x - x_i)/(x_j - x_i)$, the inner product becomes

$$\int \frac{x_j - x}{x_j - x_i} (y' - 3) \, dx + \int \frac{x - x_i}{x_j - x_i} (y' - 3) \, dx = 0$$

Recall from Eq. 16-6 that within each element,

$$\frac{d\psi}{dx} = \frac{dN_1}{dx} \psi_1 + \frac{dN_2}{dx} \psi_2$$

Then,

$$\frac{dN_1}{dx} = \frac{-1}{x_j - x_i} = -\frac{1}{1} = -1 \quad \text{and} \quad \frac{dN_2}{dx} = \frac{1}{x_j - x_i} = \frac{1}{1} = 1$$

Substitution of these into the two integrals yields

$$\int \frac{x_j - x}{x_j - x_i} (-y_i + y_j - 3) \, dx + \int \frac{x - x_i}{x_j - x_i} (-y_i + y_j - 3) \, dx = 0$$

These must be applied to the three finite elements. Recall that $x_1 = 0$, $x_2 = 1$, $x_3 = 2$, and $x_4 = 3$. Then we have the following equations:

Element I

$$\int \frac{1-x}{1-0}(-y_1 + y_2 - 3)\, dx + \int \frac{x-0}{1-0}(-y_1 + y_2 - 3)\, dx = 0$$

Element II

$$\int \frac{2-x}{2-1}(-y_2 + y_3 - 3)\, dx + \int \frac{x-1}{2-1}(-y_2 + y_3 - 3)\, dx = 0$$

Element III

$$\int \frac{3-x}{3-2}(-y_3 + y_4 - 3)\, dx + \int \frac{x-2}{3-2}(-y_3 + y_4 - 3)\, dx = 0$$

These three equations become

$$(-y_1 + y_2 - 3)\left[\int_0^1 (1-x)\, dx + \int_0^1 x\, dx\right] = 0$$

$$(-y_2 + y_3 - 3)\left[\int_1^2 (2-x)\, dx + \int_1^2 (x-1)\, dx\right] = 0$$

$$(-y_3 + y_4 - 3)\left[\int_2^3 (3-x)\, dx + \int_2^3 (x-2)\, dx\right] = 0$$

Upon simplification, these equations become

$$-y_1 + y_2 - 3 = 0$$

$$-y_2 + y_3 - 3 = 0$$

$$-y_3 + y_4 - 3 = 0$$

Now, according to the initial condition, $x_1 = 0$ at $y_1 = -5$. Hence it follows from the first equation that $y_2 = 3 - 5 = -2$. Substitution of this into the second equation yields $y_3 = 3 - 2 = 1$. Then, from the third equation, $y_4 = 4$.

Since the analytical solution is $y = 3x - 5$, Galerkin method gives the exact solution for the problem.

For some numerical solutions of differential equations, the spacing between nodes can influence the accuracy of the result. This happens if the solution changes so rapidly in time or space that the shape functions cannot adequately interpolate the value.

Any time a numerical method is used, care must be taken so that changes in the grid or node spacing do not change the results. This is called *grid independence*. This is a required condition for the computed results to be reliable.

Galerkin method may be applied in the same manner to more complicated problems having more variables and higher derivatives. However, as the level of difficulty increases, additional complications occur. These complications include the presence of second and higher-order derivatives, nonlinear terms, and the need for numerical integration.

Let us apply the Galerkin method to the Navier-Stokes equations to illustrate these procedures.

Application of Galerkin Method to the Navier-Stokes Equations

The Navier-Stokes equations, when combined with the equation of continuity, describe fluid flow at any Reynolds number. In the following unsteady Navier-Stokes equations, we use the Einstein summation, wherein subscripts specify repeating of the variable i and j, usually taking a value of 1 to 2 for two-dimensional problems:

$$\frac{\partial U_i}{\partial t} + U_j \frac{\partial U_i}{\partial x_j} = \frac{-1}{\rho} \frac{\partial P^*}{\partial x_i} + \frac{\partial}{\partial x_j}\left[v\left(\frac{\partial U_i}{\partial x_j} + \frac{\partial U_j}{\partial x_i} \right) \right] \tag{16-12}$$

in which U_i are the components of velocity; P^* is the sum of pressure P, a surface force, and γh, the gravity body force; ρ is the mass density of the fluid; v is the kinematic viscosity; and γ is the specific weight of fluid.

The two-dimensional continuity equation for incompressible fluids is

$$\frac{\partial U_1}{\partial x_1} + \frac{\partial U_2}{\partial x_2} = 0 \tag{16-13}$$

Application of Galerkin method to these equations results in the following integral equations. The quadratic weighting function is used for the inner product involving the steady Navier-Stokes equations. In the case illustrated here, the 8-node quadratic shape function is used to interpolate u, h, and v. The 4-node linear weighting function (M_r) is used for the inner product involving the continuity equation, all other weighting functions use the 8-node quadrilateral (N_r).

Steady Navier-Stokes equations.   For i = 1:

$$\iint N_r \left[U_1 \frac{\partial U_1}{\partial x_1} + U_2 \frac{\partial U_1}{\partial x_2} + \frac{1}{\rho} \frac{\partial}{\partial x_1}(P + \gamma h) \right.$$
$$\left. - \frac{\partial}{\partial x_1}\left(2v \frac{\partial U_1}{\partial x_1} \right) - \frac{\partial}{\partial x_2}\left(v\left(\frac{\partial U_1}{\partial x_2} + \frac{\partial U_2}{\partial x_1} \right) \right) \right] dx_1 \, dx_2 = 0, \qquad r = 1 \text{ to } 8 \tag{16-14}$$

For $i = 2$:

$$\iint N_r \left[U_1 \frac{\partial U_2}{\partial x_1} + U_2 \frac{\partial U_2}{\partial x_2} + \frac{1}{\rho} \frac{\partial}{\partial x_2}(P + \gamma h) \right.$$
$$\left. - \frac{\partial}{\partial x_1}\left(v\left(\frac{\partial U_2}{\partial x_1} + \frac{\partial U_1}{\partial x_2} \right) \right) - \frac{\partial}{\partial x_2}\left(2v \frac{\partial U_2}{\partial x_2} \right) \right] dx_1 \, dx_2 = 0, \qquad r = 1 \text{ to } 8 \tag{16-15}$$

Continuity Equation

$$\iint M_r \left[\frac{\partial U_1}{\partial x_1} + \frac{\partial U_2}{\partial x_2} \right] dx_1 \, dx_2 = 0, \qquad r = 1 \text{ to } 4 \tag{16-16}$$

Reduction of Derivatives to First Order

The equations that result from the Galerkin process may contain derivatives higher than first order. Since the shape functions are continuous at the element boundaries in the variable but not in its derivatives, any second- and higher-order derivatives must be manipulated to first order. For example, the part of the Navier-Stokes equations that includes second-order derivatives is as follows.

For $i = 1$:

$$\int \int N_r \left\{ \frac{\partial}{\partial x_1} \left(2v \frac{\partial U_1}{\partial x_1} \right) + \frac{\partial}{\partial x_2} \left[v \left(\frac{\partial U_1}{\partial x_2} + \frac{\partial U_2}{\partial x_1} \right) \right] \right\} dx_1 \, dx_2 \tag{16-17}$$

$$r = 1 \text{ to } 8$$

Recall that the Green-Gauss theorem (or integration by parts) is given as

$$\int u \, dv = uv - \int v \, du \tag{16-18}$$

It can be applied to Eq. 16-17.

For i = 1:

$$\int N_r \left(2v \frac{\partial U_1}{\partial x_1} \right) dx_2 + \int N_r \left[v \left(\frac{\partial U_1}{\partial x_2} + \frac{\partial U_2}{\partial x_1} \right) \right] dx_1$$

$$- \int \int \frac{\partial N_r}{\partial x_1} \left(2v \frac{\partial U_1}{\partial x_1} \right) dx_1 \, dx_2 - \int \int \frac{\partial N_r}{\partial x_2} \left[v \left(\frac{\partial U_1}{\partial x_2} + \frac{\partial U_2}{\partial x_1} \right) \right] dx_1 \, dx_2, \qquad r = 1 \text{ to } 8 \tag{16-19}$$

The Green-Gauss theorem has been applied to the two terms of Eq. 16-17 to reduce their order. This results in naturally occurring gradient boundary conditions (the single integral). This operation is also done on the pressure and gravity terms, which also contribute necessary boundary conditions.

The following integral equations result from the Galerkin method for the steady Navier-Stokes equations (for $i = 1$) and the continuity equation.

Steady Navier-Stokes Equation ($i = 1$)

$$\int \int \left\{ N_r \left[U_1 \frac{\partial U_1}{\partial x_1} + U_2 \frac{\partial U_1}{\partial x_2} \right] - \frac{\partial N_r}{\partial x_1} \frac{1}{\rho} (P + \gamma h) \right.$$

$$\left. + \frac{\partial N_r}{\partial x_1} \left(2v \frac{\partial U_1}{\partial x_1} \right) + \frac{\partial N_r}{\partial x_2} \left[v \left(\frac{\partial U_1}{\partial x_2} + \frac{\partial U_2}{\partial x_1} \right) \right] \right\} dx_1 \, dx_2 \tag{16-20}$$

$$= \int N_r \left[\frac{-1}{\rho} (P + \gamma h) + 2v \frac{\partial U_1}{\partial x_1} \right] dx_2 + \int N_r \left[v \left(\frac{\partial U_1}{\partial x_2} + \frac{\partial U_2}{\partial x_1} \right) \right] dx_1, \qquad r = 1 \text{ to } 8$$

in which h is height of the point in question above the datum.

Continuity Equation

$$\int \int M_r \left(\frac{\partial U_1}{\partial x_1} + \frac{\partial U_2}{\partial x_2} \right) dx_1 \, dx_2 = 0, \qquad r = 1 \text{ to } 4 \tag{16-21}$$

The next step is to substitute expressions incorporating the shape functions for the variables. For example,

$$u_1 = N_1 U_1(1) + \cdots + N_8 U_1(8) \tag{16-22}$$

in which $U_1(s)$ is the value of the variable at local node s. The derivatives are approximated as

$$\frac{\partial U_1}{\partial x_j} = \frac{\partial N_1}{\partial x_j} U_1(1) + \cdots + \frac{\partial N_8}{\partial x_j} U_1(8) \tag{16-23}$$

All the variables and nonconstant knowns in these equations can be approximated in terms of shape functions and derivatives of shape functions. Upon substitution, a system of nonlinear equations is obtained with the velocity components and pressure at the nodes as the unknowns. The boundary conditions become the known or force vector. These equations are assembled for all the elements. This requires that global node numbers are used. Figure 16-6(b) shows the local node numbers for an eight-node, two-dimensional element. Figure 16-7 shows the global node numbers for a small problem.

Numerical Integration

Since quadratic shape functions use local coordinates, Gauss quadrature (numerical integration) and a coordinate transformation must be used. Gauss quadrature calculates an integral between the limits of $x = -1$ to $x = 1$ by evaluating the function $f(x)$ at intermediate values of x. These values are then multiplied by the appropriate weights and summed. Additional points can be sampled if acceptable accuracy is needed. For solutions of the Navier-Stokes equations, three-point quadrature is sufficient.

For three-point Gauss quadrature, the sampling points are at $x = -0.774597, 0,$ 0.774597, with the following weights, respectively: 5/9, 8/9, and 5/9.

When three-point Gauss quadrature is applied to a two-dimensional problem (Fig. 16-9) the function is evaluated at nine points, and the function value at each of the nine locations is multiplied by the product of two weights. Figure 16-9 also shows a local coordinate system with its ordinates ξ and η. Note that the eight velocity nodes are numbered in order, so that the four pressure nodes are local node numbers 1, 3, 5, and 7. See Carnahan, Luther, and Wilkes (1969) or other numerical analysis texts for further information on Gauss quadrature.

Coordinate Transformations

When local coordinates are used in an integration, a correction for the coordinate transformation must be made during the integration. In general, for two-dimensional

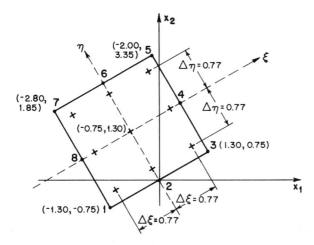

Figure 16-9 Local and global coordinates and Gauss integration points.

integration using local coordinates, we may write

$$\int \int f(x_1, x_2)\, dx_1\, dx_2 = \int \int f(x_1, x_2)|\det J|\, d\xi\, d\eta \qquad (16\text{-}24)$$

in which J is the Jacobian of the coordinate transformation (Segerlind 1976, 273)

$$J = \begin{pmatrix} \dfrac{\partial x_1}{\partial \xi} & \dfrac{\partial x_2}{\partial \xi} \\[2ex] \dfrac{\partial x_1}{\partial \eta} & \dfrac{\partial x_2}{\partial \xi} \end{pmatrix} \qquad (16\text{-}25)$$

and $|\det J|$ means the absolute value of the determinant of the matrix J. In order to complete the integration, $f(x_1, x_2)$ must be transformed into $f(\xi, \eta)$. The shape functions N_r are already in the local coordinates, but the partial derivatives of the shape function must still be transformed. This is accomplished with the following relationship:

$$\begin{pmatrix} \dfrac{\partial N}{\partial x_1} \\[2ex] \dfrac{\partial N}{\partial x_2} \end{pmatrix} = [J]^{-1} \begin{pmatrix} \dfrac{\partial N}{\partial \xi} \\[2ex] \dfrac{\partial N}{\partial \eta} \end{pmatrix} \qquad (16\text{-}26)$$

where $[J]^{-1}$ is the inverse of J.

Application to the Continuity Equation

The following procedure shows the numerical integration of the two-dimensional continuity equation.

$$\int \int \mathbf{M} \left\{ \frac{\partial u_i}{\partial x_i} \right\} dx_1\, dx_2 = \int \int \mathbf{MB}u \Big| \det J \Big| W(\xi, \eta)\, d\xi_i\, d\eta \qquad (16\text{-}27)$$

in which $W(\xi, \eta)$ is the Gauss weight; $\mathbf{B}$ is the matrix of derivatives of the quadratic shape function; and $\mathbf{M}$ is the vector of the linear shape functions:

$$\mathbf{M} = \frac{1}{4}\left[(1-\xi)(1-\eta) \quad (1+\xi)(1-\eta) \quad (1+\xi)(1+\eta) \quad (1-\xi)(1+\eta) \right]$$

$$\mathbf{B} = \left[\frac{\partial N_1}{\partial x_1} \cdots \frac{\partial N_8}{\partial x_1} \quad \frac{\partial N_1}{\partial x_2} \cdots \frac{\partial N_8}{\partial x_2} \right]$$

$$\mathbf{u} = \{ U_1(1), \ldots U_1(8) \quad U_2(1), \ldots U_2(8) \}^T$$

$$\left\{ \frac{\partial u_i}{\partial x_i} \right\} = \frac{\partial U_1}{\partial x_1} + \frac{\partial U_2}{\partial x_2}$$

The superscript T denotes transpose.

$W(\xi, \eta)$ is assigned according to the value of the local coordinates. All these variables are in terms of the spatial coordinates and the local coordinates. Since three-point integration is used, nine locations within each element are evaluated for each of the preceding terms and are multiplied by the product of their respective weights. The sum of these operations is the value of the integral for the element.

Boundary Conditions

Due to the elliptic nature of the equations, boundary conditions must be specified for all variables. The two typical boundary conditions are the specification of the variable and its gradients. Known boundary conditions are applied where the value of the variable is known a priori. Such a value would correspond to a place on the boundary where the velocity is known from experiment or because there is no flow across the boundary. The gradient boundary conditions for the Navier-Stokes equations are

$$\text{Normal}: \quad -\frac{P + \gamma h}{\rho} + 2v\frac{\partial U_n}{\partial x_n}$$

$$\text{Traction}: \quad v\left(\frac{\partial U_n}{\partial x_t} + \frac{\partial U_t}{\partial x_n} \right)$$

(16-28)

in which n indicates normal to the boundary, t indicates tangent to the boundary, and v is the kinematic viscosity.

These gradients represent the shear stress at the boundary. They arise in the integrals that result from integration by parts (Eq. 16-19).

The inlet velocities are either known or a pressure gradient is specified by the normal gradient using Eq. 16-28. The wall boundary conditions use either a known velocity (usually zero) or a zero-traction gradient, which means zero stress at the boundary.

The boundary conditions for the longitudinal velocity at a free surface can be zero-traction gradient. This condition is only approximately true for turbulent flow, since a free surface produces an inflection in the horizontal velocity profile. The normal velocity at the surface is zero.

While solving the Navier-Stokes equations, do not under- or overspecify the known velocity boundary conditions. It is not a good practice to specify known velocity at the outflow boundaries. This can lead to impossible demands due to violations of the continuity equation by setting inflow $\neq$ outflow. The better alternative is to set zero-valued normal velocity gradients along the outflow boundary by using Eq. 16-28. This assumes that the flow has sufficient distance to become well developed at the outlet, so be sure that the domain is large enough in the direction of flow.

Problems can arise when specifying boundary conditions where surfaces meet. For example, make sure that the inlet velocity of an open channel at the free surface is compatible with zero-stress boundary conditions along the free surface.

So far we have not discussed the boundary conditions for pressure. In setting up the matrix equations for an element, the row representing the continuity equation is adjacent to the pressure in the unknowns. As a result, setting the pressure at a node deletes the continuity equation at that node. Many researchers have indicated that specifying the pressures, especially at inlets and outlets, leads to poor convergence and sometimes strange results (Chung 1978). Jackson (1984) points out that the effect of deleting continuity at a point is to collect all round-off errors at that point. Pressure anomalies or discontinuities should be expected at that location. For these reasons the pressure is usually set to zero at one wall node where all velocities are zero or in the interior of the flow (Gresho, Lee, and Sani 1980; Schamber and Larock 1981). Any other boundary conditions for pressure are included in "normal" stress conditions.

Solution of Equations

Since the equations are nonlinear, they are solved by an iterative method. One such method is the Newton-Raphson method, presented in Chapter 6. The Newton-Raphson method has the advantage of rapid convergence, but it requires initial estimates of variables that are "close" to the root. If they are not close enough, the calculation procedure may diverge.

The Jacobian matrix $\mathbf{D}$ obtained during the solution of the Navier-Stokes equations is asymmetric due to the nonlinear convection terms, is nonpositive definite, (since the continuity equation does not contain pressure as a variable), is not diagonally dominant, and may contain zeros on the diagonal (Olson 1977). It is nonsingular and solvable if proper boundary conditions and starting values are specified.

Relaxation Factors

In certain cases, Newton's method could oscillate around a solution. To prevent this and to improve the convergence rate, a relaxation factor λ may be used. This factor is used to reduce the calculated corrections. The equation using the relaxation factor is

$$\begin{bmatrix} x_1 \\ x_2 \end{bmatrix}^{n+1} = \begin{bmatrix} x_1 \\ x_2 \end{bmatrix}^{n} + \lambda [D]^{-1} \begin{bmatrix} f_1 \\ f_2 \end{bmatrix}^{n} \qquad (16\text{-}29)$$

in which the superscript represents the number of iterations.

Experience with the Navier-Stokes equations indicates that λ should be less than 0.5.

16-4 Solution of Navier-Stokes Equations

In this section we will describe how to solve two modifications of the Navier-Stokes equations. The two classes of problems presented are turbulent flow and shallow-water (or depth-averaged) flow.

Turbulent Flows

The most widely used method to calculate turbulent flows is the two equation (k-ϵ) turbulence model.

The k-ϵ turbulence model. This model introduces two transport equations for k and ϵ. These equations are similar in form to the Navier-Stokes equations and are solved in a similar manner. This model requires a modification of the Navier-Stokes equations that introduces a turbulent viscosity term. The new equations are called the *Reynolds equations of motion.*

The steady Reynolds equations of motion are shown here using the Einstein summation, wherein subscripts specify repeating of the variable i and j, taking values of 1 to 2 for two-dimensional flows.

$$\overline{U}_j \frac{\partial \overline{U}_i}{\partial x_j} = \frac{-1}{\rho} \frac{\partial \overline{P}}{\partial x_i} - g \frac{\partial h}{\partial x_i} + \frac{\partial}{\partial x_j} \left[\nu \left(\frac{\partial \overline{U}_i}{\partial x_j} + \frac{\partial \overline{U}_j}{\partial x_i} \right) - \overline{u'_i u'_j} \right] \quad (16\text{-}30)$$

The components of mean velocities are the $\overline{U}_i$ variables, and $\overline{P}$ is the mean pressure. These equations use a variable substitution introduced by Reynolds. He separated velocity and pressure into their mean and fluctuating parts:

$$\begin{aligned} U_i &= \overline{U}_i + u'_i \\ P &= \overline{P} + p' \end{aligned} \quad (16\text{-}31)$$

in which the overbar denotes time averaging and the prime indicates the fluctuating part. The last term in the Reynolds equations of motion are Reynolds stresses. These are modeled using the Bousinesq approximation:

$$-\overline{u'_i u'_j} = \nu_T \left(\frac{\partial \overline{U}_i}{\partial x_j} + \frac{\partial \overline{U}_j}{\partial x_i} \right) - \frac{2}{3} k \delta_{ij} \quad (16\text{-}32)$$

in which ν_T is the turbulent viscosity, δ_{ij} is Kroneker's delta, and k is turbulent kinetic energy per unit mass of the fluctuating velocities:

$$k = \frac{1}{2} \overline{u'_i u'_i} \quad (16\text{-}33)$$

The equation of continuity for incompressible fluids is

$$\frac{\partial \overline{U}_i}{\partial x_i} = 0 \quad (16\text{-}34)$$

The steady transport equations for k and ϵ are

$$\overline{U}_i \frac{\partial k}{\partial x_i} = \frac{\partial}{\partial x_i}\left(\frac{\nu_T}{\sigma_k}\frac{\partial k}{\partial x_i}\right) + P_k - \epsilon \tag{16-35}$$

$$\overline{U}_i \frac{\partial \epsilon}{\partial x_i} = \frac{\partial}{\partial x_i}\left(\frac{\nu_T}{\sigma_\epsilon}\frac{\partial \epsilon}{\partial x_i}\right) + C_{\epsilon 1}\frac{\epsilon}{k}P_k - C_{\epsilon 2}\frac{\epsilon^2}{k} \tag{16-36}$$

in which the σ's represent the turbulent Prandtl numbers for diffusion of k and ϵ and P_k is production of kinetic energy by shear, given by

$$P_k = \nu_T \left(\overline{\frac{\partial U_i}{\partial x_j}} + \overline{\frac{\partial U_j}{\partial x_j}}\right)\overline{\frac{\partial U_i}{\partial x_j}} \tag{16-37}$$

The terms to the left of the equal sign in the turbulence equations are the convection terms. The next term is the diffusion term. The remaining terms are source and sink terms. Once k and ϵ are determined, they are used to calculate a new turbulent viscosity at each node using the relationship

$$\nu_T = \frac{C^* k^2}{\epsilon} \tag{16-38}$$

The turbulence model has introduced five constants. Their values originally proposed by Launder and Spalding (1972) are given in Table 16-1. These additional two equations (and two unknowns) significantly increase the computation time.

Table 16-1 Values of constants in the $k - \epsilon$ model

C^*	$C_{\epsilon 1}$	$C_{\epsilon 2}$	σ_k	σ_ϵ
0.09	1.44	1.9	1.0	1.3

Two excellent introductions to the k-ϵ turbulence models are Nallasamy (1985), who provides a thorough literature review of recent research, and Rodi (1980), who presents the theory and finite difference applications of the model.

Recent researchers who have applied the finite element method to turbulent flow include Schamber and Larock (1981), who used the k-ϵ turbulence model to flow in a sedimentation basin. Devantier and Larock (1986) added sediment transport and variable water density to a similar flow. Taylor, Thomas, and Morgan (1981) and Autret, Grandotto, and Dekeyser (1987) modeled flow over a backward step. Sharma and Carey (1986) studied heat transfer in a boundary layer. They employed a variable transformation that helped convergence but required an extremely fine grid. Kim and Schetz (1989) analyzed propeller-driven flow over an air foil.

Boundary conditions.   For all variables in turbulent flow, the boundary conditions are chosen to place the calculations within the turbulent flow region. See Finnie and Jeppson (1991) for a discussion of turbulent variable boundary conditions.

Flow under a sluice gate.   Figure 16-10(a) presents the finite-element grid for a complicated turbulent flow problem ($\mathbf{R}_e = 179,000$). This problem represents flow under a sluice gate (Finnie and Jeppson 1991). The computed and measured velocities and pressures on the sluice gate are compared in Fig. 16-10(b) and (c). These results indicate that the combination of finite-element method and two-equation turbulence model yields reasonable results. Because of the large number of unknowns (1851) this

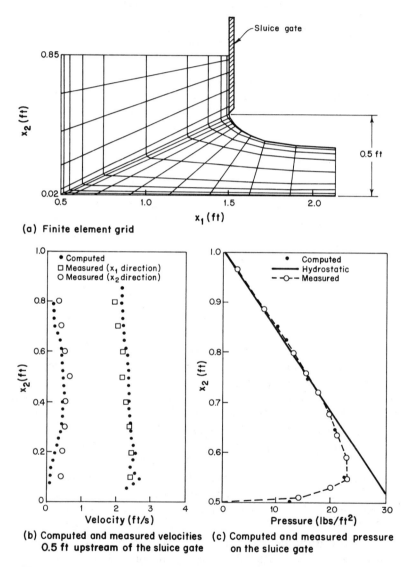

Figure 16-10   Flow under a sluice gate (After Finnie and Jeppson, 1991).

problem was calculated on a supercomputer. The calculation took approximately 1/2 h of computing time.

Methods to promote convergence.   To obtain convergence for the solution of the turbulence equations can be a very difficult problem. If the finite element grid is too coarse, the variables oscillate during the calculations, which may lead to negative values. Subsequent iterations spread the negative values until the solution diverges. Previous finite-element turbulence modelers resorted to adding artificial viscosity or adding false diffusion (Polansky, Lamb, and Crawford 1984). This is done by artificially increasing the viscosity in the diffusion terms (second-order derivatives) or by employing unequally weighted shape functions (called *upwinding*) (Benim 1990). Both of these procedures reduce the accuracy of the results.

Devantier and Larock (1986) reported similar convergence problems and solved them by removing some of the nonlinear parts of the k and ϵ equations and adding them back in as the calculations progress.

For the sluice-gate problem, the following procedures were employed to obtain convergence for the Newton-Raphson method:

1. Relaxation was applied to the correction.
2. Artificial diffusion was added by multiplying the second derivative terms by a constant. This constant was slowly divided out.
3. The following variable substitution was employed in the k-ϵ equations: $k = q^2$ and $\epsilon = r^2$. This substitution was made and the equations for q and r were solved.

The q-r model used in this problem is a simplification of Smith's (1984a, b) q-f turbulence model. Actually, q-r is just a variable substitution and is not a new turbulence model. As such, it does not require additional proof of its model assumptions (which are the same as the k-ϵ model) and needs no additional calibration of the five constants given in Table 16-1. The reason for using the q-r variable substitution is that it is a simple way to avoid or reduce the need for upwinding or related techniques.

The sluice-gate problem is one of the few turbulence solutions where the permanent addition of artificial diffusion is not used in the Reynolds equations of motion or the turbulence equations.

Additional research is needed to improve convergence of finite element methods to minimize or eliminate the use of artificial viscosity.

Shallow-Water Equations

The solution of shallow-water equations by the finite-difference method was discussed in Chapters 14 and 15. In this section, we present some results of finite-element solutions of these equations. Galerkin's method can be applied to the shallow-water equations in the same manner as it was applied to the Navier-Stokes equations. The most significant difference is that a method to integrate the time dependent term must now be developed. Lee and Froehlich (1986) present a good discussion of the different available methods.

Integration of the time-dependent term.   Either explicit or implicit finite-difference methods can be applied to the time derivatives. The rest of the terms are evaluated using finite elements. These different time-stepping methods include Euler, the trapezoidal rule, Adams-Moulton predictor-correctors, Runge-Kutta, and split-step schemes.

These schemes differ in the evaluation of the time derivatives (time $i - 1, i$, or $i + 1$) and how they combine time derivatives at the current and future time steps.

The finite-element method can also be used for the time derivatives. Its disadvantage is that more work is required than the finite differences approach. Some authors perform a "lumping" operation on the equations, which reduces accuracy but improves the convergence of the equations.

The first example is taken from Thompson (1990), who solved the shallow-water equations to analyze two-dimensional steady flow in a hydraulic jump. Typical results and the grid of nine node elements are shown in Fig. 16-11. Satisfactory agreement was found between measured results and the calculations, but the calculated front of the jump was not as steep as measured. A hydraulic jump provides a demanding test of a solution method due to the rapid rise in the surface. Artifical viscosity must be added for the solution to converge. It is believed that less artificial viscosity results in a steeper front and a better prediction. The calculations began as a subcritical open channel flow. The upstream boundary conditions were changed over time until the hydraulic jump resulted. So while the result is steady, it was developed with the unsteady equations.

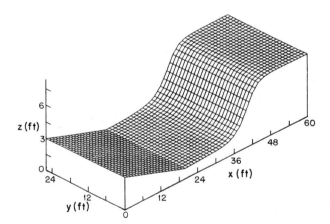

Figure 16-11 Hydraulic jump in a rectangular channel (After Thompson, 1990).

Katopodes (1984) also solved flow in a hydraulic jump and introduced an artificial diffusion scheme that can be selectively applied.

The second example is taken from Dhatt et al. (1986), who simulated flow in the St. Lawrence River. They compared results using a number of different triangular elements and obtained acceptable agreement between a 4-3 (four velocity nodes, three pressure nodes) and a 6-3 element. The grid is presented in Fig. 16-12(a), and the steady results are presented in Fig. 16-12(b).

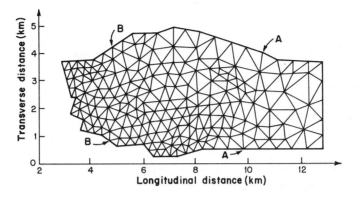

(a) Finite element grid (374 elements)

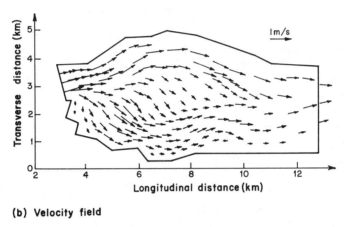

(b) Velocity field

Figure 16-12   Steady flow in St. Lawrence River (After Dhatt et al. 1986).

These two examples show the type of shallow-water flow problems that can be solved using the finite-element method. Further development of the finite-element method is needed to improve accuracy and efficiency of the calculations.

16-5 Summary

An introduction to the finite-element method was presented in this chapter. Procedures for the discretization of the solution domain were described and one- and two-dimensional shape functions were discussed. Transformation of the governing differential equations was outlined. The application of Galerkin method to solve the Navier-Stokes equation was discussed. A brief description of the k-ϵ turbulence model was presented. Additional work remains to be done to make this method an easy-to-use and computationally efficient design tool.

Problems

16-1. Calculate the shape function at nodes i, j, k and the value of ψ_a at $x_a = 1.0$ for the following one-dimensional quadratic element: For $x_i = 0$, $\psi_i = -3$; for $x_j = 1.5, \psi_j = -1$; and for $x_k = 3, \psi_k = 2.5$.

16-2. For one-dimensional quadratic element show that if $x_a = x_j$, then $N_j = 1.0$ and $N_i = N_k = 0.0$.

16-3. Apply Galkerkin's method to solve

$$\frac{dy}{dx} = x + 1$$

from $x = 1$ to $x = 2.5$, with the initial conditions $y = 1.5$ at $x = 1$. Use three linear, one-dimensional shape functions.

REFERENCES

Autret, A.; Grandotto, M.; and Dekeyser, I.   1987. Finite element computation of a turbulent flow over a backward-facing step, *International Jour. for Numerical Methods in Fluids,* 7: 89–102.

Benim, A. C.   1990. Finite element analysis of confined turbulent swirling flows, *International Jour. for Numerical Methods in Fluids,* 11, no. 6: 697–718.

Carnahan, B.; Luther, H. A.; and Wilkes, J. O.   1969. *Applied numerical methods,* John Wiley, New York, NY.

Chapra, S. C., and Canale, R. P.   1988. *Numerical methods for engineers,* 2nd ed. McGraw-Hill, New York, NY.

Chung, T. J.   1978. *Finite element analysis in fluid dynamics,* McGraw-Hill, New York, NY.

Devantier, B. A., and Larock, B. E.   1986. Modeling a recirculating density-driven turbulent flow. *International Jour. for Numerical Methods in Fluids.* 6: 241–53.

Dhatt, G.; Soulaimani, A.; Ouellet, Y.; and Fortin, M.   1986. Development of new triangular elements for free surface flows, *International Jour. for Numerical Methods in Fluids.* 6: 895–911.

Finnie, J. I., and Jeppson, R. W.   1991. Solving turbulent flows using finite elements, *Jour. Hydraulic Engineering,* Amer. Soc. Civil Engrs., 117, no 11: 1513–30.

Finnie, J. I.   1987. An application of the finite element method and two equation $(k\text{-}\epsilon)$ turbulence model to two- and three-dimensional fluid flow problems governed by the Navier-Stokes equations, *Ph. D. thesis,* Utah State University, Logan, Utah.

Gresho, P. M., and Lee, R. L.   1981. Don't suppress the wiggles -they're telling you something, *Computers and Fluids,* 9:223–253.

Gresho, P. M., Lee, R. L., and Sani R. L.   1980. On the time dependent solution of the incompressible Navier-Stokes equations in two and three dimensions, In *Recent Advances in Numerical Methods in Fluids,* edited by C. Taylor and K. Morgan, vol 1. Pineridge Press Limited, Swansea, United Kingdom.

Huyakorn, P. S., Taylor, C., Lee, R. L., and Gresho, P. M.   1978. A comparison of various mixed-interpolation finite elements in the velocity-pressure formulation of the Navier-Stokes equations, *Comput. Fluids,* 6:25.

Jackson, C. P. 1984. The effect of the choice of the reference pressure location in numerical modeling of incompressible flow, *International Journal for Numerical Methods in Fluids,* vol 4:147–158.

Katopodes, N.D. 1984. A dissipative Galerkin scheme for open channel flow, *Jour of Hydraulic Engineering,* Amer Soc Civ Engrs., 110, no. 4:450–466.

Kim, S. J., and Schetz, J. A. 1989. Finite element analysis of the flow of a propeller on a slender body with a two-equation turbulence model, *Proc. Seventh Inter Conference on Finite Element Methods in Flow Problems,* pp. 1541–50.

Launder, B. E., and Spalding, D. B. 1972. *Mathematical Models of Turbulence,* Academic Press, London, England, p. 169.

Lee, J. K. and Froelich, D. C. 1986. Review of literature on the finite element solution of the equations of two-dimensional surface-water flow in the horizontal plane, *Circular: 1009* U. S. Geological Survey, Denver, CO. 60 pp.

Nallasamy, M. 1985. A critical evaluation of various turbulence models as applied to internal fluid flows, *Technical Paper 2474,* NASA, Springfield, VA.

Olson, M. D. 1977. Comparison of various finite element solution methods for the Navier-Stokes equations, *Proc.* First Inter. Conf. on Finite Elements in Water Resources, edited by W. G. Gray, G. F. Pinder, and C. A. Brebbia, Pentech Press Limited, Plymouth, Devon, England, 4.185–4.203.

Polansky, G. F., Lamb, J. P., and Crawford, M. E. 1984. A finite element analysis of incompressible turbulent backstep flow with heat transfer, *Second Aerospace Sciences Meeting,* AIAA Paper No. 84-0178.

Rodi, W. 1980. Turbulence models and their applications in hydraulics, International Association for Hydraulic Research, Delft, the Netherlands.

Saez, A. E., and Carbonell, R. G. 1985. On the performance of quadrilateral finite elements in the solution to the stokes equations in periodic structures, *Int. J. Numer. Methods Fluids,* 5:601–614.

Schamber, D. R., and B. E. Larock. 1981. Numerical analysis of flow in sedimentation basins, *Jour. Hydraulics Division,* Amer. Soc. Civil Engrs., 107, no 5:575–91.

Segerlind, L. J. 1976. *Applied finite element analysis,* John Wiley, New York, NY.

Sharma, M., and Carey, G. F. 1986. Turbulent boundary-layer analysis using finite elements, *International Jour. for Numerical Methods in Fluids,* 6:769–87.

Smith, R. M. 1984a. On the finite-element calculation of turbulent flow using the k-ϵ model, *International Jour. for Numerical Methods in Fluids*, 4:303–19.

————. 1984b. A practical method of two-equation turbulence modeling using finite elements, *International Jour. for Numerical Methods in Fluids,* 4:321–36.

Spiegel, Murray R. 1968. *Mathematical handbook of formulas and tables, Schaum's outline series,* McGraw-Hill, New York, NY.

Taylor, C.; Thomas, C. E.; and Morgan, K. 1981. Modeling flow over a backward-facing step using the F.E.M. and the two-equation model of turbulence, *International Jour. for Numerical Methods in Fluids,* 1:295–304.

Thompson, J. 1990. Numerical modeling of irregular hydraulic jumps, *Proc. Hydraulic Engineering Conference,* Amer. Soc. Civil Engrs., 749–54.

Zienkiewicz, O. C., and Taylor R. L. 1989. *The finite element method,* 4th ed. McGraw-Hill, London.

17 Special Topics

Flooding of Interstate Highway 10 near Slidell, Louisiana during a 200-year flood on the Pearl River. (Courtesy of U.S. Geological Survey, after Lee and Froehlich, 1989).

17-1 Introduction

In this chapter, we discuss a number of special topics, to which we apply concepts discussed in the previous chapters. First, we discuss rating curve at a channel cross section during steady and unsteady flow conditions. Then, we describe different methods for flood routing. This is followed by a discussion of the aggradation and degradation of channel bottom due to the imbalance between the actual amount of sediment in the flow and the carrying capacity of flow in a channel.

17-2 Rating Curve

A rating curve describes a relationship between the water level or stage at a channel cross section with the rate of discharge at that section. In Chapter 4, we presented several empirical resistance formulas for steady-uniform flow that relate the channel discharge with the parameters of the channel. We may express these formulas in a general form as

$$Q_n = kAR^m\sqrt{S_o} \tag{17-1}$$

in which Q_n = rate of discharge if the flow is uniform; k = a coefficient; A = flow area; R = hydraulic radius; S_o = channel-bottom slope; and m = an exponent that depends upon the formula used. For example, for the Manning formula, $k = C_o/n$, and $m = \frac{2}{3}$; and for the Chezy formula, k = Chezy C and $m = \frac{1}{2}$. Similar to Eq. 17-1, a resistance formula for unsteady, nonuniform flow may be written as

$$Q = kAR^m\sqrt{S_f} \tag{17-2}$$

in which S_f = slope of the energy-grade line for discharge Q. If AR^m is a monotonically increasing function of flow depth y (this is the case for regular channel cross sections), then Eq. 17-1 plots as a dotted line, as shown in Fig. 17-1. This is a single-valued relationship between Q and y. Let us now discuss how unsteadiness with respect to time and nonuniformity with respect to distance modify this relationship.

It follows from Eqs. 17-1 and 17-2 that

$$Q = Q_n\sqrt{\frac{S_f}{S_o}} \tag{17-3}$$

Substitution of expression for S_f from the momentum equation (Eq. 12-17) into this equation yields

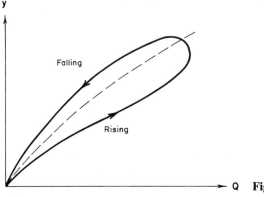

Figure 17-1 Rating curve.

$$Q = Q_n\sqrt{1 - \frac{1}{S_o}\left(\frac{\partial y}{\partial x} + \frac{V}{g}\frac{\partial V}{\partial x} + \frac{1}{g}\frac{\partial V}{\partial t}\right)} \qquad (17\text{-}4)$$

Let us consider a channel reach in which inflow and stage are increasing with time. Thus, the flow depth and velocity at the upper end of the reach are higher than that at the lower end, i.e., the flow depth and flow velocity decrease with distance and the flow velocity increases with time. In other words, $\partial y/\partial x$ and $\partial V/\partial x$ are both negative and $\partial V/\partial t$ is positive. Normally, the last term of Eq. 17-4 representing the local acceleration is small as compared to the other two terms representing the convective acceleration. Therefore, it follows from Eq. 17-4 that Q during a rising stage is greater than Q_n for a given value of y. By proceeding similarly, we can show that Q is less than Q_n for the same flow depth for a falling stage. Thus the rating curve has hysteresis, as shown in Fig. 17-1. Note that the difference between the discharges during rising and falling stages is due to unsteadiness and nonuniformity of the flow depth. The larger the hysteresis, the more pronounced is the effect of these terms. In such situations it becomes necessary to include these terms in the analysis.

The shape and the magnitude of a flood wave may change as it travels in a body of water due to friction and due to storage effects. The storage may be produced by pondage or by cross-sectional changes in the natural channels. The wave is elongated and its magnitude is reduced due to storage. Such a reduction many times may be much more than that due to friction. In such situations, it is possible to simplify the analysis by disregarding the frictional effects.

17-3 Flood Routing

The computation of the height and velocity of a flood wave as it propagates in a body of water is referred to as *flood routing*. The body of water may be a lake, reservoir, channel, stream, etc. For this purpose, we may solve the continuity and momentum equations derived in Chapter 11 by using various numerical schemes presented in Chapters 12 through 14. For two-dimensional flows, the governing equations and the numerical methods outlined in Chapter 15 may be utilized. Such computations have been referred to as *hydraulic routing*, and the models developed for this purpose are called *dynamic models*. In many situations, however, some terms of the governing equations are smaller than the other terms. For example, the inertial terms for a typical flood wave are negligible as compared to the other terms and may be dropped without introducing significant errors. We will call such analysis procedures *approximate methods*.

Depending upon the terms included in the analysis, the approximate procedures for flood routing have been given different names. For example, the continuity equation solved simultaneously with a simplified form of the momentum equation is called *hydrologic routing*. If the simplified momentum equation is the steady-uniform equation, the routing procedure is called *kinematic routing*, and if an additional term for the slope of the water surface is included, the routing is called *diffusion routing*. By approximating the complex relationships between the storage capacity of a channel length and the inflow and outflow, several coefficient methods have been developed.

An approximate flood-routing procedure yields satisfactory results if the simplifying assumptions on which it is based are valid; i.e., the terms of the governing equations excluded in its development are negligible. These methods have the advantage that they are simple to apply and that they do not require detailed data for the channel geometry. However, an improper application may yield totally incorrect results. We briefly discuss some of these approximate procedures in this Chapter.

17-4 Reservoir Routing

Lakes, reservoirs, ponds, and detention basins act as storage facilities. We will refer to each of them as a reservoir. The water level in such a storage facility may be considered horizontal. This simplifies the analysis significantly since dynamic effects are neglected and we need to consider only the continuity equation. According to this equation, the difference between the inflow and outflow is equal to the rate of change of volume of water stored in the reservoir. The stored volume is called *storage, S*.

We may write the continuity equation relating the inflow, I, outflow, O, and the rate of change of storage, S, in the reservoir (Fig. 17-2) as

$$\frac{dS}{dt} = I - O \tag{17-5}$$

The inflow at different times is given as an inflow hydrograph; and the storage and outflow are specified as functions of the water level in the reservoir. To route a flood wave through a reservoir, we integrate this equation. For this purpose, any of the numerical schemes presented in Chapter 6 may be used. However, we will discuss in the following paragraphs a procedure initially developed for hand calculations.

A finite-difference approximation of Eq. 17-5 may be written as

$$\frac{\Delta S}{\Delta t} = \bar{I} - \bar{O} \tag{17-6}$$

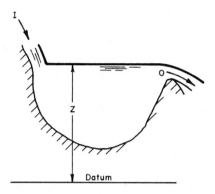

Figure 17-2 Definition sketch.

in which $\bar{I}$ and $\bar{O}$ indicate the mean values during the time interval Δt. This interval is referred to as the *routing interval*. Let us designate the variables at the beginning of a routing interval by superscript k and the values at the end of the interval by superscript $k+1$. In addition, let us assume the variation of different variables is linear during the time interval. Then Eq. 17-6 becomes

$$\frac{S^{k+1} - S^k}{\Delta t} = \frac{1}{2}(I^k + I^{k+1}) - \frac{1}{2}(O^k + O^{k+1}) \tag{17-7}$$

By rearranging the known and unknown terms of this equation, we obtain

$$I^k + I^{k+1} + \frac{2S^k}{\Delta t} - O^k = O^{k+1} + \frac{2S^{k+1}}{\Delta t} \tag{17-8}$$

Now, if O and S are functions of the water level, z, in the reservoir, then we may say that O is a function of $2S/\Delta t$ as well. We can utilize this fact to solve Eq. 17-8 as follows.

1. We select a value for the routing interval and plot a curve between O and $(2S/\Delta t) + O$.

2. At the beginning of any routing interval, we know the values of water level in the reservoir, inflow I^k and outflow O^k. When we start the calculations, the values of these variables are specified; later, they are computed during the previous time interval. In addition, we know I^{k+1} from the inflow hydrograph. For known O^k, we first read $(2S/\Delta t) + O$ from the curve between $(2S/\Delta t) + O$ and O. Then by subtracting $2O^k$ from this value, we compute $(2S/\Delta t) - O$. Now, we can determine the left-hand side of Eq. 17-8.

3. For the value computed in step 2, which is also equal to $(2S^{k+1}/\Delta t) + O^{k+1}$ (from Eq. 17-8), we read the value of O^{k+1} from the curve between $(2S/\Delta t) + O$ and O.

4. We repeat steps 1 to 3 for the next time interval and continue this process until the computations for the desired period are done.

17-5 Channel Routing

In the previous section, we assumed that the water surface in the reservoir always remains horizontal although its level may change if the inflow is not equal to the outflow. Also, the storage and outflow were assumed as functions of water level. Hence, we could say that the storage is a function of the outflow. These assumptions are valid for a channel, if the flow in the channel reach is uniform. However, for non-uniform flow, the storage depends upon the inflow and outflow (Fig. 17-3). The storage in a channel reach may be divided into *prism storage* where S is proportional to O and *wedge storage* where S is proportional to the difference between inflow and outflow. Based on these assumptions, a procedure was presented in 1938 to do flood-routing studies on the Muskingum river by U.S. Army Corps of Engineers. Nowadays, this procedure is commonly called *Muskingum routing*.

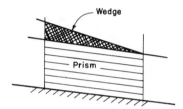

Figure 17-3 Prism and wedge storage.

In this method, the storage in a channel reach is expressed as a function of inflow and outflow as

$$S = KO + KX(I - O) \tag{17-9}$$

in which K and X are constants. Note that for dimensional reasons, K has the units of time and X is dimensionless. We shall show later in Sec. 17-8 that K is the time for a wave to travel from one end of the reach to the other.

By using the notation of the previous section, we may write Eq. 17-9 for the storage at time k as

$$S^k = K[XI^k + (1 - X)O^k] \tag{17-10}$$

Similarly, for the storage at time $k + 1$ may be written as

$$S^{k+1} = K[XI^{k+1} + (1 - X)O^{k+1}] \tag{17-11}$$

By substituting these equations into Eq. 17-8 and simplifying the resulting equation, we obtain

$$O^{k+1} = C_o I^{k+1} + C_1 I^k + C_2 O^k \tag{17-12}$$

in which

$$C_o = \frac{\Delta t - 2KX}{\Delta t + 2K(1-X)}$$

$$C_1 = \frac{\Delta t + 2KX}{\Delta t + 2K(1-X)} \qquad (17\text{-}13)$$

$$C_2 = \frac{-\Delta t + 2K(1-X)}{\Delta t + 2K(1-X)}$$

We may use Eq. 17-12 for flood routing through a channel reach if we know the values of K and X. These may be determined from the observed flow records as discussed in the following paragraph; or their values may be computed from the expressions derived by a rigorous analysis in Sec. 17-8.

We may solve Eq. 17-10 (Roberson, et al. 1988) for K as

$$K = \frac{0.5\Delta t[(I^k + I^{k+1}) - (O^k + O^{k+1})]}{X(I^{k+1} - I^k) + (1-X)(O^{k+1} - O^k)} \qquad (17\text{-}14)$$

For an observed hydrograph, we plot the numerator of this equation as the ordinate and the denominator as the abcissa for different time intervals and different assumed values of X, say 0.1, 0.2, 0.3, etc. The value of X that gives the plot close to a straight line is the X value to use and the slope of the graph is the value of K.

17-6 Kinematic Routing

In kinematic routing, we solve the continuity equation, Eq. 12-4, simultaneously with an approximate form of the momentum equation. This approximate form is obtained by neglecting the local and convective acceleration terms of the momentum equation. The remaining terms represent the resistance equation for steady, uniform flow. In other words, we consider the flow to be steady for momentum conservation but take into consideration the effects of unsteadiness by an increase or decrease in the flow depth.

We may write the resistance equation in a form similar to Eq. 17-1, i.e.,

$$Q = f(A) \qquad (17\text{-}15)$$

or

$$A = F(Q) \qquad (17\text{-}16)$$

Hence, applying the chain rule we may write

$$\frac{\partial A}{\partial t} = \frac{\partial A}{\partial Q}\frac{\partial Q}{\partial t} \qquad (17\text{-}17)$$

or

$$\frac{\partial A}{\partial t} = \frac{\partial Q}{\partial t}\frac{dA}{dQ}\bigg|_{x=x_o} \qquad (17\text{-}18)$$

Substituting this equation into Eq. 12-4, assuming $q_l = 0$, and simplifying the resulting equation, we obtain

$$\frac{\partial Q}{\partial t} + a \frac{\partial Q}{\partial x} = 0 \tag{17-19}$$

in which $a = dQ/dA$. Since this is a kinematic model, the wave is called a *kinematic wave*.

It follows from the expression for a that it has dimensions of L/T. Thus, it represents a velocity. The following discussion will show that a is the velocity of a flood wave. Note that this expression is different from the one we derived in Chapters 12 and 13 for the absolute velocity of a disturbance as $V \pm c$. However, extensive field measurements of the propagation of the crest of flood waves confirm this relationship (Seddon 1900).

Equation 17-19 is a first-order partial differential equation with Q as the dependent variable and x and t as the independent variables. It describes the movement of a flood wave in terms of the rate of discharge. If a is constant, then Eq. 17-19 is linear. D'Alembert presented a general solution of this linear equation as

$$Q = f(x - at) \tag{17-20}$$

In this solution, we assume that the partial derivatives of f with respect to x and t exist. By taking the partial derivative of Eq. 17-20 with respect to x and t and substituting them into Eq. 17-19, we can prove that Eq. 17-20 represents the general solution of Eq. 17-19. At $t = 0$, $Q = f(x)$ represents the initial conditions. This curve describes the variation of discharge Q with distance x. The solution at time t_1 is $f(x - at_1)$, and at time t_2 it is $f(x - at_2)$. Let us assume that an observer is traveling at velocity a in the downstream direction. To this observer, this curve always appears as $f(x)$. We may draw the same conclusion by considering a moving coordinate system such that $\xi = x - at$. Then, $Q = f(\xi)$; the shape of the solution curve is always $Q = f(\xi)$, and it is independent of time t.

This discussion shows that a flood hydrograph in kinematic routing travels in the positive x-direction at velocity a; the shape of the hydrograph does not change and its peak does not attenuate (Fig. 17-4). However, note that these conclusions are based on the assumption that a is constant. If this is not the case—i.e., $dQ/dA \neq$ constant—then Eq. 17-19 becomes nonlinear and the wave shape may change due to nonlinear effects as it propagates in the channel. The change in the wave shape depends upon the variation of a with Q. The positive front of a wave steepens with distance if a increases with an increase in Q and the front flattens if a decreases with an increase in Q. Sometimes, the wave front may steepen so much that it forms an almost vertical front; this is referred to as a *kinematic shock*.

A kinematic model is based on the solution of Eq. 17-19. The solution may be analytical or numerical. In a numerical solution the wave height and shape may be modified as it propagates. This modification is purely due to the characteristics of the numerical method and does not represent simulation of the actual physical phenomenon. The wave modification may be in the form of reduction of the wave height, change in shape, or a combination of the two. The reduction in the height is called *dissipation*, the change in the shape is referred to as *dispersion*, and the combination of dissipation and dispersion is called *diffusion*.

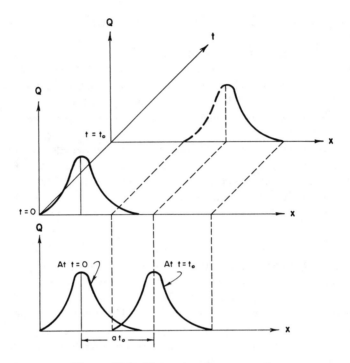

Figure 17-4 Kinematic wave propagation.

The modification of the wave shape depends upon the numerical method employed and the value of the Courant number, $C_n = a\Delta t/\Delta x$. To illustrate this point, let us assume that the wave velocity, a, is constant; i.e., the slope of the curve between Q and A is constant. Let us now use the Lax scheme to numerically integrate Eq. 17-19. Then, by using the notation outlined in Chapter 14 (p. 311), we get

$$Q_i^{k+1} = \tfrac{1}{2}(Q_{i+1}^k + Q_{i-1}^k) - \tfrac{1}{2}a\frac{\Delta t}{\Delta x}(Q_{i+1}^k - Q_{i-1}^k) \tag{17-21}$$

Rearrangement of the terms of this equation yields

$$Q_i^{k+1} = \tfrac{1}{2}(1 - C_n)Q_{i+1}^k + \tfrac{1}{2}(1 + C_n)Q_{i-1}^k \tag{17-22}$$

It is clear from this equation that $Q_i^{k+1} = Q_{i-1}^k$ if $C_n = 1$. Thus, if the wave peak was at computational node $(i - 1)$ at time t_o, then it will be at node i at time $t_o + \Delta t$. Also note that the shape of the wave is not modified as it propagates from one node to the next. However, if the value of C_n is not 1, then the wave shape is modified as it travels. For example, if $C_n = 0.8$, then $Q_i^{k+1} = 0.1Q_{i+1}^k + 0.9Q_{i-1}^{k+1}$. In other words, the wave shape is modified and the wave peak is attenuated while traveling from time k to time $k + 1$. According to the governing equation of a kinematic model, a flood wave travels without modifying its shape and with no attenuation. However, numerical calculations with C_n different than 1 may result in attenuating the wave peaks and in modifying the wave shapes. This is mainly due to the limitations of the numerical solution and is not due to the simulation of actual dissipation.

Applicability criterion.   For the applicability of the kinematic model to overland flow, Woolhiser and Liggett (1967) developed the following criterion:

$$K_f = \frac{S_o L_o}{y_n \mathbf{F}_r} \geq 20 \tag{17-23}$$

in which K_f = kinetic flow number; S_o = bottom slope; L_o = length of the overland flow plane; y_n = normal depth; and $\mathbf{F}_r$ = Froude number corresponding to uniform flow. Morris and Woolhiser (1980) state that it is also necessary for low Froude number flows that $K_f \mathbf{F}_r^2 \geq 5$, in addition to the preceding criterion.

By using an analytical solution of the linearized equations, Ponce et al. (1978) showed that the accuracy of the computed results for a sinusoidal perturbation of mean flow is within 95 percent accurate after one period of propagation if the dimensionless wave period

$$T_w = \frac{T S_o V_o}{y_o} > 171 \tag{17-24}$$

in which T = wave period; V_o = reference mean velocity; and y_o = reference flow depth. The wave period T may be taken as twice the time of rise of the flood wave.

17-7 Diffusion Routing

In the diffusion routing, we solve a simplified form of the momentum equation with the continuity equation. The simplified form of the momentum equation includes the convective acceleration term representing the spatial change in the flow depth as well as the source terms but neglects the temporal derivative term as well as the convective acceleration terms due to spatial change in the flow velocity. Thus, the momentum equation with these simplifications becomes

$$S_f = S_o - \frac{\partial y}{\partial x} \tag{17-25}$$

We showed in Chapter 4 that any resistance formula may be written in the form $Q = K S_f^{1/2}$, where K is the conveyance factor. Substitution of this expression into Eq. 17-25 yields

$$\frac{Q^2}{K^2} = S_o - \frac{\partial y}{\partial x} \tag{17-26}$$

Let us eliminate y from this equation and the continuity equation (Eq. 12-4) so that the resulting equation describes the variation of Q with respect to x and t. To do this, let us first differentiate Eq. 17-26 with respect to t and Eq. 12-4 with respect to x assume

$q_l = 0$ and then eliminate $\partial^2 y / \partial x \partial t$ from the resulting equations. Differentiation of Eq. 17-26 with respect to t gives

$$\frac{2Q}{K^2}\frac{\partial Q}{\partial t} - \frac{2Q^2}{K^3}\frac{\partial K}{\partial t} = -\frac{\partial^2 y}{\partial x \partial t} \tag{17-27}$$

Differentiating Eq. 12-4 with respect to x, noting that $\partial A / \partial x = B \partial y \partial x$, and dividing throughout by B yields

$$\frac{\partial^2 y}{\partial x \partial t} = -\frac{1}{B}\frac{\partial^2 Q}{\partial x^2} \tag{17-28}$$

By eliminating $\partial^2 y / \partial x \partial t$ from Eqs. 17-27 and 17-28, we obtain

$$\frac{1}{B}\frac{\partial^2 Q}{\partial x^2} - \frac{2Q}{K^2}\frac{\partial Q}{\partial t} + \frac{2Q^2}{K^3}\frac{\partial K}{\partial t} = 0 \tag{17-29}$$

Based on the chain rule, we may write that $\partial K / \partial t = (\partial K / \partial A)(\partial A / \partial t)$. Substitution for $\partial A / \partial t$ from Eq. 12-4 into this expression gives

$$\frac{\partial K}{\partial t} = \frac{dK}{dA}\bigg|_{x=x_o}\left(-\frac{\partial Q}{\partial x}\right) \tag{17-30}$$

To simplify the derivation, let us obtain the derivative dK/dA from the expression $Q = K\sqrt{S_o}$ for uniform flow instead of from the general expression $Q = K\sqrt{S_f}$. Then, it follows from Eq. 17-30 that

$$\frac{\partial K}{\partial t} = \frac{1}{\sqrt{S_o}}\frac{dQ}{dA}\left(-\frac{\partial Q}{\partial x}\right) \tag{17-31}$$

By eliminating $\partial K / \partial t$ from Eqs. 17-29 and 17-31, noting that $Q = K\sqrt{S_o}$ and $a = dQ/dA$, multiplying throughout by $K^2/(2BQ)$, and simplifying the resulting equation, we obtain

$$D\frac{\partial^2 Q}{\partial x^2} - \left(\frac{\partial Q}{\partial t} + a\frac{\partial Q}{\partial x}\right) = 0 \tag{17-32}$$

in which $D = Q/(2BS_o)$.

A comparison of Eqs. 17-19 and 17-32 shows that other than the first term, the remaining equation is the same as the equation for the kinematic model. The first term of this equation represents the diffusion of a flood wave as it travels in the channel. By using the coefficients D and a determined from the observed hydrographs, the attenuation of a flood wave due to storage and friction may be included in the analysis.

Applicability. The following criterion (Ponce et al. 1978) may be used for the applicability of the diffusion model:

$$K_w = \mathbf{F}_r T S_o \sqrt{\frac{g}{y_o}} \geq 30 \tag{17-33}$$

in which $\mathbf{F}_r$ = reference flow Froude number and the other variables are as defined for the kinematic model.

17-8 Muskingum-Cunge Routing

In Sec. 17-5 we discussed the Muskingum routing and presented a procedure to determine the coefficients K and X from the observed flood hydrographs. Cunge (1969) derived expressions for these coefficients from a finite-difference approximation of the kinematic wave equation, Eq. 17-19. We outline this procedure in this section.

Let us substitute the following finite-difference approximations into Eq. 17-19

$$\frac{\partial Q}{\partial x} = \frac{(O^k + O^{k+1}) - (I^k + I^{k+1})}{2\Delta x}$$

$$\frac{\partial Q}{\partial t} = \frac{\alpha(I^{k+1} - I^k) + (1 - \alpha)(O^{k+1} - O^k)}{\Delta t}$$

(17-34)

in which α is the weighting coefficient for the time derivative. The simplification of the resulting equation yields Eq. 17-12 except that the following expressions for different coefficients are obtained instead of those given in Eq. 17-13:

$$C_o = \frac{0.5\Delta t - \alpha\Delta x/a}{0.5\Delta t + (1 - \alpha)\Delta x/a}$$

$$C_1 = \frac{0.5\Delta t + (\alpha\Delta x/a)}{0.5\Delta t + (1 - \alpha)\Delta x/a}$$

(17-35)

$$C_2 = \frac{-0.5\Delta t + (1 - \alpha)\Delta x/a}{0.5\Delta t + (1 - \alpha)\Delta x/a}$$

Note that for $\alpha = X$ and $\Delta x/a = K$, these expressions reduce to those given in Eq. 17-13. Thus, $K = \Delta x/a$, which is the travel time for a flood wave to propagate through the reach.

We can show that if $\alpha = 0.5$ and $a\Delta t/\Delta x = 1$ in the Muskingum-Cunge routing, then a wave does not attenuate and does not change shape as it is propagated through a channel reach. This result is similar to that obtained by the kinematic routing.

17-9 Aggradation and Degradation of Channel Bottom

In this section we present a mathematical model to simulate the aggradation and degradation of the channel bottom.

Introduction

The channel bottom may aggrade or degrade if the balance among water discharge, sediment flow and the channel shape is disturbed. Such a disturbance may be due to natural or other factors, such as the construction of a dam, change in the sediment supply rate, lowering of the channel bottom, migration of knickpoints, etc. A reliable, estimation of the bed aggradation and degradation becomes necessary in river control engineering and water-management projects.

Several experimental studies have been conducted to investigate the short- and long-term bed-level changes in alluvial channels. Lane and Borland (1954) conducted experiments to study riverbed scour during floods. Brush and Wolman (1960) measured the time variation of bed levels in a laboratory channel due to the migration of knickpoints or the points of abrupt change in the longitudinal profile. Newton (1951) obtained laboratory data for degradation due to sediment diminution and Soni et al. (1980) studied aggradation due to sediment overloading. Begin et al. (1981) experimentally studied degradation of alluvial channels in response to lowering of the channel bottom. Suryanarayana (1969) obtained laboratory data for the degradation of alluvial channels downstream of a dam.

A number of analytical solutions have been developed by simplifying the governing equations describing the aggradation and degradation processes. Soni et al. (1980) used a linear diffusion model to predict the transient-bed profiles due to sediment overloading. Jain (1981) pointed out an error in their boundary conditions and presented an analytical solution utilizing more appropriate boundary conditions. His computed results compared satisfactorily with the experimental data. Begin et al. (1981) used a diffusion model to compute longitudinal profiles produced by base-level lowering. Gill (1983a, b) solved the linear diffusion equation describing the aggradation and degradation process by the Fourier series and by the error-function methods. Jaramillo and Jain (1984) developed a nonlinear parabolic partial differential equation, solved it by the method of residuals, and compared their computed results with the available experimental data. Zhang and Kahawita (1987) and Gill (1987) presented nonlinear solutions for aggradation and degradation that compared better with the experimental data than the linear solutions.

The linear and nonlinear parabolic models are based on the assumption of quasi-steady water flow. This assumption may not be valid during floods or during other unsteady flow conditions. This may not also be valid even if the discharge is constant if the slope of the channel bottom changes during the period of interest. Therefore, the complete unsteady water-flow equations along with the sediment continuity equation are solved by numerical techniques. Holly (1986), Dawdy and Vanoni (1986), and Cunge et al. (1980) reviewed the literature on the numerical simulation of alluvial hydraulics. Lu and Shen (1986) tested several numerical aggradation and degradation models by comparing the computed results with Suryanarayana's laboratory data (1969).

One-dimensional, unsteady sediment transport models may be classified into two categories: the uncoupled models, in which the water-flow equations and sediment continuity equation are uncoupled during a given time step, and the quasi-steady flow models, in which the energy equation is solved along with the sediment continuity equation. Lyn (1987) used perturbation techniques to identify multiple time scales in the governing equations and suggested that complete coupling between the full unsteady water-flow equations and the sediment continuity equation is desirable in cases where the conditions are rapidly changing at the boundaries. He proposed to use the Preissmann linearized implicit scheme for a simultaneous solution of the governing equations. Park and Jain (1986) used this procedure in their unsteady, uncoupled model for the analysis of the aggradation due to sediment overloading. To avoid instability, they had to iterate the solution whenever the spatial gradient of change in bed level became too large.

In this section, a one-dimensional, unsteady, coupled deformable bed model (Bhallamudi and Chaudhry 1991) is presented. The complete St. Venant equations for water flow and the sediment continuity equation are solved simultaneously by the MacCormack explicit scheme (MacCormack 1969). The computed results are compared with the experimental data to verify the model.

Governing Equations

The following equations describe unsteady flow in a wide rectangular channel with deformable bed (Fig. 17-5):

Continuity equation for water

$$\frac{\partial h}{\partial t} + \frac{\partial q}{\partial x} = 0 \tag{17-36}$$

Momentum equation for water

$$\frac{\partial q}{\partial t} + \frac{\partial}{\partial x}\left(\frac{q^2}{h} + \frac{1}{2}gh^2\right) + gh\frac{\partial z}{\partial x} + ghS_f = 0 \tag{17-37}$$

Continuity equation for sediment

$$\frac{\partial}{\partial t}\left[(1-p)z + \frac{q_s h}{q}\right] + \frac{\partial q_s}{\partial x} = 0 \tag{17-38}$$

in which q = water discharge per unit width; h = flow depth; z = channel bottom elevation; q_s = unit sediment discharge, and p = porosity of the bed layer. The sediment discharge may be estimated by an empirical power function of the flow velocity

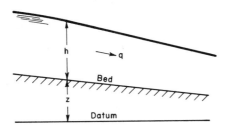

Figure 17-5 Definition sketch.

$$q_s = a \left(\frac{q}{h}\right)^b \tag{17-39}$$

in which a and b are empirical constants whose values depend upon the sediment properties. Note that this relationship for the sediment discharge is used here mainly for simplicity. More elaborate and complex relationships may easily be included in the model since the governing equations are solved by an explicit numerical scheme.

Experimental investigations of Soni et al. (1977) show that the Manning n under nonuniform conditions in an aggrading channel is smaller than its value for uniform flow. The resistance law for uniform flow may be used under nonuniform conditions, provided local friction slope is used instead of the bottom slope. The applicability of the steady, uniform sediment transport formula under nonuniform conditions is questionable and needs further investigations.

Numerical Scheme

Equations 17-36 through 17-38 are a set of nonlinear hyperbolic partial differential equations and closed-form solutions are available only for idealized cases. We solve these equations numerically by using the MacCormack scheme. As we discussed in Chapters 14 and 15, this scheme is simple to implement and captures shocks without any special treatment, and the incorporation of general empirical equations for roughness and sediment discharge is easy.

By using the forward finite differences for the spatial partial derivatives in the predictor part and the backward finite differences in the corrector part, we obtain the following equations.

Predictor

$$h_i^* = h_i^k - \frac{\Delta t}{\Delta x}\left(q_{i+1}^k - q_i^k\right)$$

$$q_i^* = q_i^k - \frac{\Delta t}{\Delta x}\left\{\frac{\left(q_{i+1}^k\right)^2}{h_{i+1}^k} - \frac{\left(q_i^k\right)^2}{h_i^k} + \frac{g}{2}\left[\left(h_{i+1}^k\right)^2 - \left(h_i^k\right)^2\right]\right\}$$

$$\qquad - gh_i^k \frac{\Delta t}{\Delta x}\left(z_{i+1}^k - z_i^k\right) - gh_i^k \Delta t \frac{\left(q_i^k n\right)^2}{\left(h_i^k\right)^{3.33}}$$

$$z_i^* = z_i^k + \frac{1}{1-p}\left[\left(\frac{q_s h}{q}\right)_i^k - \left(\frac{q_s h}{q}\right)_i^*\right] - \frac{\Delta t}{(1-p)\,\Delta x}\left[(q_s)_{i+1}^k - (q_s)_i^k\right]$$

$$(q_s)_i^* = a\left(\frac{q_i^*}{h_i^*}\right)^b$$

$$\tag{17-40}$$

in which the superscript * indicates values at the end of the predictor part.

Corrector

$$h_i^{**} = h_i^* - \frac{\Delta t}{\Delta x}\left[q_i^* - q_{i-1}^*\right]$$

$$q_i^{**} = q_i^* - \frac{\Delta t}{\Delta x}\left[\frac{\left(q_i^*\right)^2}{h_i^*} - \frac{\left(q_{i-1}^*\right)^2}{h_{i-1}^*} + \frac{g}{2}\left\{\left(h_i^*\right)^2 - \left(h_{i-1}^*\right)^2\right\}\right]$$

$$- gh_i^*\frac{\Delta t}{\Delta x}\left(z_i^* - z_{i-1}^*\right) - gh_i^*\Delta t\frac{\left(q_i^*n\right)^2}{\left(h_i^*\right)^{3.33}}$$

$$z_i^{**} = z_i^* + \frac{1}{1-p}\left[\left(\frac{q_s h}{q}\right)_i^* - \left(\frac{q_s h}{q}\right)_i^{**}\right] - \frac{\Delta t}{(1-p)\,\Delta x}\left[(q_s)_i^* - (q_s)_{i-1}^*\right]$$

$$(q_s)_i^{**} = a\left(\frac{q_i^{**}}{h_i^{**}}\right)^b$$

$$(17\text{-}41)$$

in which the superscript ** denotes the value of the variable after the corrector step.

Now, the values of the unknowns at time level $k + 1$ (i.e., at the end of time interval Δt) are given by

$$h_i^{k+1} = \tfrac{1}{2}\left(h_i^k + h_i^{**}\right)$$
$$q_i^{k+1} = \tfrac{1}{2}\left(q_i^k + q_i^{**}\right) \qquad (17\text{-}42)$$
$$z_i^{k+1} = \tfrac{1}{2}\left(z_i^k + z_i^{**}\right)$$

By using the preceding algorithm, the values of h, q, and z at the new time level $k + 1$ are determined at every interior node ($i = 2, ..., N$). The values of the dependent variables h, q, and z at the boundary nodes 1 and $N + 1$ are determined by using the boundary conditions. For subcritical flow conditions, it can be shown by using the characteristic theory that two boundary conditions at the upstream boundary and one condition at the downstream boundary have to be specified. The values of the dependent variables which are not specified through boundary conditions may be determined from the characteristic equations. Their values may also be determined by interpolation from the known values at the interior nodes (Roache 1972). The inclusion of these boundary conditions into the finite-difference scheme is problem specific and is discussed for each problem later.

Stability

For stability, the MacCormack scheme has to satisfy the Courant-Friedrichs-Lewy (CFL) condition. Since the water waves travel at a much higher velocity than the bed transients, this condition is given by the following equation:

$$C_n = \frac{(q/h + \sqrt{gh})\,\Delta t}{\Delta x} \leq 1 \qquad (17\text{-}43)$$

in which C_n is the Courant number. Equation 17-43 has to be satisfied at every grid point for the scheme to be stable.

Artificial viscosity

Numerical oscillations near the steep wave fronts may be dampened by introducing artificial viscosity. For this purpose, the Jameson procedure discussed in Chapter 14 may be utilized. Of the several cases studied, smoothing was required only in the case of knickpoint migration.

Computational procedure

Let us say the values of h_i^k, q_i^k, and z_i^k are known at the known time level k at all the grid points ($i = 1, \ldots, N + 1$) of a channel divided into N reaches and we want to determine their values at the unknown time level, $k + 1$. The known values are the initial conditions at $t = 0$ if the computations are just starting; otherwise, they are the computed values during the previous time interval.

The values at time level $k + 1$ may be computed as follows:

1. The values of h_i^*, q_i^*, and z_i^* at the interior nodes ($i = 2, \ldots, N$) are computed by using Eqs. 17-40 and their values at the boundaries ($i = 1$ and $i = N + 1$) are computed by using the appropriate boundary conditions.

2. Now, h_i^{**}, q_i^{**}, and z_i^{**} at the interior nodes ($i = 2, \ldots, N$) are determined from Eqs. 17-41 and their values at the boundaries from the specified boundary conditions.

3. Then, h, q, and z at the end of time interval Δt, i.e., h_i^{k+1}, q_i^{k+1}, and z_i^{k+1}, are determined by using Eq. 17-42.

4. The values determined in step 3 are modified if necessary to dampen the high-frequency oscillations by using the Jameson procedure for artificial viscosity presented in Chapter 14.

5. The values of h_i^k, q_i^k and z_i^k for the next time interval are set equal to h_i^{k+1}, q_i^{k+1}, and z_i^{k+1}; the time interval, Δt, for the next step is determined from Eq. 17-43; and the procedure is repeated until simulation for the specified time is done.

The flow equations and the sediment continuity equation are coupled in this procedure because it uses a two-level predictor-corrector approach. Strictly speaking, there is no coupling during the predictor part. However, the predicted values of the bed elevation are used to determine the correct values of discharge in Eq. 17-37. Similarly, the predicted values of h and q are used to determine q_s and to evaluate the spatial derivative term in Eq. 17-38. Therefore, each dependent variable computed at the end of the time step takes into account the changes in all the other variables. In this sense, this procedure may be called *coupled*. On the other hand, coupling is not achieved if Eq. 17-38 is solved after completely solving Eqs. 17-36 and 17-37, i.e., after corrector step for Eqs. 17-36 and 17-37. This later approach is similar to the *uncoupled implicit method*. According to Park and Jain (1986), this approach results in numerical instabilities whenever the gradient of bed level became too large. They iterated the solution during each time step for numerical stability. In addition to its simplicity, the computational procedure presented herein does not require iterations.

Applications

To illustrate the application of this model, the computed results are compared with experimental results for two cases. For the comparisons for other cases, see Bhallamudi and Chaudhry (1991).

Aggradation Due to Sediment Overloading

Test results on the aggradation process (Soni et al. 1980) in a laboratory channel were compared with the computed results. The channel was 0.2 m wide and 30 m long. The sand forming the bed and the injected material had a mean diameter of 0.32 mm. From the uniform-flow measurements, the following values were computed: $a = 1.45 \times 10^{-3}$; $b = 5.0$; Manning $n = 0.022$; and the porosity of the sediment bed layer, $p = 0.4$. For the results presented here, the initial water discharge, $q_o = 0.020$ m^2/s; the uniform flow depth, $h_o = 0.05$ m; the initial bed slope, $S_o = 0.00356$; and the equilibrium sediment discharge, $q_{so} = 4$.

In the mathematical model, uniform unit discharge, uniform flow depth, and the initial bed elevations were specified at every node as the initial conditions. The transient state was initiated by increasing the rate of sediment discharge at the upstream end by Δq_s. As mentioned earlier, one boundary condition at the downstream end and two boundary conditions at the upstream end were specified. The upstream boundary representing constant discharge was implemented by specifying $q(0, t) = q_o$ for $t \geq 0$. However, the inclusion of the second boundary $q_s(0, t) = q_{so} + \Delta q_s$ was not as straightforward as the former; and it had to be translated into an equation so that the bed elevation at the upstream end could be calculated. This was achieved by assuming a fictitious node upstream of node 1 and specifying the sediment discharge at that node equal to $q_{so} + \Delta q_s$. By using the sediment continuity equation and applying the backward finite difference on the spatial differential term, we obtain

$$\left[(1-p)z + \frac{q_s h}{q}\right]_1^{k+1} = \left[(1-p)z + \frac{q_s h}{q}\right]_1^k + \frac{\Delta t}{\Delta x}\left[\left(q_{so} + \Delta q_s\right) - \left(q_s\right)\right]_1^k \quad (17\text{-}44)$$

The left-hand side of Eq. 17-44 and, therefore, z at the unknown time level $k + 1$ can be calculated, since the terms on the right-hand side are known for the time level k. The flow depth at node 1 is determined from the characteristic equation by using the known discharge at the boundary. The constant-depth downstream boundary condition was specified by $h(N\Delta x, t) = h_o$ for $t \geq 0$. The discharge and the bed elevation at the downstream end are determined by extrapolation from the values at the interior nodes. Note that the constant-depth boundary condition is valid for long channels in which bed transients do not reach the downstream end within the period for which conditions are computed. This boundary condition is not valid for flood flows and for short channels. The boundary values in these cases may be evaluated by the characteristic method along with a rating curve between the discharge and depth as the specified boundary condition.

The mathematical model was used on a 50-m-long channel, which was divided into 50 reaches ($\Delta x = 1.0$ m). The computational time step, Δt, was selected so that the Courant condition for stability ($C_n = 0.9$) was satisfied. Figure 17-6 compares the computed transient bed and water surface profiles at $t = 40$ min with the measured values. The "measured" points here correspond to the average transient profiles. Averaging of the actual data was required because of the presence of ripples and dunes on the bed. As can be seen, the mathematical model satisfactorily simulates the aggradation of the channel bed as well as the transient water surface profile.

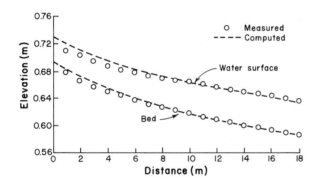

Figure 17-6 Comparison of computed and measured bed and water-surface profiles.

Knickpoint Migration

A *knickpoint* is defined as an abrupt change in the longitudinal bottom profile of a channel (Fig. 17-7). In a channel with nonerodible bed, a knickpoint remains intact indefinitely. However, in channels flowing over an erodible bed, the knickpoints are obliterated as they migrate upstream. Brush and Wolman (1960) explained the migration of knickpoints as a result of erosion potential (hS_f) becoming maximum at the point where slope changes. Referring to Fig. 17-7, the flow is subcritical on the upstream side of the knickpoint and supercritical on the downstream side. The boundary shear ($\tau_o = \gamma h S_f$) is maximum at the break in the bottom slope, it decreases downstream as the flow depth decreases, and it also decreases on the upstream side of the knickpoint as the energy-grade line flattens, even though flow depth is higher. Since the sediment transport is directly proportional to the shear stress, the higher shear stress at the knickpoint results in more material being carried away from this location than from the upstream or downstream reaches. Therefore, the knickpoint migrates upstream, the eroded material deposits downstream, and, finally, the oversteepened reach flattens.

The experimental data obtained by Brush and Wolman (1960) in a 15.8-m-long and 1.2-m-wide flume was compared with the computed results. Before the experiment,

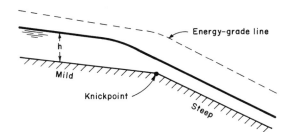

Figure 17-7 Definition sketch for knickpoint migration.

a 0.21-m-wide, trapezoidal channel with rounded corners was molded in noncohesive sands with a median size of 0.67 mm. A fall of approximately 0.0305 m was provided at a distance of 10.8 m from the upstream end to simulate the oversteepened reach or the knickpoint. The slope of the channel upstream and downstream of this point was approximately equal to 0.00125. Water was then turned on ($q_o = 0.0028$ m²/s, $h_o = 0.0305$ m) and the bed levels were recorded at successive times.

To simulate this experiment by the mathematical model, the channel was divided into 52 reaches ($\Delta x = 0.3048$ m). The initial and boundary conditions were included as in the previous case; the Courant number was 0.85; and the Jameson procedure (see Chapter 14) was used to add artificial viscosity. The sediment discharge was determined from Eq. 17-39, a value of $b = 4.2$ was used, and the coefficient a was assumed as a function of the energy-grade line, S_e. At the start of the computations for each new time step, the slope of the energy-grade line at every grid point was computed using the backward finite difference and the values of q, h, and z at the known time level:

$$S_e = \frac{1}{\Delta x}\left[\left(z_{i-1} + h_{i-1} + \frac{q_{i-1}^2}{2gh_{i-1}^2}\right) - \left(z_i + h_i + \frac{q_i^2}{2gh_i^2}\right)\right] \qquad (17\text{-}45)$$

Then the value of a was determined from $a = S_e^{1.71}$. The exponents in Eq. 17-39 were estimated from the sediment transport measurements (Brush and Wolman 1960). The conditions were identical in both the experiments except for the channel slope, and the difference in the sediment transport rates could be related to the slope of the energy grade line by a power law.

Figure 17-8 compares the computed and measured results at $t = 2.67$ h. The computed bed profile satisfactorily matches with the measured bed profile despite the inherent uncertainty in the sediment transport equation. As can be seen from Fig. 17-8, the downstream bed level in the experiment is higher than that predicted by the model, although the volume of predicted upstream erosion is almost equal to the deposition on the downstream side. Channel widening on the downstream side of the knickpoint was observed in the experiments and the larger deposition in the experiments might be due to extra sediment provided by the erosion of the bank.

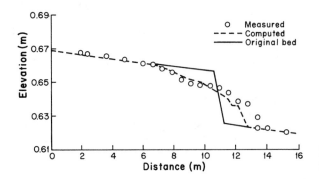

Figure 17-8 Bed level variation due to knickpoint migration.

Problems

17-1. Compute the outflows from a reservoir for the following data by using a routing interval of 5 min.

    i. Spillway outflow = $50H^{1.5}$, where H = head above the spillway crest in ft.

    ii. The reservoir has vertical sides and the surface area is 300,000 ft^2.

    iii. The inflow increases linearly from zero at $t = 0$ to 500 ft^3/sec at $t = 15$ min and then linearly decreases to 100 ft^3/sec in 10 min, after which the inflow remains constant at this value.

    iv. The reservoir is at the spillway crest level at time $t = 0$.

17-2. Write a computer program to route an inflow hydrograph through a reservoir. Assume the data for the inflow hydrograph are specified at discrete times and the reservoir surface area and the outflow through the spillway at specified elevations are given. Use a second-order accurate finite-difference scheme to integrate the governing equation and parabolic interpolation to determine values from the stored data.

17-3. For the following data, compute the outflow from a detention pond until time $t = 20$ min.

    i. Spillway crest level = El. 10 ft.

    ii. Spillway discharge (in ft^3/sec) = $100(z - 10)^{1.5}$, z = water level, in ft.

    iii. The pond surface area at El. 0 is 200,000 ft^2 and it linearly increases to 300,000 ft^2 at El. 40.

    iv. Inflow for $t < 10$ min is $5t$, where t is in seconds and inflow remains constant at 3000 ft^3/sec for $t > 10$ min.

    v. The pond level at time $t = 0$ is at El. 8 ft.

17-4. Prove that if $\alpha = 0.5$ and $\alpha \Delta t / \Delta x = 1$ in the Muskingum-Cunge routing, a wave does not attenuate as it is routed through a channel reach. [*Hint:* Determine C_o, C_1, and C_2 for these values and then show that a flood wave is not attenuated].

17-5. By expanding Eq. 17-12 in a Taylor series and comparing it with the diffusion equation, show that

$$X = \frac{1}{2} - \frac{Q_0}{2 S_0 a B \Delta x}$$

17-6. Route the flood hydrograph of Prob. 17-1 through a channel reach using kinematic routing for different values of C_n. Use Lax, MacCormack, and Preissmann schemes and compare their results with the exact solution.

17-7. By using the Lax and MacCormack schemes, study the effects of different values of the dispersion coefficient, D, in Eq. 17-32 on the computed results. Route the hydrgraph of Prob. 17-1.

REFERENCES

Begin, Z. B., Meyer, D. F., and Schumm, S. A. 1981. Development of longitudinal profiles of alluvial channels in response to base-level lowering, *Earth Surface Processes and Land Forms,* 6, no 1:49–68.

Bhallamudi, S. M., and Chaudhry, M. H. 1991. Numerical modeling of aggradation and degradation in alluvial channels, *Jour. Hydraulic Engineering,* Amer. Soc. Civil Engrs., 117, no. 9: 1145–64.

Brush, L. M. and Wolman, M. G. 1960. Knickpoint behavior in noncohesive material: A laboratory study, *Bulletin of the Geological Society of America,* 71: 59–74.

Cunge, J. A. 1969. On the subject of a flood propagation computation method (Muskingum method), *Jour. Hydraulic Research,* Inter. Assoc. Hydraulic Research, 7, no. 2: 205–30.

Cunge, J. A.; Holly, F. M., Jr; and Verwey, A. 1980. *Practical aspects of computational river hydraulics,* Pitman, London.

Dawdy, D. R. and Vanoni, V. A. 1986. Modeling alluvial channels, *Water Resources Research,* 22, no. 9: 71S–81S.

Fennema, R. J., and Chaudhry, M. H. 1990. Numerical solution of 2-D free-surface flows: Explicit methods, *Jour. Hydraulic Engineering,* Amer. Soc. Civil Engrs., 116, no. 8: 1013–34.

Gill, M. A. 1983a. Diffusion model for aggrading channels, *Jour. Hydraulic Research,* Inter. Assoc. Hydraulic Research, 21, no. 5: 355–67.

———. 1983b. Diffusion model for degrading channels, *Jour. Hydraulic Research,* Inter. Assoc. Hydraulic Research, 21, no. 5: 369–78.

———. 1987. Nonlinear solution of aggradation and degradation in channels, *Jour. Hydraulic Research,* Inter. Assoc. Hydraulic Research, 25, no. 5: 537–47.

Hayami, S. 1951. On the propagation of flood waves. *Bulletin of the Disaster Prevention Research Institute,* Disaster Prevention Research Institute, (Kyoto, Japan) 1, no. 1: 1–16.

Holly, F. M., Jr. 1986. Numerical simulation in alluvial hydraulics, *5th Congress of the Asian and Pacific Regional Division of International Association for Hydraulic Research* Seoul, Korea, August 18–21.

Hromadka, T. V., and DeVries, J. J. 1988. Kinematic wave and computational error, *Jour. Hydraulic Engineering,* Amer. Soc. Civil Engrs., 114, (2): 207–17 (Discussions and closure: 116 no. 2: 278–89).

Jain, S. C. 1981. River bed aggradation due to overloading, *Jour. of Hydraulic Division,* Amer. Soc. Civil Engrs., 107, no. 1: 120–24.

Jain, S. C. and Park, I. 1989. Guide for estimating riverbed degradation, *Jour. Hydraulic Engineering,* Amer. Soc. Civil Engrs., 115, no. 3: 356–66.

Jameson, A.; Schmidt, W.; and Turkel, E. 1981. Numerical solutions of the Euler equations by finite volume methods using Runge-Kutta time-stepping schemes, *AIAA 14th Fluid And Plasma Dynamics Conference,* Palo Alto, California, AIAA-81-1259.

Jaramillo, W. F. and Jain, S. C. 1984. Aggradation and degradation of alluvial-channel beds, *Jour. Hydraulic Engineering,* Amer. Soc. Civil Engrs., 110, no. 8: 1072–85.

Katopodes, N. D. 1982. On zero-inertia and kinematic waves, *Jour. Hydraulics Division, Amer. Soc. Civil Engrs.,* 108, no. 11: 1380–87.

Lane, E. W. and Borland, W. M. 1954. River-bed scour during floods, *Trans.,* Amer. Soc. Civil Engrs., 119: 1069–79.

Lee, J. K., and Froehlich, D. C. 1989. Two-dimensional finite-element hydraulic modeling of bridge crossings, Report FHWA-RD-88-146, U.S. Department of Transportation, McLean, VA.

Lighthill, M. J., and Whitham, G. B. 1955. On kinematic waves I: Flood movement in long rivers, *Proc., Royal Society* (London) A229: 281–316.

Lu, J-Y., and Shen, H. W. 1986. Analysis and comparisons of degradation models, *Jour. Hydraulic Engineering,* Amer. Soc. Civil Engrs., 112 no. 4: 281–99.

Lyn, D. A. 1987. Unsteady sediment transport modeling, *Jour. Hydraulic Engineering,* Amer Soc. Civil Engrs., 113 no. 1: 1–15.

MacCormack, R. W. 1969. The effect of viscosity in hypervelocity impact cratering, *American Institute of Aeronautics and Astronautics* Paper 69-354.

Morris, E. M., and Woolhiser, D. A. 1980. Unsteady one-dimensional flow over a plane: Partial equilibrium and recession hydrograph, *Water Resources Research,* 16, no. 2: 355–60.

Newton, C. T. 1951. An experimental investigation of bed degradation in an open channel, *Trans.,* Boston Soc. Civil Engrs., pp. 28–60.

Park, I., and Jain, S. C. 1986. River-bed profiles with imposed sediment load, *Jour. Hydraulic Engineering,* Amer. Soc. Civil Engrs., 112, no. 4: 267–79.

Ponce, V. M. 1990. *Engineering hydrology,* Prentice Hall, Englewood Cliffs, NJ.

———— 1991. The kinematic wave controversy, *Jour. Hydraulic Engineering,* Amer. Soc. Civil Engrs., 117, no. 4: 511–25.

————; Li, R. M.; and Simons, D. B. 1978. Applicability of kinematic and diffusion models, *Jour. Hydraulic Engineering,* Amer. Soc. Civil Engrs., 104, no. 3: 353–60.

————, and Theurer, F. D. 1982. Accuracy criteria in diffusion routing, *Jour. Hydraulic Engineering,* Amer. Soc. Civil Engrs., 108, no. 6: 747–57.

Roache, P. J. 1972. *Computational fluid dynamics,* Hermosa Publishers, Albuquerque, NM.

Roberson, J. A., Cassidy, J. J., and Chaudhry, M. H., 1988, *Hydraulic Engineering,* Houghton Mifflin, Boston.

Seddon, J. A. 1900. River hydraulics, *Trans.,* Amer. Soc. Civil Engrs., pp. 179–229.

Soni, J. P.; Garde, R. J.; and Raju, K. G. 1980. Aggradation in streams due to overloading, *Jour Hydraulic Division,* Proc. Amer. Soc. Civil Engrs., 106, no. 1: 117–32.

Soni, J. P.; Garde, R. J.; Raju, K. R.; and Kittur, G. J. 1977. Nonuniform flow in aggrading channels, *Jour. Waterways Port Coastal Ocean Division,* Amer Soc. Civil Engrs., 103, no. 3 (August): 321–33.

Suryanarayana, B. 1969. Mechanics of degradation and aggradation in a laboratory flume, Ph.D. diss., Colorado State University, Ft. Collins, Colorado.

U.S. Army Corps of Engineers. 1985. *HEC-1, Flood hydrograph package: users' manual.* Hydrologic Engineering Center, Davis, Calif.

Woolhiser, D. A., and Liggett, J. A. 1967. Unsteady one-dimensional flow over a plane — the rising hydrograph, *Water Resources Research,* 3, no. 3: 753–71.

Zhang, H., and Kahawita, R. 1987. Nonlinear model for aggradation in alluvial channels, *Jour. Hydraulic Engineering,* Amer. Soc. Civil Engrs., 113, no. 3: 353–69.

A Appendix

```
C
C          COMPUTATION OF ENERGY AND MOMENTUM COEFFICIENT
C          FOR A RECTANGULAR CHANNEL USING TRAPEZOIDAL RULE
C
C          ****************** NOTATION *************************
C
C          ALPHA = KINETIC ENERGY COEFFICIENT;
C          BETA  = MOMENTUM COEFFICIENT;
C          B0 = CHANNEL BOTTOM WIDTH;
C          N = NUMBER OF POINTS IN THE VELOCITY PROFILE;
C          S = SLOPE OF CHANNEL SIDES, S: HORIZONTAL TO 1 VERTICAL;
C          V = ARRAY CONTAINING THE VELOCITIES AT KNOWN DEPTHS;
C          Y = ARRAY FOR WATER DEPTH AT WHICH VELOCITIES ARE GIVEN;
C          VM = MEAN FLOW VELOCITY
C          *********************************************************
C
           DIMENSION Y(20),V(20),F(20),VALUE(3)
           READ(5,*) B0,S
           READ(5,*) N,(Y(I),V(I),I = 1,N)
           DO 10 J = 1,3
             I = 1
  20         IF(I.LE.N) THEN
                F(I) = V(I)**J*(2*S*Y(I) + B0)
                I = I+1
                GO TO 20
             END IF
           CALL TRAP (N,Y,F,XINT)
           VALUE(J) = XINT
  10       CONTINUE
           AR = (B0+S*Y(N))*Y(N)
           VM = VALUE(1)/AR
           BETA = VALUE(2)/(AR*VM**2)
           ALPHA = VALUE(3)/(AR*VM**3)
           WRITE(6,40) VM,B0,S
           WRITE(6,50) ALPHA, BETA
```

```
40    FORMAT(3X,'MEAN FLOW VELOCITY = ',F6.2/,3X,'CHANNEL BOTTOM WIDTH
     1 = ',F6.2,/3X,'CHANNEL LATERAL SLOPE = ',F6.2//)
50    FORMAT(2X,' * VELOCITY DISTRIBUTION COEFFICIENTS *'/,
     1 2X,'KINETIC ENERGY COEFFICIENT = ',F6.2/,2X,
     2 'MOMENTUM COEFFICIENT = ',F6.2)
      STOP
      END
C
C       TRAPEZOIDAL RULE SUBROUTINE
C
      SUBROUTINE TRAP(N,Y,F,SUM)
      DIMENSION Y(20),F(20)
      SUM = 0.0
      DO 10 I = 1,N-1
       DY = Y(I+1) - Y(I)
       SUM = SUM + 0.5*DY*(F(I+1)+F(I))
10    CONTINUE
      RETURN
      END
```

*************** INPUT DATA ********************************

```
10,1
7,0.0,0.0,0.2,3.87,0.4,4.27,0.6,4.53,0.8,4.72,1.0,4.87,1.2,5.0
```

*************** PROGRAM OUTPUT ****************************

```
TRAPEZOIDAL RULE
MEAN FLOW VELOCITY =    4.19
CHANNEL BOTTOM WIDTH =  10.00
CHANNEL LATERAL SLOPE =    1.00

* VELOCITY DISTRIBUTION COEFFICIENTS *
  KINETIC ENERGY COEFFICIENT =   1.19
  MOMENTUM COEFFICIENT =   1.09
```

B-1 Bisection Method

```
C      COMPUTATION OF CRITICAL DEPTH IN A TRAPEZOIDAL CHANNEL
C       USING BISECTION METHOD
C
C      ******************* NOTATION **********************
C
C      ALPHA = VELOCITY-HEAD COEFFICIENT;
C      BO = CHANNEL-BOTTOM WIDTH, IN M;
C      G = ACCELERATION DUE TO GRAVITY, IN M/S2;
C      Q = DISCHARGE, IN M/S;
C      THETA = SLOPE OF CHANNEL BOTTOM, IN DEGREES;
C      TOLER = SPECIFIED TOLERANCE;
C      S = SLOPE OF CHANNEL SIDES, S: HORIZONTAL TO 1 VERTICAL;
C      YC = CRITICAL DEPTH, IN M;
C      YL = INITIAL LOW ESTIMATE FOR CRITICAL DEPTH;
C      YR = INITIAL HIGH ESTIMATE FOR CRITICAL DEPTH.
C      ********************************************************
C
       READ(5,*) Q,BO,S,YL,YR,G,ALPHA,THETA,TOLER
       WRITE (6,6)
6      FORMAT (5X,'BISECTION METHOD'/)
       WRITE(6,20) Q,BO,S,YL,YR,G,ALPHA,THETA
20     FORMAT(3X,'DISCHARGE =',F8.2,' M3/S'/3X,'BOTTOM WIDTH =',F5.2,
      1  ' M'/3X,'SIDE SLOPE =',F4.2/3X,'INITIAL ESTIMATE FOR YL =',
      2  F5.2,' M'/3X,'INITIAL ESTIMATE FOR YR =',F5.2,' M'/
      3  3X,'G =',F5.2,' M/S2'/3X,'ALPHA =',F5.2/3X,'THETA =',
      4  F5.2,' DEG'/)
       THETA = 0.0349*THETA
       CONST=(Q/COS(THETA))/SQRT(G/ALPHA)
       DBDY=2.*S
       K=0
```

```
30      K=K+1
        IF (K.GT.50) GO TO 80
        B = BO+2.*S*YI
        A= 0.5*(BO+B)*YI
        F=A**1.5/SQRT(B) - CONST
        IF (F.LT.0.0) YL=YI
        IF (F.GT.0.0) YR=YI
        IF (ABS(F).LE.TOLER) GO TO 60
        YI=0.5*(YL+YR)
        GO TO 30
60      YC=0.5*(YL+YR)
        WRITE(6,70) YC
70      FORMAT(5X,'CRITICAL DEPTH =',F6.2,' M')
        GO TO 90
80      WRITE(6,85)
85      FORMAT(10X,'ITERATIONS FAILED')
90      STOP
        END
```

```
*********** INPUT DATA ***************

1000.,10.,2.0,.10,10.,9.81,1.,0.0,0.001


********** PROGRAM OUTPUT *************

BISECTION METHOD
DISCHARGE = 1000.00 M3/S
BOTTOM WIDTH =10.00 M
SIDE SLOPE =2.00
INITIAL ESTIMATE FOR YL = .10 M
INITIAL ESTIMATE FOR YR =10.00 M
G = 9.81 M/S2
ALPHA = 1.00
THETA =   .00 DEG

CRITICAL DEPTH =  6.65 M
```

B-2 Newton-Raphson Method

```
C       COMPUTATION OF CRITICAL DEPTH IN A TRAPEZOIDAL CHANNEL
C        USING NEWTON-RAPHSON METHOD
C
C       ******************* NOTATION ************************
C
C       ALPHA = VELOCITY-HEAD COEFFICIENT;
C       BO = CHANNEL-BOTTOM WIDTH;
```

```
C       G = ACCELERATION DUE TO GRAVITY;
C       Q = DISCHARGE;
C       S = SLOPE OF CHANNEL SIDES, S: HORIZONTAL TO 1 VERTICAL;
C       THETA = CHANNEL-BOTTOM SLOPE, IN DEGREES;
C       TOLER = SPECIFIED TOLERANCE;
C       YC = CRITICAL DEPTH;
C       YI = INITIAL ESTIMATE FOR CRITICAL DEPTH.
C       ********************************************************
C
        READ(5,*) Q,BO,S,YI,G,ALPHA,THETA,TOLER
        WRITE (6,6)
6       FORMAT (5X,'NEWTON-RAPHSON METHOD'/)
        WRITE(6,20) Q,BO,S,YI,G,ALPHA,THETA,TOLER
20      FORMAT(3X,'DISCHARGE =',F8.2,' M3/S'/3X,'BOTTOM WIDTH =',F5.2,
     1    ' M'/3X,'SIDE SLOPE =',F4.2/3X,'INITIAL ESTIMATE FOR YC =',
     2     F5.2,' M'/3X,'G =',F5.2,' M/S2'/3X,'ALPHA =',F5.2/
     3     3X,'CHANNEL BOTTOM SLOPE =',F5.2,' DEGREES'/
     4     3X,'SPECIFIED TOLERANCE FOR CONVERGENCE OF ITERATIONS =',
     5      F6.3,' M'/)
        THETA=0.034906*THETA
        CONST=(Q/SQRT(COS(THETA)))/SQRT(G/ALPHA)
        DBDY=2.*S
        K=0
30      K=K+1
        IF (K.GT.50) GO TO 80
        B = BO+2.*S*YI
        A= 0.5*(BO+B)*YI
        F=A**1.5/SQRT(B)-CONST
        FD=1.5*SQRT(A*B) - 0.5*((A/B)**1.5)*DBDY
        DY=F/FD
        YI=YI-DY
        IF (ABS(DY).LE.TOLER) GO TO 60
        GO TO 30
60      WRITE(6,70) YI
70      FORMAT(5X,'CRITICAL DEPTH =',F6.2,' M')
        GO TO 90
80      WRITE(6,85)
85      FORMAT(10X,'ITERATIONS FAILED')
90      STOP
        END
```

```
********** INPUT DATA *************

1000.,10.,2.0,5.0,9.81,1.,0.0,0.001

********** PROGRAM OUTPUT *********

NEWTON-RAPHSON METHOD
DISCHARGE = 1000.00 M3/S
```

```
BOTTOM WIDTH =10.00 M
SIDE SLOPE =2.00
INITIAL ESTIMATE FOR YC = 5.00 M
G = 9.81 M/S2
ALPHA = 1.00
CHANNEL BOTTOM SLOPE =  .00 DEGREES
SPECIFIED TOLERANCE FOR CONVERGENCE OF ITERATIONS =  .001 M

 CRITICAL DEPTH = 6.65 M
```

C Appendix

C-1 Bisection Method

```
C       COMPUTATION OF NORMAL DEPTH IN A TRAPEZOIDAL CHANNEL
C       USING BISECTION METHOD
C
C       ********************** NOTATION ************************
C
C       Q = DISCHARGE, IN m3/s;
C       BO = CHANNEL BOTTOM WIDTH, IN m;
C       S  = SLOPE OF CHANNEL SIDES, S: HORIZONTAL TO 1 VERTICAL;
C       YL = INITIAL LOW ESTIMATE FOR NORMAL DEPTH, IN m;
C       YH = INITIAL HIGH ESTIMATE FOR NORMAL DEPTH, IN m;
C       YNORM = NORMAL DEPTH, IN m;
C       AR = ARITHMETICAL FUNCTION FOR FLOW AREA;
C       HR = ARITHMETICAL FUNCTION FOR HYDRAULIC RADIUS;
C       CMAN = MANNING'S COEFFICIENT;
C       SO = CHANNEL BOTTOM SLOPE;
C       ****************************************************
C
        AR(BO,S,Y) = (BO + S*Y)*Y
        HR(BO,S,Y) = ((BO + S*Y)*Y)/(BO + 2*Y*SQRT(1+S*S))
        READ(5,*) Q,CMAN,BO,S,SO
        READ(5,*) YL,YH
        WRITE(6,5) YL,YH
        C1 = (CMAN*Q)/SQRT(SO)
10      YNEW = (YL+YH)/2
        FL = AR(BO,S,YL)*HR(BO,S,YL)**0.6667 - C1
        FYN = AR(BO,S,YNEW)*HR(BO,S,YNEW)**0.6667 - C1
        IF(FL*FYN .LT. 0.0) THEN
          ERR = ABS((YH-YNEW)/YNEW)
          YH = YNEW
```

```
     ELSE
       ERR = ABS((YL-YNEW)/YNEW)
       YL = YNEW
     END IF
     IF(ERR.GT. 5E-03) GO TO 10
     YNORM = YNEW
     WRITE(6,20) Q,CMAN,BO,S,SO,YNORM
  5  FORMAT(2X, 'BISECTION METHOD',/2X,'INITIAL ESTIMATE FOR YL = ',
    1 F6.2,' m',/2X,'INITIAL ESTIMATE FOR YH = ',F6.2,' m')
 20  FORMAT(2X,'DISCHARGE = ',F6.2,'m3/s',/2X
    1,'MANNING N = ',F6.4,/2X,'CHANNEL BOTTOM WIDTH = ',F6.2,' m'/2X
    2,'SIDE SLOPE = ',F5.2,/2X,'CHANNEL BOTTOM SLOPE = ',F6.4,//2X,
    3'NORMAL DEPTH = ',F6.2,' m')
     STOP
     END
```

***** INPUT DATA ******

```
30,0.013,10.0,2.0,0.001
0.5,10
```

****** PROGRAM OUTPUT ******

```
BISECTION METHOD
INITIAL ESTIMATE FOR YL = .50 m
INITIAL ESTIMATE FOR YH = 10.00 m
DISCHARGE =   30.00m3/s
MANNING N =   .0130
CHANNEL BOTTOM WIDTH =  10.00 m
SIDE SLOPE =  2.00
CHANNEL BOTTOM SLOPE =  .0010

NORMAL DEPTH =   1.09 m
```

C-2 Newton-Raphson Method

```
C     COMPUTATION OF NORMAL DEPTH IN A TRAPEZOIDAL CHANNEL
C     USING NEWTON-RAPHSON METHOD
C
C     **************** NOTATION ***************************
C
C     Q = DISCHARGE, m3/S;
C     BO = CHANNEL BOTTOM WIDTH, m;
C     S = SLOPE OF CHANNEL SIDES, S HORIZONTAL TO 1 VERTICAL;
C     Y = INITIAL ESTIMATE FOR NORMAL DEPTH, m;
C     YNORM = NORMAL DEPTH, m;
C     AR = STATEMENT FUNCTION FOR FLOW AREA;
```

```
C      HR = STATEMENT FUNCTION FOR HYDRAULIC RADIUS;
C      BW = STATEMENT FUNCTION FOR CHANNEL TOP WIDTH;
C      CMAN = MANNING'S COEFFICIENT;
C      SO = CHANNEL BOTTOM SLOPE;
C      TOL = TOLERANCE FOR ITERATIONS
C      **********************************************************
C
       AR(B0,S,Y) = (B0 + S*Y)*Y
       HR(B0,S,Y) = ((B0 + S*Y)*Y)/(B0 + 2*Y*SQRT(1+S*S))
       BW(B0,S,Y) = B0 + 2*S*Y
       READ(5,*) Q,CMAN,B0,S,SO
       READ(5,*) Y,TOL
       WRITE(6,5) Y,TOL
       C1 = (CMAN*Q)/SQRT(SO)
       C2 = 2*SQRT(1 + S*S)
       ERR = 1.0E06
10     IF(.NOT.(ERR.LT.TOL)) THEN
          FY = AR(B0,S,Y)*HR(B0,S,Y)**0.6667 - C1
          DFDY = 1.6667*BW(B0,2,Y)*HR(B0,S,Y)**0.6667 - 0.6667*HR(B0,S,Y)
     1          **1.6667*C2
          YNEW = Y - FY/DFDY
          ERR = ABS((YNEW-Y)/YNEW)
          Y = YNEW
          GO TO 10
       END IF
       WRITE(6,20) Q,CMAN,B0,S,SO,YNEW
  5    FORMAT(2X,'NEWTON RAPHSON METHOD',/2X,'INITIAL ESTIMATE FOR Y =
     1 ',F6.2,' m',/2X,'TOLERANCE = ',F6.4)
 20    FORMAT(2X,'DISCHARGE = ',F6.2,'m3/s',/2X
     1 ,'MANNING N = ',F6.4,/2X,'CHANNEL BOTTOM WIDTH = ',F6.2,' m'/2X
     2 ,'SIDE SLOPE = ',F5.2,/2X,'CHANNEL BOTTOM SLOPE = ',F6.4,//2X,
     3'NORMAL DEPTH = ',F6.2,' m')
       STOP
       END
```

****** INPUT DATA ******

```
30,0.013,10.0,2.0,0.001
0.5,0.0001
```

**** PROGRAM OUTPUT ****

```
NEWTON RAPHSON METHOD
INITIAL ESTIMATE FOR Y =  .50 m
TOLERANCE =  .0001
```

```
DISCHARGE =  30.00m3/s
MANNING N =  .0130
CHANNEL BOTTOM WIDTH =  10.00 m
SIDE SLOPE =  2.00
CHANNEL BOTTOM SLOPE =  .0010

NORMAL DEPTH =    1.09 m
```

D Appendix

D-1 Direct Step Method

```
C      COMPUTATION OF WATER-SURFACE PROFILE BY USING
C         DIRECT STEP METHOD
C      DOWNSTREAM FLOW DEPTH SPECIFIED; COMPUTATIONS PROCEED IN THE
C         UPSTREAM DIRECTION.
C
C      ******************** NOTATION ********************************
C      A=FLOW AREA;
C      B=TOP WATER-SURFACE WIDTH;
C      BO=CHANNEL-BOTTOM WIDTH;
C      E = SPECIFIC ENERGY;
C      G = ACCELERATION DUE TO GRAVITY;
C      MN = MANNING N;
C      P = WETTED PERIMETER;
C      R = HYDRAULIC RADIUS;
C      Q = DISCHARGE;
C      N = NUMBER OF FLOW DEPTHS FOR WHICH DISTANCES ARE TO BE COMPUTED;
C      S = CHANNEL-SIDE SLOPE, S HORIZONTAL : 1 VERTICAL;
C      SO = CHANNEL-BOTTOM SLOPE;
C      SF = SLOPE OF ENERGY GRADE LINE;
C      X = DISTANCE ALONG CHANNEL BOTTOM, POSITIVE IN THE DOWNSTREAM
C            DIRECTION;
C      Y = FLOW DEPTH;
C      YD = DEPTH AT DOWNSTREAM END.
C      ************************************************************
C
       REAL MN,MN2
       DIMENSION Y(50)
       AR(YY)=YY*(BO+S*YY)
       WP(YY)=BO+2.*YY*SQRT(1.+S*S)
       READ (5,*) BO,S,SO,MN,Q,G
       WRITE (6,6)
6      FORMAT (5X,'DIRECT STEP METHOD'/)
       WRITE(6,10) BO,S,SO,MN,Q,G
```

```
10     FORMAT(5X,'B =',F5.1,' M'/5X,'S =',F4.1/5X,'SO =',F6.4/
    1   5X,'N =',F5.3/5X,'Q =',F10.3,' M3/S'/5X,'G =',F5.3/)
       MN2=MN*MN
       READ (5,*) N,(Y(I),I=1,N)
       X1=0.0
       A1=AR(Y(1))
       P1=WP(Y(1))
       R1=A1/P1
       V1=Q/A1
       E1 = Y(1)+(V1*V1)/(2.*G)
       SF1=(MN2*V1*V1)/(R1**1.333)
       WRITE(6,14) Y(1),X1
14     FORMAT(5X,'Y',10X,'X'//F8.2,F10.1)
       DO 30 I = 2,N
       A2=AR(Y(I))
       P2=WP(Y(I))
       R2=A2/P2
       V2=Q/A2
       SF2=(MN2*V2*V2)/(R2**1.333)
       E2=Y(I)+(V2*V2)/(2.*G)
       SFA=0.5*(SF1+SF2)
       DE=E2-E1
       DEN=SO-SFA
       DX=DE/DEN
       X2=X1+DX
       WRITE(6,20) Y(I),X2
20     FORMAT(F8.2,F10.1)
       X1=X2
       E1=E2
       SF1=SF2
30     CONTINUE
       STOP
       END
```

```
 ************ INPUT DATA *****************

 20.,2.,.001,.013,900.,9.81
 14,11.,10.,9.,8.5,8.,7.5,7., 6.75,6.5,6.25,6.,5.9,5.8,5.7

************ PROGRAM OUPUT **************

DIRECT STEP METHOD

B = 20.0 M
S = 2.0
SO = .0010
N = .013
Q =   900.000 M3/S
G =9.810
```

Y	X
11.00	.0
10.00	-998.6
9.00	-1999.3
8.50	-2500.2
8.00	-3002.9
7.50	-3508.7
7.00	-4019.8
6.75	-4278.1
6.50	-4539.8
6.25	-4806.7
6.00	-5082.0
5.90	-5195.5
5.80	-5311.9
5.70	-5432.5

D-2 Standard Step Method

```
C       COMPUTATION OF WATER-SURFACE PROFILE BY USING
C          STANDARD STEP METHOD
C       DOWNSTREAM FLOW DEPTH SPECIFIED, COMPUTATIONS PROCEED IN
C       THE UPSTREAM DIRECTION.
C       THE ENERGY EQUATION IS SOLVED USING NEWTON-RAPHSON METHOD
C
C       ************** NOTATION *************************************
C       A=FLOW AREA;
C       B=TOP WATER-SURFACE WIDTH;
C       BO=CHANNEL-BOTTOM WIDTH;
C       P = WETTED PERIMETER;
C       Q = DISCHARGE;
C       MN = MANNING'S N;
C       S = CHANNEL-SIDE SLOPE, S HORIZONTAL : 1 VERTICAL;
C       SO = CHANNEL-BOTTOM SLOPE;
C       X = DISTANCE ALONG CHANNEL BOTTOM, POSITIVE IN THE DOWNSTREAM
C            DIRECTION;
C       Y = FLOW DEPTH
C       YD = DEPTH AT DOWNSTREAM END.
C       ************************************************************
C
        REAL MN
        DIMENSION X(100)
        AR(YY)=YY*(BO+S*YY)
        WP(YY)=BO+2.*YY*SQRT(1.+S*S)
        READ (5,*) BO,S,SO,MN,Q,YD,ZD,G,ALPHA
        READ (5,*) N,(X(I),I=1,N)
        WRITE(6,5)
```

```
 5      FORMAT(5X,'STANDARD STEP METHOD'/)
        WRITE(6,10) BO,S,SO,MN,Q,YD,ZD,G,ALPHA
10      FORMAT(5X,'B = ',F5.1,' M'/5X,'S = ',F4.2/5X,'SO = ',F6.4/
     1    5X,'N = ',F5.3/5X,'Q = ',F9.2,' M3/S'/5X,'YD = ',F8.3,' M'/
     2    5X,'DOWNSTREAM INVERT LEVEL = ',F8.2,' M'/
     3    5X,'G = ',F6.2,' M/S2'/5X,'ALPHA = ',F6.2/)
        Q2=Q*Q
        ALQ=ALPHA*Q2/(2.*G)
        ALQ2=2.*ALQ
        QN2=(MN*Q)**2
        Y1=YD
        Z1=ZD
        DPDY=2.*SQRT(1.+S*S)
        A1=AR(Y1)
        P1=WP(Y1)
        R1=A1/P1
        B1=BO+2.*S*Y1
        SF1=QN2/(A1*A1*R1**1.3333)
        WRITE(6,15)
15      FORMAT(6X,'X',10X,'Y')
        WRITE(6,20) X(1),Y1
        DX=X(2)-X(1)
        DO 30 I = 2,N
        DX=X(I)-X(I-1)
        Z2=Z1-SO*DX
        H1=Z1+Y1+ALQ/(A1*A1)
        YE= Y1
        KK=0
16      KK=KK+1
        AE=AR(YE)
        PE=WP(YE)
        RE=AE/PE
        VE = Q/AE
        VHE=ALPHA*VE*VE/(2.*G)
        HE=Z2+YE+VHE
        SFE=QN2/(AE*AE*RE**1.333)
        SFA=0.5*(SF1+SFE)
        HF=DX*SFA
        H2E=H1-HF
        BE=BO+2.*S*YE
        DRDY=BE/PE - RE*DPDY/PE
        F=YE+ALQ/(AE*AE)+0.5*SFE*DX+Z2-H1+0.5*SF1*DX
        FD=1.- ALQ2*BE/(AE**3) - DX*SFE*(BE/AE + .6667*DRDY/RE)
        DY=F/FD
        Y=YE-DY
        IF (ABS(DY).LE.0.0001) GO TO 18
        YE=Y
        IF (KK.GT.150) GO TO 100
        GO TO 16
18      WRITE(6,20) X(I),Y
20      FORMAT(F10.1,F10.3)
        SF1=SFE
        Z1=Z2
        Y1 = Y
30      CONTINUE
        GO TO 200
100     WRITE (6,150) I,Y
```

```
150    FORMAT (5X,'ITERATIONS FAILED'/5X, 'SECTION NO. =',I3/
    1  5X,'DEPTH =',F8.2,' M')
200    STOP
       END
```

                  INPUT DATA

```
40.,2.,.001,.035,4000.,15.,0.,9.81,1.
4,0.,-1000.,-2000.,-4000.
```

                  PROGRAM OUTPUT

STANDARD STEP METHOD

```
B =   40.0 M
S = 2.00
SO =   .0010
N =   .035
Q =   4000.00 M3/S
YD =   15.000 M
DOWNSTREAM INVERT LEVEL = .00 M
G =   9.81 M/S2
ALPHA =   1.00
```

```
    X         Y
      .0    15.000
 -1000.0    14.843
 -2000.0    14.699
 -4000.0    14.509
```

D-3 Improved Euler Method

```
C      COMPUTATION OF WATER-SURFACE PROFILE BY USING
C        IMPROVED EULER METHOD
C      DOWNSTREAM DEPTH SPECIFIED, COMPUTATIONS PROCEED IN
C      THE UPSTREAM DIRECTION
C
C      ****************** NOTATION ******************************
C      A=FLOW AREA;
C      B=TOP WATER-SURFACE WIDTH;
C      BO=CHANNEL-BOTTOM WIDTH;
C      P = WETTED PERIMETER;
```

```
C      Q = DISCHARGE;
C      MN = MANNING'S N;
C      S = CHANNEL-SIDE SLOPE, S HORIZONTAL : 1 VERTICAL;
C      SO = CHANNEL-BOTTOM SLOPE;
C      X = DISTANCE ALONG CHANNEL BOTTOM, POSITIVE IN THE DOWNSTREAM
C            DIRECTION;
C      Y = FLOW DEPTH
C      YD = DEPTH AT DOWNSTREAM END.
C      *************************************************************
C
       REAL MN
       DIMENSION X(100)
       AR(Y)=Y*(BO+S*Y)
       WP(Y)=BO+2.*Y*SQRT(1.+S*S)
       READ (5,*) BO,S,SO,MN,Q,YD
       READ (5,*) N,(X(I),I=1,N)
       WRITE (6,6)
6      FORMAT (5X,'IMPROVED EULER METHOD'/)
       WRITE(6,10) BO,S,SO,MN,Q,YD
10     FORMAT(5X,'B =',F5.1,' M'/5X,'S =',F5.1/5X,'SO =',F8.4/
      1    5X,'N =',F6.3/5X,'Q =',F9.2,' M3/S'/5X,'YD =',F7.3,' M'/)
       Q2=Q*Q
       QN2=(MN*Q)**2
       Y=YD
       WRITE(6,15)
15     FORMAT(6X,'X',10X,'Y')
       WRITE(6,20) X(1),Y
       DO 30 I = 2,N
       DX=X(I)-X(I-1)
       A=AR(Y)
       P=WP(Y)
       R=A/P
       SF1=QN2/(A*A*R**1.333)
       B=BO+2.*S*Y
       DY1=(SO-SF1)/(1-(B*Q2)/(9.81*A**3))
       Y2=Y+DY1*DX
       A=AR(Y2)
       P=WP(Y2)
       R=A/P
       SF2=QN2/(A*A*R**1.333)
       B=BO+2.*S*Y2
       DY2=(SO-SF2)/(1.-(B*Q2)/(9.81*A**3))
       Y= Y+0.5*(DY1+DY2)*DX
```

```
      WRITE(6,20) X(I),Y
20    FORMAT(F10.1,F10.3)
30    CONTINUE
      STOP
      END
```

```
****** INPUT DATA *********

40.,2.,0.001,.035,4000.,15.
4,0.0,-1000.,-2000.,-4000.
```

```
****** PROGRAM OUTPUT *****

IMPROVED EULER METHOD

B = 40.0 M
S =  2.0
SO =   .0010
N =  .035
Q =  4000.00 M3/S
YD = 15.000 M

    X          Y
     .0      15.000
-1000.0      14.846
-2000.0      14.727
-4000.0      14.579
```

D-4 Simultaneous Solution Approach

```
C     SIMULTANEOUS SOLUTION OF SERIES CHANNELS USING NEWTON-RAPHSON
C     METHOD.  TRAPEZOIDAL CHANNEL.
C     UPSTREAM FLOW DEPTH SPECIFIED.
C     ENGLISH OR SI UNITS CAN BE SPECIFIED.
C
```

```
C       ********************* NOTATION ***************************
C
C       A = FLOW AREA;
C       ALPHA = KINETIC ENERGY COEFFICIENT;
C       BOT = CHANNEL BOTTOM WIDTH;
C       CHELEV = UPSTREAM CHANNEL BOTTOM ELEVATION;
C       CHL = CHANNEL LENGTH;
C       CMAN = MANNING'S COEFFICIENT;
C       C0 = CONSTANT FOR MANNING'S EQUATION;
C       DF = JACOBIAN MATRIX;
C       EQ = CHANNEL REACH EQUATIONS;
C       ELEV = WATER SURFACE ELEVATION
C       G = ACCELERATION OF GRAVITY;
C       HR = HYDRAULIC RADIUS;
C       MAXI = MAXIMUM NUMBER OF ITERATIONS;
C       NCHAN = NUMBER OF CHANNELS;
C       NEQ = TOTAL NUMBER OF EQUATIONS;
C       NR = NUMBER OF REACHES;
C       PER = WETTED PERIMETER;
C       Q0 =  CHANNEL DISCHARGE;
C       S0 = CHANNEL BOTTOM SLOPE;
C       TOL = TOLERANCE FOR CONVERGENCE OF ITERATIONS
C       TOP = TOP WIDTH;
C       UNIT = VARIABLE FOR UNITS SYSTEM;
C       Y = FLOW DEPTH;
C       Y0 = INITIAL DEPTH ESTIMATE;
C       YUP = FLOW DEPTH AT UPSTREAM END;
C       Z0 = LATERAL SLOPE, Z0 HORIZONTAL TO 1 VERTICAL
C       ***********************************************************
C
        CHARACTER*15 UNIT
        COMMON/GROUP1/ DF(100,100),F(100),X(100),NEQ
        COMMON  MAXI,G,TOL,YUP,C0,FLAG,KOUNT,NITER,Y0,Q0,S0(5),CHL(5),
       1CMAN(5),BOT(5),Z0(5),NR(5),CHELEV(5),ALPHA(5),NCHAN,DX(5),DPDY(5),
       2Y(5,100),ELEV(5,100),A(5,100),HR(5,100),PER(5,100),TOP(5,100),
       3DRDY(5,100),EQ(5,100),ITER
        FLAG = 0.0
        READ(5,5) NCHAN, MAXI,G,TOL,YUP,Y0,Q0
        DO 10 I = 1, NCHAN
        READ(5,20) S0(I),CHL(I),CMAN(I),BOT(I),Z0(I)
        READ(5,15) NR(I),CHELEV(I), ALPHA(I)
  10    CONTINUE
        IF(G.LE.11) THEN
          C0 = 1.0
          UNIT = 'SI'
```

```
      ELSE
        CO = 1.49
        UNIT = 'ENGLISH'
      END IF
   DO WHILE(ITER.LE.MAXI)
        L = 1
        F(L) = 0.0
        X(L) = 0.0
   DO WHILE(L.LE.100)
        LL = 1
        DO WHILE(LL.LE.100)
        DF(L,LL) = 0.0
        LL = LL+ 1
        END DO
        L = L+1
        END DO
        CALL REACH
        CALL TRIMAT
        CALL NEWVAL
   END DO
   IF (KOUNT.GE.NEQ) THEN
        WRITE(6,16) UNIT
        WRITE(6,21) NCHAN,NITER, TOL,G,Y0
        I = 1
        DO WHILE (I.LE.NCHAN)
         WRITE(6,30) I, SO(I),CHL(I),CMAN(I),BOT(I),Z0(I),DX(I),Q0,
     *                   ALPHA(I)
         WRITE(6,35)
         J = 1
         XLENGTH = -DX(I)
          DO WHILE (J.LE.NR(I)+1)
             XLENGTH = XLENGTH + DX(I)
             WRITE(6,40)J,XLENGTH,Y(I,J),ELEV(I,J)
          J = J+1
          END DO
        I = I+1
        END DO
   ELSE
   WRITE(6,50)
   END IF
5     FORMAT(I2,I3,F6.3,F7.4,3F7.2)
 20   FORMAT(5F10.5)
 15   FORMAT(I3,F10.3,F5.2)
16    FORMAT(/5X,'UNITS = ', A10)
21    FORMAT(5X,'NUMBER OF CHANNELS = ', I3,/5X,
     *'NUMBER OF ITERATIONS PERFORMED = ',I3,/5X,
     *'TOLERANCE = ',F10.6,/5X,'ACCELERATION OF GRAVITY = ', F6.2/
     *5X,'INITIAL VALUE FOR ITERATIONS = ',F6.2)
```

```
30      FORMAT(/5X,'** CHANNEL NUMBER ',I2,1X,'***',/5X,'BOTTON SLOPE =
       *',F10.5,/5X,'CHANNEL LENGTH = ',F8.2,/5X,'MANNING"S COEFFICIENT
       *= ',F7.5,/5X,'BOTTON WIDTH = ',F6.2,/5X,'LATERAL SLOPE = 1H:',
       *F4.1,'V',/5X,'LENGTH INTERVAL FOR COMPUTATIONS = ',F7.2/,5X,
       *'FLOW =',F8.2,/5X,'KINETIC ENERGY COEF = ',F5.2)
35      FORMAT(/6X,'SEC',7X,'DIST',7X,'DEPTH',7X,'ELEV')
40      FORMAT(6X,I3,5X,F7.2,5X,F6.3,6X,F6.2)
50      FORMAT(10X,'**** MAXIMUM NUMBER OF ITERATIONS EXCEEDED ***')
        STOP
        END
C       ****************************************************************
C          SUBROUTINE REACH
C          COMPUTE CHANNEL PROPERTIES AND CHANNEL EQUATIONS
C       ****************************************************************
        SUBROUTINE REACH
        COMMON/GROUP1/ DF(100,100),F(100),X(100),NEQ
        COMMON  MAXI,G,TOL,YUP,CO,FLAG,KOUNT,NITER,Y0,Q0,S0(5),CHL(5),
       1CMAN(5),BOT(5),Z0(5),NR(5),CHELEV(5),ALPHA(5),NCHAN,DX(5),DPDY(5),
       2Y(5,100),ELEV(5,100),A(5,100),HR(5,100),PER(5,100),TOP(5,100),
       3DRDY(5,100),EQ(5,100),ITER
        I = 1
        J = 1
        DO WHILE(I.LE.NCHAN)
         DX(I) = CHL(I)/NR(I)
         DPDY(I) = 2*SQRT(1+Z0(I)**2)
         IF(FLAG.GE.90.0) GO TO 10
        DO WHILE (J.LE.NR(I)+1)
           Y(I,J) = Y0
           ELEV(I,J) = CHELEV(I) - DX(I)*S0(I)*(J-1)
           A(I,J) = (Y0*Z0(I)+BOT(I))*Y0
           PER(I,J) = BOT(I)+2*Y0*SQRT(1+Z0(I)**2)
           HR(I,J) = A(I,J)/PER(I,J)
           TOP(I,J) = BOT(I)+2*Z0(I)*Y0
           DRDY(I,J) = TOP(I,J)/PER(I,J)-(A(I,J)/PER(I,J)**2)*DPDY(I)
           J = J+1
        END DO
10         K=1
           J = 1
        DO WHILE(J.LE.NR(I))
        EQ(I,K) = Y(I,J+1)-Y(I,J)+ELEV(I,J+1)-ELEV(I,J)+
       *    (ALPHA(I)/(2*G))*(Q0*ABS(Q0)/A(I,J+1)**2 -
       *    (ABS(Q0)*Q0)/A(I,J)**2)+0.5*DX(I)*((Q0*
       * ABS(Q0)*CMAN(I)**2)/(CO**2*A(I,J+1)**2*HR(I,J+1)**1.333)+
       *(Q0*ABS(Q0)*CMAN(I)**2)/(CO**2*A(I,J)**2*HR(I,J)**1.333))
        J = J+1
        K = K+1
        END DO
```

```
             J = 1
              I = I+1
              END DO
              FLAG = 100
             NEQ = 2
             I = 1
             J = 1
             K = 1
C                               **** CHANNEL REACH EQUATIONS ****
100      DO WHILE(K.LE.NR(I))
           F(NEQ) = -EQ(I,K)
C
           DF(NEQ,NEQ-1) = -1+Q0*ABS(Q0)*((ALPHA(I)*TOP(I,J))/
     *     (G*A(I,J)**3)-((2*CMAN(I)**2*DX(I))/(3*C0**2*A(I,J)**2*HR(I,J)
     *     **2.33))*DRDY(I,J)-(CMAN(I)**2*TOP(I,J)*DX(I))/(C0**2*A(I,J)**3
     *     *HR(I,J)**1.33))
C
           DF(NEQ,NEQ) = 1-Q0*ABS(Q0)*((ALPHA(I)*TOP(I,J+1))
     *     /(G*A(I,J+1)**3)+((2*CMAN(I)**2*DX(I))/(3*C0**2*A(I,J+1)**2
     *     *HR(I,J+1)**2.33))*DRDY(I,J+1)+(CMAN(I)**2*TOP(I,J+1)*DX(I))/
     *           (C0**2*A(I,J+1)**3*HR(I,J+1)**1.33))
         J = J+1
         K = K + 1
         NEQ = NEQ+1
      END DO
C                               ***  BOUNDARY CONDITIONS ***
C                               *** SERIES JUNCTION ***
      IF (I.EQ.NCHAN) GO TO 200
        F(NEQ) = (ELEV(I,NR(I)+1)+Y(I,NR(I)+1)+(Q0*
     *  ABS(Q0))/(2*G*A(I,NR(I)+1)**2) - ELEV(I+1,1)-Y(I+1,1)
     *  -(Q0* ABS(Q0))/(2*G*A(I+1,1)**2))
        DF(NEQ,NEQ-1) = -1+(Q0*ABS(Q0)*TOP(I,NR(I)+1))/
     *                      (G*A(I,NR(I)+1)**3)
        DF(NEQ,NEQ) =  1 - (Q0*ABS(Q0)*TOP(I+1,1))/(G*A(I+1,1)**3)
        I = I + 1
        NEQ = NEQ+1
        J = 1
        K = 1
        IF( I.LE.NCHAN) GO TO 100
C                               *****  END CONDITIONS ******
200   NEQ = NEQ -1
        F(1) = -(Y(1,1)-YUP)
      DF(1,1) = 1
      RETURN
      END
```

```
C       ****************************************************************
C           SUBROUTINE NEWVAL: UPDATE VALUES FOR NEXT ITERATION
C       ****************************************************************
        SUBROUTINE NEWVAL
        COMMON/GROUP1/ DF(100,100),F(100),X(100),NEQ
        COMMON  MAXI,G,TOL,YUP,CO,FLAG,KOUNT,NITER,YO,QO,SO(5),CHL(5),
       1CMAN(5),BOT(5),ZO(5),NR(5),CHELEV(5),ALPHA(5),NCHAN,DX(5),DPDY(5),
       2Y(5,100),ELEV(5,100),A(5,100),HR(5,100),PER(5,100),TOP(5,100),
       3DRDY(5,100),EQ(5,100),ITER
        I = 0
        K = 1
100     I = I+1
        J = 1
        DO WHILE(J.LE.NR(I)+1)
        Y(I,J) = Y(I,J) + X(K)
            A(I,J) = (Y(I,J)*ZO(I)+BOT(I))*Y(I,J)
            PER(I,J) = BOT(I)+2*Y(I,J)*SQRT(1+ZO(I)**2)
            HR(I,J) = A(I,J)/PER(I,J)
            TOP(I,J) = BOT(I)+2*ZO(I)*Y(I,J)
            DRDY(I,J) = TOP(I,J)/PER(I,J)-(A(I,J)/PER(I,J)**2)*DPDY(I)
        K = K+1
        J = J+1
        END DO
        IF( I.LT.NCHAN) GO TO 100
        IOK = 0
        KOUNT = 1
        DO WHILE (KOUNT.LE.NEQ)
        IF(ABS(X(KOUNT)).LE.TOL) IOK = IOK +1
            KOUNT = KOUNT + 1
        END DO
        IF (IOK.LT.NEQ) THEN
            ITER = ITER + 1
        ELSE
            NITER = ITER
            ITER = MAXI +1
        END IF
            RETURN
        END


C       ****************************************************************
C       SUBROUTINE TRIMAT: TRIDIAGONAL MATRIX ALGORITHM
C       ****************************************************************
        SUBROUTINE TRIMAT
        COMMON/GROUP1/ C(100,100),F(100),X(100),N
         DIMENSION U(100,100),Z(100)
        REAL LL(100,100)
```

```
LL(1,1) = C(1,1)
U(1,2) = C(1,2)/LL(1,1)
I = 2
DO WHILE(I.LE.N-1)
LL(I,I-1) = C(I,I-1)
LL(I,I) = C(I,I)-LL(I,I-1)*U(I-1,I)
U(I,I+1) = C(I,I+1)/LL(I,I)
I = I+1
END DO
LL(N,N-1) = C(N,N-1)
LL(N,N) = C(N,N)-LL(N,N-1)*U(N-1,N)
Z(1) = F(1)/LL(1,1)
I = 2
DO WHILE(I.LE.N)
Z(I) = (F(I)-LL(I,I-1) * Z(I-1))/LL(I,I)
I = I+1
END DO
X(N) = Z(N)
I = N-1
DO WHILE (I.GE.1)
X(I) = Z(I) - U(I,I+1)*X(I+1)
I = I-1
END DO
RETURN
END
```

********* INPUT DATA *********

```
2,25,9.81,0.0001,8.0,8.0,399.5
0.0001,2000.0,0.020,10.0,1.5
20,100.0,1.0
0.0003,2000.0,0.018,9.00,1.0
20,99.8,1.0
```

********* PROGRAM OUTPUT *****

```
UNITS = SI
NUMBER OF CHANNELS =          2
NUMBER OF ITERATIONS PERFORMED =    7
TOLERANCE =   .000100
ACCELERATION OF GRAVITY =   9.81
INITIAL VALUE FOR ITERATIONS =    8.00

** CHANNEL NUMBER  1 ***
BOTTON SLOPE =        .00010
CHANNEL LENGTH = 2000.00
MANNING"S COEFFICIENT =  .02000
```

```
BOTTON WIDTH =   10.00
LATERAL SLOPE = 1H: 1.5V
LENGTH INTERVAL FOR COMPUTATIONS = 100.00
FLOW =   399.50
KINETIC ENERGY COEF = 1.00
```

SEC	DIST	DEPTH	ELEV
1	.00	8.000	100.00
2	100.00	7.980	99.99
3	200.00	7.960	99.98
4	300.00	7.940	99.97
5	400.00	7.919	99.96
6	500.00	7.898	99.95
7	600.00	7.877	99.94
8	700.00	7.855	99.93
9	800.00	7.833	99.92
10	900.00	7.810	99.91
11	1000.00	7.787	99.90
12	1100.00	7.763	99.89
13	1200.00	7.739	99.88
14	1300.00	7.714	99.87
15	1400.00	7.689	99.86
16	1500.00	7.663	99.85
17	1600.00	7.637	99.84
18	1700.00	7.610	99.83
19	1800.00	7.583	99.82
20	1900.00	7.555	99.81
21	2000.00	7.526	99.80

```
** CHANNEL NUMBER   2 ***
BOTTON SLOPE =     .00030
CHANNEL LENGTH =  2000.00
MANNING"S COEFFICIENT =   .01800
BOTTON WIDTH =    9.00
LATERAL SLOPE = 1H: 1.0V
LENGTH INTERVAL FOR COMPUTATIONS =   100.00
FLOW =   399.50
KINETIC ENERGY COEF = 1.00
```

SEC	DIST	DEPTH	ELEV
1	.00	7.259	99.80
2	100.00	7.220	99.77
3	200.00	7.180	99.74
4	300.00	7.138	99.71
5	400.00	7.094	99.68
6	500.00	7.047	99.65
7	600.00	6.997	99.62

8	700.00	6.944	99.59
9	800.00	6.888	99.56
10	900.00	6.828	99.53
11	1000.00	6.763	99.50
12	1100.00	6.692	99.47
13	1200.00	6.616	99.44
14	1300.00	6.531	99.41
15	1400.00	6.438	99.38
16	1500.00	6.332	99.35
17	1600.00	6.210	99.32
18	1700.00	6.067	99.29
19	1800.00	5.888	99.26
20	1900.00	5.643	99.23
21	2000.00	5.126	99.20

E Appendix

```
C          COMPUTATION OF UNSTEADY FREE-SURFACE FLOWS BY LAX DIFFUSIVE
C          EXPLICIT SCHEME IN A TRAPEZOIDAL CHANNEL
C          CONSTANT FLOW DEPTH ALONG THE CHANNEL IS USED AS INITIAL
C          CONDITION
C          TRANSIENT CONDITIONS ARE PRODUCED BY THE SUDDEN CLOSURE
C          OF A DOWNSTREAM GATE
C
C          *********************** NOTATION **************************
C
C          AR = STATEMENT FUNCTION FOR FLOW AREA
C          B0 = CHANNEL BOTTOM WIDTH
C          C = CELERITY
C          CMN = MANNING'S COEFFICIENT
C          CHL = CHANNEL LENGTH
C          G = ACCELERATION OF GRAVITY
C          HR = STATEMENT FUNCTION FOR HYDRAULIC RADIUS
C          IPRINT = COUNTER FOR PRINTING RESULTS
C          NSEC = NUMBER OF CHANNEL SECTIONS
C          Q0 = INITIAL STEADY STATE DISCHARGE
C          S = CHANNEL LATERAL SLOPE; S HORIZONTAL TO 1 VERTICAL
C          S0 = CHANNEL BOTTOM SLOPE
C          TLAST = TIME FOR TRANSIENT FLOW COMPUTATION
C          TOP = STATEMENT FUNCTION FOR WATER TOP WIDTH
C          V = FLOW VELOCITY AT DIFFERENT SECTIONS
C          Y = FLOW DEPTH AT DIFFERENT SECTIONS
C          Y0 = INITIAL STEADY STATE FLOW DEPTH
C          UNITS = ALPHANUMERICAL VARIABLE FOR UNITS SYSTEM
C          **********************************************************
C
          CHARACTER*15 UNITS
          DIMENSION Y(60),V(60),YNEW(60),VNEW(60)
          AR(D) = (B0+D*S)*D
          HR(D) = (B0+D*S)*D/(B0+2*D*SQRT(1+S*S))
          TOP(D) = B0+2*D*S
```

```
READ(5,*) G,NSEC,TLAST,IPRINT
READ(5,*) CHL,B0,S,CMN,S0,Q0,Y0
T = 0
IF(G.GT.10) THEN
CMN2 = (CMN*CMN)/2.22
ELSE
CMN2 = CMN*CMN
END IF
C = SQRT(G*AR(Y0)/TOP(Y0))
C
C     STEADY STATE CONDITIONS
C
V0 = Q0/AR(Y0)
DX = CHL/NSEC
DT = DX/(V0+C)
DTDX = DT/DX
I = 1
NP1 = NSEC + 1
DO WHILE(I.LE.NP1)
V(I) = V0
Y(I) = Y0
I = I+1
END DO
IF(G.LE.10) UNITS = 'SI'
IF(G.GT.10) UNITS = 'ENGLISH'
WRITE(6,15) UNITS,CHL,B0,S,S0,CMN,Q0,Y0,NP1,TLAST
IP = IPRINT
C
C     COMPUTE TRANSIENT CONDITIONS
C
DO WHILE(T.LE.TLAST)
 IF(IP.GE.IPRINT) THEN
 WRITE(6,30) T
 WRITE(6,40) (Y(I),I = 1,NP1)
 WRITE(6,50) (V(I),I = 1,NP1)
 IP = 0
 END IF
IP = IP + 1
T = T+DT
C
C     UPSTREAM END
C
YNEW(1) = Y0
CB = SQRT(G*TOP(Y(2))/AR(Y(2)))
SF2 = CMN2*V(2)*ABS(V(2))/HR(Y(2))**1.333
CN = V(2) - CB*Y(2) + G*(S0-SF2)*DT
VNEW(1) = CN + CB*YNEW(1)
```

```
C
C       DOWNSTREAM END
C
        VNEW(NP1) = 0.0
        CA = SQRT(G*TOP(Y(NSEC))/AR(Y(NSEC)))
        SF = V(NSEC)*ABS(V(NSEC))*CMN2/HR(Y(NSEC))**1.333
        CP = CA*Y(NSEC) + G*(SO - SF)*DT+V(NSEC)
        YNEW(NP1) = CP/CA
C
C       INTERIOR NODES
C
        I = 2
        DO WHILE(I.LE.NSEC)
        HD1 = AR(Y(I-1))/TOP(Y(I-1))
        HD2 = AR(Y(I+1))/TOP(Y(I+1))
        HDI = 0.5*(HD1+HD2)
        YP1 = 0.5*(Y(I-1)+Y(I+1))
        VP1 = 0.5*(V(I-1)+V(I+1))
        HRIM1 = HR(Y(I-1))
        HRIP1 = HR(Y(I+1))
        SFIP1 = V(I+1)*ABS(V(I+1))*CMN2/HRIP1**1.333
        SFIM1 = V(I-1)*ABS(V(I-1))*CMN2/HRIM1**1.333
        SFP = 0.5*(SFIP1+SFIM1)
        DTX2 = DTDX*0.5
        YNEW(I) = YP1-DTX2*HDI*(V(I+1)-V(I-1))-DTX2*VP1*(Y(I+1)-Y(I-1))
        VNEW(I) = VP1-DTX2*G*(Y(I+1)-Y(I-1))-DTX2*VP1*(V(I+1)-V(I-1))
       1          + G*DT*(SO-SFP)
        I = I+1
        END DO
C
C       CHECK FOR STABILITY
C
        IFLAG = 0
        J = 1
        DO WHILE(J.LE.NP1 .AND. IFLAG.EQ.0)
        C = SQRT(G*AR(YNEW(J))/TOP(YNEW(J)))
        DTNEW = DX/(ABS(VNEW(J))+C)
        IF(DT.LT.0.75*DTNEW) DTNEXT = 1.15*DT
        IF(DT.GE.0.75*DTNEW) DTNEXT = DT
        J = J+1
        END DO
C
C       REDUCE DT FOR STABILITY AND RE-CALCULATE
C
        IF(IFLAG.EQ.1) THEN
        T = T-DT
        DT = 0.9*DTNEW
        IP = IP - 1
```

```
      ELSE
      DT = DTNEXT
      IP = IP + 1
      I = 1
      DO WHILE(I.LE.NP1)
      V(I) = VNEW(I)
      Y(I) = YNEW(I)
      I = I+1
        END DO
        END IF
        DTDX = DT/DX
        END DO
15    FORMAT(5X,'LAX DIFFUSIVE SCHEME',
     1 /5X,'UNITS = ',A10,/5X,'CHANNEL LENGTH = ',F7.2,
     2 /5X,'CHANNEL BOTTOM WIDTH = ',F6.2,/5X,'CHANNEL LATERAL SLOPE =
     3 ',F5.2,/5X,'CHANNEL BOTTOM SLOPE = ',F10.5,/5X,'MANNING"S n = '
     4,F10.5,/5X,'INITIAL STEADY STATE DISCHARGE = ',F7.2,/5X,
     5'UNIFORM FLOW DEPTH = ',F7.2,/5X,'NUMBER OF CHANNEL SECTIONS = ',
     6 I3,/5X,'TIME FOR WHICH TRANSIENTS WILL BE COMPUTED = ',F6.2/)
30    FORMAT(/1X,'T = ',F8.3)
40    FORMAT(1X,'Y = ',(12F6.2))
50    FORMAT(1X,'V = ',(12F6.3))
      STOP
      END
```

********** INPUT DATA *************

```
      9.81,10,350,1
      1000,10,2,0.010,0.0001,77.46,3.0
```

********** PROGRAM OUTPUT **********

```
LAX DIFFUSIVE SCHEME
UNITS = SI
CHANNEL LENGTH = 1000.00
CHANNEL BOTTOM WIDTH =   10.00
CHANNEL LATERAL SLOPE =   2.00
CHANNEL BOTTOM SLOPE =      .00010
MANNING"S n =      .01000
INITIAL STEADY STATE DISCHARGE =   77.46
UNIFORM FLOW DEPTH =         3.00
NUMBER OF CHANNEL SECTIONS =        11
TIME FOR WHICH TRANSIENTS WILL BE COMPUTED = 350.00
```

```
T =      .000
Y =    3.00   3.00   3.00   3.00   3.00   3.00   3.00   3.00   3.00   3.00   3.00
V =   1.614  1.614  1.614  1.614  1.614  1.614  1.614  1.614  1.614  1.614  1.614


T =    16.025
Y =    3.00   3.00   3.00   3.00   3.00   3.00   3.00   3.00   3.00   3.00   3.76
V =   1.614  1.614  1.614  1.614  1.614  1.614  1.614  1.614  1.614  1.614   .000


T =    32.050
Y =    3.00   3.00   3.00   3.00   3.00   3.00   3.00   3.00   3.00   3.64   3.76
V =   1.614  1.614  1.614  1.614  1.614  1.614  1.614  1.614  1.614   .321   .000


T =    48.076
Y =    3.00   3.00   3.00   3.00   3.00   3.00   3.00   3.00   3.52   3.64   3.81
V =   1.614  1.614  1.614  1.614  1.614  1.614  1.614  1.614   .570   .321   .000


T =    64.101
Y =    3.00   3.00   3.00   3.00   3.00   3.00   3.00   3.41   3.52   3.78   3.81
V =   1.614  1.614  1.614  1.614  1.614  1.614  1.614   .784   .570   .079   .000


T =    80.126
Y =    3.00   3.00   3.00   3.00   3.00   3.00   3.32   3.41   3.72   3.78   3.83
V =   1.614  1.614  1.614  1.614  1.614  1.614   .963   .784   .181   .079   .000


T =    96.151
Y =    3.00   3.00   3.00   3.00   3.00   3.24   3.32   3.66   3.72   3.81   3.83
V =   1.614  1.614  1.614  1.614  1.614  1.111   .963   .303   .181   .026   .000


T =   112.177
Y =    3.00   3.00   3.00   3.00   3.19   3.24   3.59   3.66   3.79   3.81   3.83
V =   1.614  1.614  1.614  1.614  1.229  1.111   .440   .303   .063   .026   .000


T =   128.202
Y =    3.00   3.00   3.00   3.14   3.19   3.51   3.59   3.76   3.79   3.83   3.83
V =   1.614  1.614  1.614  1.322  1.229   .584   .440   .114   .063   .014   .000


T =   144.227
Y =    3.00   3.00   3.11   3.14   3.44   3.51   3.73   3.76   3.82   3.83   3.84
V =   1.614  1.614  1.394  1.322   .727   .584   .180   .114   .031   .014   .000


T =   160.252
Y =    3.00   3.08   3.11   3.37   3.44   3.68   3.73   3.80   3.82   3.83   3.84
V =   1.614  1.449  1.394   .864   .727   .261   .180   .053   .031   .011   .000


T =   176.278
Y =    3.00   3.08   3.31   3.37   3.63   3.68   3.78   3.80   3.83   3.83   3.85
V =   1.287  1.449   .990   .864   .357   .261   .082   .053   .022   .011   .000
```

```
T =   192.303
Y =    3.00   3.18   3.31   3.58   3.63   3.76   3.78   3.82   3.83   3.84   3.85
V =   1.287   .933   .990   .464   .357   .120   .082   .035   .022   .010   .000


T =   208.328
Y =    3.00   3.18   3.45   3.58   3.74   3.76   3.81   3.82   3.84   3.84   3.86
V =    .573   .933   .425   .464   .168   .120   .050   .035   .019   .010   .000


T =   224.353
Y =    3.00   3.23   3.45   3.64   3.74   3.80   3.81   3.83   3.84   3.85   3.86
V =    .573   .169   .425   .089   .168   .069   .050   .029   .019   .010   .000


T =   240.379
Y =    3.00   3.23   3.45   3.64   3.72   3.80   3.82   3.83   3.84   3.85   3.86
V =   -.294   .169  -.172   .089  -.031   .069   .040   .029   .019   .010   .000


T =   256.404
Y =    3.00   3.21   3.45   3.56   3.72   3.76   3.82   3.84   3.84   3.86   3.86
V =   -.294  -.565  -.172  -.300  -.031  -.059   .040   .028   .019   .010   .000


T =   272.429
Y =    3.00   3.21   3.34   3.56   3.61   3.76   3.78   3.84   3.85   3.86   3.87
V =   -.978  -.565  -.680  -.300  -.317  -.059  -.064   .028   .018   .010   .000


T =   288.454
Y =    3.00   3.14   3.34   3.42   3.61   3.64   3.78   3.80   3.85   3.86   3.87
V =   -.978-1.060  -.680  -.676  -.317  -.303  -.064  -.064   .018   .009   .000


T =   304.480
Y =    3.00   3.14   3.22   3.42   3.46   3.64   3.67   3.80   3.82   3.86   3.88
V = -1.329-1.060-1.037  -.676  -.637  -.303  -.285  -.064  -.065   .009   .000


T =   320.505
Y =    3.00   3.08   3.22   3.28   3.46   3.50   3.67   3.70   3.82   3.83   3.88
V = -1.329-1.308-1.037  -.979  -.637  -.595  -.285  -.268  -.065  -.066   .000


T =   336.530
Y =    3.00   3.08   3.14   3.28   3.33   3.50   3.54   3.70   3.73   3.83   3.81
V = -1.452-1.308-1.249  -.979  -.917  -.595  -.556  -.268  -.253  -.066   .000
```

F Appendix

```
C      COMPUTATION OF UNSTEADY FREE-SURFACE FLOWS BY MACCORMACK
C      EXPLICIT SCHEME IN A TRAPEZOIDAL CHANNEL
C      CONSTANT FLOW DEPTH ALONG THE CHANNEL IS USED AS INITIAL
C      CONDITION.
C      TRANSIENT CONDITIONS ARE PRODUCED BY THE SUDDEN CLOSURE
C      OF A DOWNSTREAM GATE
C
C      *********************** NOTATION ************************
C
C      AR = STATEMENT FUNCTION FOR FLOW AREA
C      B0 = CHANNEL BOTTOM WIDTH
C      CHL = CHANNEL LENGTH
C      CENTR = MOMENT OF FLOW AREA ABOVE THE FREE SURFACE
C      CMN = MANNING'S COEFFICIENT
C      G = ACCELERATION OF GRAVITY
C      HR = STATEMENT FUNCTION FOR HYDRAULIC RADIUS
C      IPRINT = COUNTER FOR PRINTING RESULTS
C      NSEC = NUMBER OF CHANNEL SECTIONS
C      Q0 = INITIAL STEADY STATE DISCHARGE
C      S = CHANNEL LATERAL SLOPE; S HORIZONTAL TO 1 VERTICAL
C      S0 = CHANNEL BOTTOM SLOPE
C      TLAST = TIME FOR TRANSIENT FLOW COMPUTATIONS
C      TOP = STATEMENT FUNCTION FOR WATER TOP WIDTH
C      V = FLOW VELOCITY AT DIFFERENT SECTIONS
C      Y = FLOW DEPTH AT DIFFERENT SECTIONS
C      Y0 = INITIAL STEADY STATE FLOW DEPTH
C      UNITS = ALPHANUMERICAL VARIABLE FOR UNITS SYSTEMS
C      ********************************************************
C
       CHARACTER*15 UNITS
       DIMENSION Y(60),Q(60),QP1(60),AP1(60),YNEW(60),ARNEW(60),
      1FACTOR(60),SOSFP1(60),QNEW(60)
       AR(D) = (B0+D*S)*D
       HR(D) = (B0+D*S)*D/(B0+2*D*SQRT(1+S*S))
```

```
      TOP(D) = B0+2*D*S
      CENTR(D) = D*D*(B0/2+D*S/3)
      YBACK(A) = (-B0+SQRT(B0**2+4*A*S))/(2*S)
      READ(5,*) G,NSEC,TLAST,IPRINT
      READ(5,*) CHL,B0,S,CMN,S0,Q0,Y0
      T = 0
      IF(G.GT.10) THEN
      CMN2 = (CMN*CMN)/2.22
      ELSE
      CMN2 = CMN*CMN
      END IF
C
C     STEADY-STATE CONDITIONS
C
      C = SQRT(G*AR(Y0)/TOP(Y0))
      V0 = Q0/AR(Y0)
      DX = CHL/NSEC
      DT = DX/(V0+C)
      DTDX = DT/DX
      I = 1
      NP1 = NSEC + 1
      DO WHILE(I.LE.NP1)
      Q(I) = Q0
      Y(I) = Y0
      I = I+1
      END DO
      IF(G.LE.10) UNITS = 'SI'
      IF(G.GT.10) UNITS = 'ENGLISH'
      WRITE(6,15) UNITS,CHL,B0,S,S0,CMN,Q0,Y0,NP1,TLAST
      IP = IPRINT
C
C     COMPUTE TRANSIENT CONDITIONS
C
      DO WHILE(T.LE.TLAST)
       IF(IP.GE.IPRINT) THEN
       WRITE(6,30) T
       WRITE(6,40) (Y(I),I = 1,NP1)
       WRITE(6,60) (Q(I)/AR(Y(I)),I = 1,NP1)
       IP = 0
       END IF
      IP = IP + 1
      T = T+DT
C
C     UPSTREAM END
C
      YNEW(1) = Y0
      ARNEW(1) = AR(YNEW(1))
```

```
      CB = SQRT(G*TOP(Y(2))/AR(Y(2)))
      SF2 = CMN2*Q(2)*ABS(Q(2))/(AR(Y(2))**2*HR(Y(2))**1.333)
      CN = Q(2)/AR(Y(2)) - CB*Y(2) + G*(S0-SF2)*DT
      V = CN + CB*YNEW(1)
      QNEW(1) = V*AR(YNEW(1))
C
C     DOWNSTREAM END
C
      QNEW(NP1) = 0.0
      CA = SQRT(G*TOP(Y(NSEC))/AR(Y(NSEC)))
      SF = Q(NSEC)*ABS(Q(NSEC))*CMN2/(AR(Y(NSEC))**2*HR(Y(NSEC))**1.333)
      CP = CA*Y(NSEC) + G*(S0 - SF)*DT+Q(NSEC)/AR(Y(NSEC))
      YNEW(NP1) = CP/CA
      ARNEW(NP1) = AR(YNEW(NP1))
C
C     INTERIOR NODES
C
C     PREDICTOR STEP
C
      I = 2
      DO WHILE(I.LE.NSEC)
      QVI = Q(I)**2/AR(Y(I))+ G*CENTR(Y(I))
      QVIM1 = Q(I-1)**2/AR(Y(I-1)) + G*CENTR(Y(I-1))
      SFI = CMN2*Q(I)*ABS(Q(I))/(AR(Y(I))**2*HR(Y(I))**1.333)
      SOSFI = G*AR(Y(I))*(S0-SFI)
      SFIM1 = CMN2*Q(I-1)*ABS(Q(I-1))/(AR(Y(I-1))**2*HR(Y(I-1))**1.333)
      QP1(I) = Q(I) - DTDX*(QVI -  QVIM1) + SOSFI*DT
      AP1(I) = AR(Y(I)) - DTDX*(Q(I) - Q(I-1))
      IF(S.LE.0.) THEN
      YP1 = AP1(I)/B0
      ELSE
      YP1 = YBACK(AP1(I))
      END IF
      SFP1 = CMN2*QP1(I)*ABS(QP1(I))/(AP1(I)**2*HR(YP1)**1.333)
      SOSFP1(I) = G*AP1(I)*(S0-SFP1)
      FACTOR(I) = QP1(I)**2/AR(YP1) + G*CENTR(YP1)
      I = I+1
      END DO
      QP1(NP1) = 0.0
      FACTOR(NP1) =  G*CENTR(YNEW(NP1))
C
C     CORRECTOR STEP
C
      I = 2
      DO WHILE(I.LE.NSEC)
      QP2 = Q(I) - DTDX*(FACTOR(I+1)-FACTOR(I))+SOSFP1(I)*DT
      AP2 = AR(Y(I)) - DTDX*(QP1(I+1) - QP1(I))
      QNEW(I) = (QP1(I) + QP2)*0.5
```

```
      ARNEW(I) = (AP1(I)+AP2)*0.5
      IF(S.LE.0) THEN
      YNEW(I) = ARNEW(I)/B0
      ELSE
      YNEW(I) = YBACK(ARNEW(I))
      END IF
      I = I+1
      END DO
C
C     CHECK FOR STABILITY
C
      IFLAG = 0
      J = 1
      DO WHILE(J.LE.NP1 .AND. IFLAG.EQ.0)
      C = SQRT(G*ARNEW(J)/TOP(YNEW(J)))
      V = QNEW(J)/ARNEW(J)
      DTNEW = DX/(ABS(V)+C)
      IF(DT.LT.0.75*DTNEW) DTNEXT = 1.15*DT
      IF(DTNEW.GE.0.75*DT) DTNEXT = DT
      J = J+1
      END DO
C
C     REDUCE DT FOR STABILITY AND RECALCULATE
C
      IF(IFLAG.EQ.1) THEN
      T = T-DT
      DT = 0.9*DTNEW
      IP = IP - 1
      ELSE
      DT = DTNEXT
      IP = IP + 1
      I = 1
      DO WHILE(I.LE.NP1)
      Q(I) = QNEW(I)
      Y(I) = YNEW(I)
      I = I+1
        END DO
        END IF
        DTDX = DT/DX
        END DO


15    FORMAT(5X,'MACCORMACK SCHEME',
     1 /5X,'UNITS = ',A10,/5X,'CHANNEL LENGTH = ',F7.2,
     2 /5X,'CHANNEL BOTTOM WIDTH = ',F6.2,/5X,'CHANNEL LATERAL SLOPE =
     3 ',F5.2,/5X,'CHANNEL BOTTOM SLOPE = ',F10.5,/5X,'MANNING"S n = '
     4,F10.5,/5X,'INITIAL STEADY STATE DISCHARGE = ',F7.2,/5X,
     5'UNIFORM FLOW DEPTH = ',F7.2,/5X,'NUMBER OF CHANNEL SECTIONS = ',
     6 I3,/5X,'TIME FOR WHICH TRANSIENTS WILL BE COMPUTED = ',F6.2)
```

```
30    FORMAT(/2X,'T = ',F8.3)
40    FORMAT(2X,'Y = ',(12F6.2))
60    FORMAT(2X,'V = ',(12F6.3))
      STOP
      END


  ********** INPUT DATA **********

  9.81,10,350,1
  1000,10,2,0.010,0.0001,77.46,3.00

  ********* PROGRAM OUTPUT ********

    MACCORMACK SCHEME
    UNITS = SI
    CHANNEL LENGTH = 1000.00
    CHANNEL BOTTOM WIDTH =   10.00
    CHANNEL LATERAL SLOPE =   2.00
    CHANNEL BOTTOM SLOPE =       .00010
    MANNING"S n =      .01000
    INITIAL STEADY STATE DISCHARGE =    77.46
    UNIFORM FLOW DEPTH =           3.00
    NUMBER OF CHANNEL SECTIONS =          11
    TIME FOR WHICH TRANSIENTS WILL BE COMPUTED = 350.00


T =       .000
Y =    3.00  3.00  3.00  3.00  3.00  3.00  3.00  3.00  3.00  3.00  3.00
V =    1.614 1.614 1.614 1.614 1.614 1.614 1.614 1.614 1.614 1.614 1.614

T =     16.025
Y =    3.00  3.00  3.00  3.00  3.00  3.00  3.00  3.00  3.00  3.28  3.76
V =    1.614 1.614 1.614 1.614 1.614 1.614 1.614 1.614 1.614  .987  .000

T =     32.050
Y =    3.00  3.00  3.00  3.00  3.00  3.00  3.00  3.00  3.12  3.50  3.76
V =    1.614 1.614 1.614 1.614 1.614 1.614 1.614 1.614 1.313  .599  .000

T =     48.076
Y =    3.00  3.00  3.00  3.00  3.00  3.00  3.00  3.05  3.33  3.65  3.81
V =    1.614 1.614 1.614 1.614 1.614 1.614 1.614 1.496  .860  .318  .000

T =     64.101
Y =    3.00  3.00  3.00  3.00  3.00  3.00  3.02  3.19  3.55  3.73  3.82
V =    1.614 1.614 1.614 1.614 1.614 1.614 1.570 1.168  .461  .144  .000

T =     80.126
Y =    3.00  3.00  3.00  3.00  3.00  3.01  3.10  3.41  3.70  3.77  3.81
V =    1.614 1.614 1.614 1.614 1.614 1.598 1.393  .709  .208  .067  .000
```

```
T =    96.151
Y =    3.00   3.00   3.00   3.00   3.00   3.04   3.27   3.63   3.76   3.80   3.82
V =   1.614  1.614  1.614  1.614  1.608  1.517  1.007   .331   .079   .039   .000


T =   112.177
Y =    3.00   3.00   3.00   3.00   3.02   3.15   3.50   3.75   3.79   3.81   3.83
V =   1.614  1.614  1.614  1.612  1.574  1.270   .537   .123   .037   .021   .000


T =   128.202
Y =    3.00   3.00   3.00   3.01   3.08   3.36   3.69   3.79   3.80   3.81   3.83
V =   1.614  1.614  1.613  1.598  1.444   .818   .211   .038   .029   .015   .000


T =   144.227
Y =    3.00   3.00   3.00   3.03   3.22   3.60   3.78   3.80   3.81   3.82   3.83
V =   1.614  1.613  1.608  1.538  1.111   .369   .059   .026   .023   .015   .000


T =   160.252
Y =    3.00   3.00   3.01   3.12   3.46   3.75   3.80   3.80   3.81   3.82   3.84
V =   1.613  1.611  1.581  1.341   .614   .116   .017   .031   .023   .010   .000


T =   176.278
Y =    3.00   3.01   3.06   3.31   3.68   3.80   3.80   3.80   3.82   3.83   3.84
V =   1.609  1.600  1.482   .913   .226   .020   .030   .030   .021   .010   .000


T =   192.303
Y =    3.00   3.03   3.18   3.57   3.79   3.80   3.80   3.81   3.82   3.83   3.84
V =   1.587  1.552  1.194   .415   .050   .017   .038   .031   .017   .011   .000


T =   208.328
Y =    3.00   3.09   3.42   3.75   3.80   3.80   3.80   3.81   3.82   3.83   3.84
V =   1.499  1.383   .687   .119   .006   .041   .039   .028   .018   .009   .000


T =   224.353
Y =    3.00   3.24   3.66   3.80   3.80   3.79   3.81   3.82   3.83   3.84   3.85
V =   1.197   .943   .238   .011   .031   .047   .040   .025   .017   .010   .000


T =   240.379
Y =    3.00   3.44   3.76   3.80   3.79   3.80   3.81   3.82   3.83   3.84   3.85
V =    .452   .276   .005   .004   .052   .049   .034   .026   .016   .010   .000


T =   256.404
Y =    3.00   3.49   3.73   3.77   3.79   3.80   3.81   3.82   3.83   3.84   3.85
V =   -.588  -.350  -.146   .015   .050   .045   .033   .025   .017   .009   .000


T =   272.429
Y =    3.00   3.37   3.60   3.72   3.78   3.80   3.82   3.83   3.84   3.85   3.86
V =  -1.317  -.814  -.337  -.056   .029   .035   .033   .025   .017   .009   .000
```

```
T =    288.454
Y =     3.00   3.20   3.44   3.63   3.75   3.80   3.82   3.83   3.84   3.85   3.86
V = -1.540-1.128  -.600  -.234  -.046   .018   .028   .025   .017   .009   .000

T =    304.480
Y =     3.00   3.08   3.29   3.51   3.68   3.77   3.82   3.83   3.85   3.86   3.87
V = -1.519-1.319  -.885  -.485  -.196  -.038   .015   .021   .017   .009   .000

T =    320.505
Y =     3.00   3.02   3.18   3.39   3.58   3.71   3.80   3.83   3.85   3.86   3.87
V = -1.463-1.425-1.130  -.746  -.404  -.160  -.031   .010   .014   .009   .000

T =    336.530
Y =     3.00   3.01   3.10   3.28   3.46   3.63   3.75   3.81   3.85   3.86   3.87
V = -1.445-1.473-1.306  -.979  -.632  -.337  -.133  -.028   .004   .007   .000
```

G Appendix
_____

```
C     COMPUTATION OF UNSTEADY, FREE-SURFACE FLOWS BY PREISSMANN
C     IMPLICIT SCHEME IN A TRAPEZOIDAL CHANNEL
C     CONSTANT FLOW DEPTH ALONG THE CHANNEL IS SPECIFIED AS
C     INITIAL CONDITION.
C     TRANSIENT CONDITIONS ARE PRODUCED BY THE SUDDEN CLOSURE
C     OF A DOWNSTREAM GATE.
C
C     *********************** NOTATION ***********************
C
C      ALPHA = WEIGHTING COEFFICIENT
C      AR = STATEMENT FUNCTION FOR FLOW AREA
C      B0 = CHANNEL BOTTOM WIDTH
C      C = CELERITY
C      CENTR = MOMENT OF FLOW AREA
C      CMN = MANNING'S COEFFICIENT
C      CHL = CHANNEL LENGTH
C      G = ACCELERATION OF GRAVITY
C      HR = STATEMENT FUNCTION FOR HYDRAULIC RADIUS
C      IPRINT = COUNTER FOR PRINTING RESULTS
C      MAXITER = MAXIMUM NUMBER OF ITERATIONS
C      NSEC = NUMBER OF CHANNEL SECTIONS
C      Q0 = INITIAL STEADY STATE DISCHARGE
C      S = CHANNEL LATERAL SLOPE
C      S0 = CHANNEL BOTTOM SLOPE
C      TLAST = TIME FOR TRANSIENT FLOW COMPUTATION
C      TOL = TOLERANCE FOR INTERATIONS
C      TOP = STATEMENT FUNCTION FOR WATER TOP WIDTH
C      V = FLOW VELOCITY
C      Y = FLOW DEPTH
C      UNITS = ALPHANUMERIC VARIABLE FOR UNITS SYSTEM
C     *********************************************************
C
      CHARACTER*15 UNITS
      DIMENSION Y(60),V(60),C1(60),C2(60),DF(60),EQN(124,125)
```

467

```fortran
      AR(D) = (B0+D*S)*D
      HR(D) = (B0+D*S)*D/(B0+2*D*SQRT(1+S*S))
      TOP(D) = B0+2*D*S
      CENTR(D) = D*D*(B0/2+D*S/3)
      DCENDY(D)= D*(B0+D*S)
      READ(5,*) G,NSEC,TLAST,IPRINT
      READ(5,*) CHL,B0,S,CMN,S0,Q0,Y0,ALPHA,TOL,MAXITER
      T = 0
      IF(G.GT.10) THEN
      CMN2 = (CMN*CMN)/2.22
      ELSE
      CMN2 = CMN*CMN
      END IF
C
C     STEADY STATE CONDITIONS
C
      C = SQRT(G*AR(Y0)/TOP(Y0))
      V0 = Q0/AR(Y0)
      DX = CHL/NSEC
      DT = DX/(V0+C)
      DTX2 = 2*DT/DX
      YRES = Y0
      I = 1
      NP1 = NSEC + 1
      DO WHILE(I.LE.NP1)
      V(I) = V0
      Y(I) = Y0
      I = I+1
      END DO
      IF(G.LE.10) UNITS = 'SI'
      IF(G.GT.10) UNITS = 'ENGLISH'
      WRITE(6,15) UNITS,CHL,B0,S,S0,CMN,Q0,Y0,NP1,TLAST
      IFLAG = 0
      IP = IPRINT
C
C     COMPUTE TRANSIENT CONDITIONS
C
      DO WHILE(T.LE.TLAST .AND. IFLAG.EQ.0)
      ITER = 0
      IF(IPRINT.EQ.IP) THEN
      IP = 0
      WRITE(6,20) T
      WRITE(6,30) (Y(I),I= 1,NP1)
      WRITE(6,40) (V(I),I= 1,NP1)
      END IF
      T = T+DT
```

```
C
C     GENERATE SYSTEM OF EQUATIONS
C
      I = 1
      DO WHILE(I.LE.NSEC)
      ARI = AR(Y(I))
      ARIP1 = AR(Y(I+1))
      C1(I)=DTX2*(1-ALPHA)*(ARIP1*V(I+1)-ARI*V(I))-ARI-ARIP1
      SF1 = ABS(V(I))*V(I)*CMN2/HR(Y(I))**1.333
      SF2 = ABS(V(I+1))*V(I+1)*CMN2/HR(Y(I+1))**1.333
      TERM1 = -DT*(1-ALPHA)*(G*ARIP1*(S0-SF2)+G*ARI*(S0-SF1))
      TERM2 = -(V(I)*ARI+V(I+1)*ARIP1)
      TERM3 = DTX2*(1-ALPHA)*(V(I+1)**2*ARIP1+G*CENTR(Y(I+1))-
     1          V(I)**2*ARI-G*CENTR(Y(I)))
      C2(I) = TERM1 + TERM2 + TERM3
      I = I+1
      END DO
      NP11 = 2*NP1 + 1
      SUM = TOL+10
  100 IF(.NOT.(SUM.LE.TOL))THEN
      I = 1
      DO WHILE(I.LE.2*NP1)
       J = 1
       DO WHILE(J.LE.NP11)
        EQN(I,J) = 0.0
        J = J+1
       END DO
      I = I+1
      END DO

      ITER = ITER+1
C
C     BOUNDARY EQUATIONS
C
      EQN(1,1) = 1.0
      EQN(1,NP11) = -(Y(1)-YRES)
      EQN(2*NP1,2*NP1) = 1.0
      EQN(2*NP1,NP11) = -V(NP1)
C
C     INTERIOR NODES
C
      I = 1
      DO WHILE(I.LE.NSEC)
      ARI = AR(Y(I))
      ARIP1 = AR(Y(I+1))
      K = 2*I
```

```
      EQN(K,NP11)=-(ARI+ARIP1+DTX2*ALPHA*(V(I+1)*ARIP1-V(I)*ARI)+C1(I))
      SF1 = ABS(V(I))*V(I)*CMN2/HR(Y(I))**1.333
      SF2 = ABS(V(I+1))*V(I+1)*CMN2/HR(Y(I+1))**1.333
      TERM1 = DTX2*ALPHA*(V(I+1)**2*ARIP1+G*CENTR(Y(I+1))-V(I)**2*ARI
     1            -G*CENTR(Y(I)))
      TERM2 = -ALPHA*DT*G*((SO-SF2)*ARIP1+(SO-SF1)*ARI)
      EQN(K+1,NP11) = -(V(I)*ARI+V(I+1)*ARIP1+TERM1+TERM2+C2(I))
      DAY1 = TOP(Y(I))
      DAY2 = TOP(Y(I+1))
      EQN(K,K-1) = DAY1*(1-DTX2*ALPHA*V(I))
      EQN(K,K) = -DTX2*ALPHA*ARI
      EQN(K,K+1) = DAY2*(1+DTX2*ALPHA*V(I+1))
      EQN(K,K+2) = DTX2*ALPHA*ARIP1
      DCDY1 = DCENDY(Y(I))
      DCDY2 = DCENDY(Y(I+1))
      DSDV1 = 2*V(I)*CMN2/HR(I)**1.333
      DSDV2 = 2*V(I+1)*CMN2/HR(I+1)**1.333
      PERI = ARI/HR(Y(I))
      TERM1 = 2*SQRT(1+S**2)*ARI - DAY1*PERI
      TERM2 = HR(Y(I))**0.333*ARI**2
      DSDY1 = 1.333*V(I)*ABS(V(I))*CMN2*TERM1/TERM2
      PERIP1 = ARIP1/HR(Y(I+1))
      TERM1 = 2*SQRT(1+S**2)*ARIP1-DAY2*PERIP1
      TERM2 = HR(Y(I+1))**0.333*ARIP1**2
      DSDY2 = 1.333*V(I+1)*ABS(V(I+1))*CMN2*TERM1/TERM2
      TERM1 = -DTX2*ALPHA*(V(I)**2*DAY1 + G*DCDY1)
      TERM2 = -G*DT*ALPHA*(SO-SF1)*DAY1
      EQN(K+1,K-1)=V(I)*DAY1+TERM1+TERM2+G*DT*ALPHA*ARI*DSDY1
      EQN(K+1,K)=ARI-DTX2*ALPHA*2*V(I)*ARI+G*DT*ALPHA*ARI*DSDV1
      TERM1 = DTX2*ALPHA*(V(I+1)**2*DAY2+G*DCDY2)
      TERM2 = -G*DT*G*(SO-SF2)*DAY2
      EQN(K+1,K+1) = V(I+1)*DAY2+TERM1+TERM2+ALPHA*DT*G*ARIP1*DSDY2
      EQN(K+1,K+2) = ARIP1+DTX2*ALPHA*2*V(I+1)*ARIP1+ALPHA*DT*G
     1               *DSDV2*ARIP1
      I = I+1
      END DO
C
C     SOLVE SYSTEM OF EQUATIONS
C
      CALL MATSOL(2*NP1,DF,EQN)
      I = 1
      SUM = 0.0
      DO WHILE(I.LE.2*NP1)
      SUM = ABS(DF(I))+SUM
      IF(MOD(I,2).EQ.1) Y(I/2+1) = Y(I/2+1)+DF(I)
      IF(MOD(I,2).EQ.0) V(I/2) = V(I/2) + DF(I)
      I = I+1
      END DO
```

```
C
C       CHECK NUMBER OF ITERATIONS
C
        IF(ITER.GT.MAXITER) THEN
        IFLAG = 1
        SUM = TOL
        END IF
        GO TO 100
        END IF
        IP = IP+1
        END DO
        IF(IFLAG.EQ.1) WRITE(6,50)
15      FORMAT(5X,'PREISSMANN SCHEME',
       1 /5X,'UNITS = ',A10,/5X,'CHANNEL LENGTH = ',F7.2,
       2 /5X,'CHANNEL BOTTOM WIDTH = ',F6.2,/5X,'CHANNEL LATERAL SLOPE =
       3 ',F5.2,/5X,'CHANNEL BOTTOM SLOPE = ',F10.5,/5X,'MANNING"S n = '
       4,F10.5,/5X,'INITIAL STEADY STATE DISCHARGE = ',F7.2,/5X,
       5'UNIFORM FLOW DEPTH = ',F7.2,/5X,'NUMBER OF CHANNEL SECTIONS = ',
       6 I3,/5X,'TIME FOR WHICH TRANSIENTS WILL BE COMPUTED = ',F6.2)
20      FORMAT(/2X,'T = ',F8.3)
30      FORMAT(2X,'Y = ',(12F6.2))
40      FORMAT(2X,'V = ',(12F6.3))
50      FORMAT(5X,'MAXIMUM NUMBER OF ITERATIONS EXCEEDED')
        STOP
        END
C       ****************************************************************
C         SIMULTANEOUS SOLUTION OF THE SYSTEM OF EQUATIONS
C
        SUBROUTINE MATSOL(N,X,A)
        DIMENSION X(60),A(124,125),NROW(60)
        I = 1
        DO WHILE(I.LE.N)
        NROW(I) = I
        I = I + 1
        END DO
        I = 1
        DO WHILE(I.LE.N-1)
         AMAX = A(NROW(I),I)
         J = I
         IP =I
         DO WHILE(J.LE.N)
          IF(ABS(A(NROW(J),I)) .GT. AMAX) THEN
           AMAX = ABS(A(NROW(J),I))
           IP = J
          END IF
          J = J+1
         END DO
        IF(ABS(AMAX) .LE. 1E-08) GO TO 100
```

```
      IF(NROW(I).NE.NROW(IP))    THEN
       NC = NROW(I)
       NROW(I) = NROW(IP)
       NROW(IP) = NC
      END IF
      J = I+1
      DO WHILE(J.LE.N)
       COEF = A(NROW(J),I)/A(NROW(I),I)

       JJ = I+1
        DO WHILE(JJ.LE.N+1)
         A(NROW(J),JJ)=A(NROW(J),JJ)-COEF*A(NROW(I),JJ)
         JJ = JJ + 1
        END DO
        J = J+1
      END DO
      I = I+1
      END DO
      IF(ABS(A(NROW(N),N)) .LE. 1E-08) GO TO 100
      X(N) = A(NROW(N),N+1)/A(NROW(N),N)
      I = N-1
      DO WHILE(I.GE.1)
       SUM = 0.0
       J = I+1
       DO WHILE(J.LE.N)
        SUM = A(NROW(I),J)*X(J) + SUM
        J = J+1
       END DO
       X(I) = (A(NROW(I),N+1) - SUM)/A(NROW(I),I)
       I = I-1
      END DO
      GO TO 200
100   WRITE(*,*) 'SINGULAR MATRIX --> NO UNIQUE SOLUTION EXISTS'
200   RETURN
      END


      *************** INPUT DATA *****************

      9.81,10,350,1
      1000,10,2,0.010,0.0001,77.46,3.0,0.8,0.001,50


      ************** PROGRAM OUTPUT ***************

      PREISSMAN SCHEME
      UNITS SYSTEM = SI
      CHANNEL LENGTH = 1000.00
      CHANNEL BOTTOM WIDTH =      10.00
```

```
CHANNEL LATERAL SLOPE =   2.00
CHANNEL BOTTOM SLOPE =      .00010
MANNING"S n =      .01000
INITIAL STEADY STATE DISCHARGE =    77.46
UNIFORM FLOW DEPTH =        3.00
NUMBER OF CHANNEL SECTIONS =        11
TIME FOR WHICH TRANSIENTS WILL BE COMPUTED = 350.00


T =     .000
Y =    3.00  3.00  3.00  3.00  3.00  3.00  3.00  3.00  3.00  3.00  3.00
V =   1.614 1.614 1.614 1.614 1.614 1.614 1.614 1.614 1.614 1.614 1.614


T =    16.025
Y =    3.00  3.00  3.00  3.00  3.00  3.00  3.00  3.00  3.00  3.03  3.79
V =   1.614 1.614 1.614 1.614 1.614 1.614 1.614 1.613 1.622 1.551  .000


T =    32.050
Y =    3.00  3.00  3.00  3.00  3.00  3.00  3.00  3.00  3.01  3.57  3.77
V =   1.614 1.614 1.614 1.614 1.614 1.614 1.612 1.621 1.598  .455  .000


T =    48.076
Y =    3.00  3.00  3.00  3.00  3.00  3.00  3.00  2.98  3.40  3.71  3.80
V =   1.614 1.614 1.614 1.614 1.614 1.613 1.612 1.655  .799  .143  .000


T =    64.101
Y =    3.00  3.00  3.00  3.00  3.00  3.01  2.96  3.26  3.62  3.77  3.80
V =   1.614 1.614 1.614 1.614 1.615 1.602 1.688 1.080  .337  .053  .000


T =    80.126
Y =    3.00  3.00  3.00  3.00  3.01  2.96  3.15  3.51  3.73  3.79  3.81
V =   1.614 1.614 1.613 1.617 1.596 1.692 1.310  .557  .140  .025  .000


T =    96.151
Y =    3.00  3.00  3.00  3.01  2.97  3.06  3.40  3.66  3.77  3.80  3.81
V =   1.614 1.613 1.618 1.595 1.674 1.485  .791  .262  .066  .016  .000


T =   112.177
Y =    3.00  3.00  3.01  2.98  3.01  3.29  3.58  3.74  3.79  3.81  3.82
V =   1.611 1.618 1.599 1.647 1.601 1.024  .417  .128  .037  .013  .000


T =   128.202
Y =    3.00  3.00  3.00  2.98  3.18  3.49  3.69  3.77  3.80  3.81  3.82
V =   1.621 1.605 1.623 1.658 1.241  .599  .214  .069  .027  .011  .000


T =   144.227
Y =    3.00  3.00  2.97  3.09  3.39  3.63  3.75  3.79  3.80  3.81  3.82
V =   1.606 1.609 1.671 1.424  .803  .327  .116  .046  .023  .010  .000
```

```
T =    160.252
Y =    3.00  2.98  3.03  3.29  3.56  3.71  3.77  3.80  3.81  3.82  3.83
V =    1.591 1.652 1.557 1.015  .467  .180  .072  .036  .021  .010  .000


T =    176.278
Y =    3.00  2.99  3.19  3.48  3.67  3.75  3.79  3.80  3.81  3.82  3.83
V =    1.655 1.632 1.217  .634  .266  .108  .052  .032  .020  .010  .000


T =    192.303
Y =    3.00  3.11  3.39  3.61  3.72  3.77  3.79  3.81  3.82  3.83  3.84
V =    1.711 1.422  .826  .373  .157  .074  .044  .030  .019  .010  .000


T =    208.328
Y =    3.00  3.29  3.55  3.69  3.76  3.78  3.80  3.81  3.82  3.83  3.84
V =    1.462 1.040  .524  .227  .103  .059  .040  .029  .019  .009  .000


T =    224.353
Y =    3.00  3.41  3.64  3.74  3.78  3.79  3.80  3.82  3.83  3.84  3.85
V =     .817  .561  .303  .151  .081  .052  .038  .028  .019  .009  .000


T =    240.379
Y =    3.00  3.41  3.65  3.75  3.79  3.80  3.81  3.82  3.83  3.84  3.85
V =     .047  .060  .085  .086  .069  .051  .038  .028  .018  .009  .000


T =    256.404
Y =    3.00  3.35  3.59  3.72  3.78  3.80  3.81  3.82  3.83  3.84  3.85
V =    -.576 -.392 -.168 -.021  .038  .047  .039  .028  .018  .009  .000


T =    272.429
Y =    3.00  3.26  3.49  3.65  3.75  3.79  3.82  3.83  3.84  3.85  3.86
V =    -.979 -.752 -.436 -.185 -.039  .021  .034  .028  .018  .009  .000


T =    288.454
Y =    3.00  3.18  3.38  3.55  3.68  3.76  3.81  3.83  3.84  3.85  3.86
V =   -1.209-1.011 -.690 -.387 -.169 -.044  .009  .022  .018  .009  .000


T =    304.480
Y =    3.00  3.12  3.28  3.45  3.60  3.71  3.78  3.82  3.84  3.85  3.86
V =   -1.330-1.184 -.906 -.598 -.337 -.154 -.048 -.001  .011  .008  .000


T =    320.505
Y =    3.00  3.08  3.20  3.36  3.51  3.64  3.74  3.80  3.83  3.85  3.87
V =   -1.393-1.293-1.077 -.795 -.520 -.297 -.142 -.052 -.011  .001  .000


T =    336.530
Y =    3.00  3.05  3.15  3.28  3.43  3.56  3.68  3.76  3.81  3.85  3.86
V =   -1.425-1.360-1.203 -.965 -.701 -.459 -.267 -.134 -.056 -.017  .000
```

Author Index

Subject Index